The Biochemical Basis of Neuropharmacology

SEVENTH EDITION

JACK R. COOPER, Ph.D.
Professor of Pharmacology
Yale University School of Medicine

FLOYD E. BLOOM, M.D.
Chairman, Department of Neuropharmacology
The Scripps Research Institute

ROBERT H. ROTH, Ph.D.
Professor of Pharmacology and Psychiatry
Yale University School of Medicine

New York   Oxford
OXFORD UNIVERSITY PRESS
1996

Oxford University Press

Oxford   New York
Athens   Auckland   Bangkok   Bombay
Calcutta   Cape Town   Dar es Salaam   Delhi
Florence   Hong Kong   Istanbul   Karachi
Kuala Lumpur   Madras   Madrid   Melbourne
Mexico City   Nairobi   Paris   Singapore
Taipei   Tokyo   Toronto

and associated companies in
Berlin   Ibadan

Published by Oxford University Press, Inc.,
198 Madison Avenue, New York, New York 10016

Library of Congress Cataloging-in-Publication Data
Cooper, Jack R., 1924–
The biochemical basis of neuropharmacology / Jack R. Cooper,
Floyd E. Bloom, Robert H. Roth. — 7th ed.
p. cm.   Includes bibliographical references and index.
ISBN 0-19-510398-X—ISBN 0-19-510399-8 (pbk.)
1. Neurochemistry.   2. Neuropharmacology.   I. Bloom, Floyd E.
II. Title.
[DNLM: 1. Neuropharmacology.   2. Nerve Tissue—chemistry.
3. Nervous System—drug effects.   4. Neurotransmitters—physiology.
QV 76.5 C777b 1996]   QP356.3.C66   1996   615′.78—dc20
DNLM/DLC   for Library of Congress   95-580

Printed in the United States of America
on acid-free paper

The Biochemical Basis of
Neuropharmacology

Preface
to the Seventh Edition

Aside from updating each chapter and including a discussion of new gaseous, neuronal modulators, nitric oxide, and carbon monoxide, we have made two major changes in this edition. The chapter on Metabolism in the Central Nervous System has been eliminated, since we feel it has outlived its usefulness. However, some of that material has been incorporated into a new chapter on the genetic basis of neurological and psychiatric diseases. With the almost biweekly discovery of genes that appear to be involved in diseases of the nervous system, this area has the potential for providing a revolutionary kind of therapy.

We would be remiss if we did not thank our editor, Jeffrey House, for his helpful suggestions as well as his exhortations to turn in our revisions on time. We also thank our colleagues who took the time to point out errors in the text—sometimes gleefully we suspect, but nevertheless we appreciate their efforts.

J.R.C.
F.E.B.
R.H.R.

Contents

The Biochemical Basis of
Neuropharmacology

1 | Introduction

Neuropharmacology can be defined simply as the study of drugs that affect nervous tissue. This, however, is not a practical definition, since a great many drugs whose therapeutic value is extraneural can affect the nervous system. For example, the cardiotonic drug digitalis will not uncommonly produce central nervous system effects ranging from blurred vision to disorientation. For our purposes we must accordingly limit the scope of neuropharmacology to those drugs specifically employed to affect the nervous system. The domain of neuropharmacology would thus include psychotropic drugs that affect mood and behavior, anesthetics, sedatives, hypnotics, narcotics, anticonvulsants, analgetics, and a variety of drugs that affect the autonomic nervous system.

Since, with few exceptions, the precise molecular mechanism of action of these drugs is unknown, and since recitations of their absorption, metabolism, therapeutic indications, and toxic liability can be found in most textbooks of pharmacology, we have chosen to take a different approach to the subject. We will concentrate on the biochemistry and physiology of nervous tissue, emphasizing neurotransmitters, and will introduce the neuropharmacologic agents where their action is related to the subject under discussion. Thus, a discussion of LSD is included in the chapter on serotonin, and a suggested mechanism of action of the antipsychotic drugs is found in Chapters 9 and 13.

It is not difficult to justify this focus on either real or proposed neurotransmitters since the drugs act at junctions rather than on the events that occur with axonal conduction or within the cell body. Except for local anesthetics, which interact with axonal membranes, all neuropharmacological agents whose mechanisms of action are to some extent documented seem to be involved primarily with synaptic events. This finding appears quite logical in view of the regulatory mechanisms in the transmission of nerve impulses. The extent to which a neuron is depolarized or hyperpolarized will depend largely on its excitatory and inhibitory synaptic inputs, and these inputs must obviously involve neurotransmitters, neuromodulators, or neurohormones. What is enormously difficult to comprehend is the contrast between the action of a drug on a simple neuron, which causes it either to fire or not to fire, and the wide diversity of central nervous system effects, including subtle changes in mood and behavior which that same drug will induce. As will become clearer in subsequent chapters, at the molecular level, an explanation of the action of a drug is often possible; at the cellular level, an explanation is sometimes possible; but at a behavioral level, our ignorance is abysmal. There is no reason to assume, for example, that a drug that inhibits the firing of a particular neuron will therefore produce a depressive state in an animal: there may be dozens of unknown intermediary reactions involving transmitters and modulators between the demonstration of the action of a drug on a neuronal system and the ultimate effect on behavior.

However, the fact that one can find compounds with a specific chemical structure to control a given pathological condition is an exciting experimental finding, since it suggests an approach that the neuropharmacologist can use to clarify normal as well as abnormal brain chemistry and physiology. For instance, the use of drugs that affect the adrenergic nervous system has uncovered basic and hitherto unknown neural properties such as the uptake, storage, and release of the biogenic amines. The recognition of the analogy between curare poisoning in animals and myasthenia gravis in humans led to the understanding of the cholinergic neuromuscular transmission problem in myasthenia gravis and to subsequent treatment with anticholinesterases.

We have already referred to neuroactive agents involved in synaptic transmission as neurotransmitters, neuromodulators, and neurohormones, so definitions are now in order. Although we can define these terms in a strict, rigid fashion, it will be apparent—as noted later—that it is an exercise in futility to apply these definitions to a neuroactive agent as a classification unless one both understands its activity and specifies its locus. Briefly, the traditional definition of a *neurotransmitter* states that the compound must be synthesized and released presynaptically; it must mimic the action of the endogenous compound that is released on nerve stimulation; and, where possible, a pharmacological identity is required where drugs that either potentiate or block postsynaptic responses to the endogenously released agent also act identically to the administered suspected neurotransmitter. Conventionally, based on the studies of ACh at the neuromuscular junction, transmitter action was thought to be a brief and highly restricted point-to-point process. If ones takes the word *modulation* literally, then a *neuromodulator* has no intrinsic activity but is only active in the face of ongoing synaptic activity where it can modulate transmission either pre- or postsynaptically. In many instances, however, a modulating agent does produce changes in conductance or membrane potential. Typically, modulatory effects involve a second messenger system. A *neurohormone* has intrinsic activity and can be released from both neuronal and nonneuronal cells and—most important to the definition—travels in the circulation to act at a site distant from its release site. Just how far a neurohormone has to travel before it loses its neurotransmitter status and becomes a neurohormone has never been decided.

We stated earlier that, while we could define these terms, it would be of little use to pigeonhole known neuroactive compounds until the site of action and the activity of the agent were specified. For example, dopamine is a certified neurotransmitter in the striatum, yet it is released from the hypothalamus and travels through the hypophyseal-portal circulation to the pituitary, where it inhibits the release of prolactin. Here it obviously fits the definition of a neurohormone. Similarly, serotonin is a neurotransmitter in the raphe nuclei, yet at the facial motornucleus it acts primarily as a neuromodulator and secondarily as a transmitter. Most peptides with their multiple activities in the brain and gut are generally considered to be neuromodulators, yet Substance P fulfills the criteria of a transmitter at sensory afferents to the dorsal horn of the spinal cord. In sum, the plethora of exceptions to the aforementioned definitions of transmitter, modulator, and hormone has generated confusion in the literature. Better to describe the activity of a neuroactive agent at a specified site rather than attempt to give a profitless definition.

The multidisciplinary aspects of pharmacology in general are particularly relevant in the field of neuropharmacology, where a "pure" neurophysiologist or neurochemist would be severely handicapped in elucidating drug action at a molecular level. Since neuropharmacology is not a specific discipline with its own technology, the neuropharmacologist should be aware of the tools that are available for the total dissection of a biological problem. These would include morphological techniques such as electron microscopy, fluorescence microscopy, and freeze-etching; immunological techniques as a basis for developing radioimmunoassays, immunocytochemistry, and monoclonal antibodies; as well as the classical electrophysiological and biochemical procedures. If the investigator is concerned with the action of psychotropic drugs, a prerequisite is some knowledge of the techniques and pitfalls of behavioral testing. Finally, with the explosion of molecular biological techniques, it is a wise student who masters this technology since the basis of many psychiatric and neurological diseases, as well as addiction and tolerance, probably lies in the regulation of gene expression (see Chapter 13). Neuropharmacologists utilize the above techniques to achieve two goals: The understanding of drug mechanisms of action and the possible development of new, selective therapeutic agents.

In science one measures something. One must know what to measure, where to measure it, and how to measure it. This sounds rather obvious, but the student should be aware that, particularly in the neural sciences, these seemingly simple tasks can be enormously difficult. For example, suppose one were interested in elucidating the presumed biochemical aberration in schizophrenia. *What* would one measure? ATP? Glucose? Ascorbic acid? Unfortunately, this problem early on had been zealously investigated by people who measured everything they could think of, generally in the blood, in their search for differences between normal individuals and schizophrenics. As could be predicted, the problem was not solved. (It may be assumed, however, that these studies produced a large population of anemic schizophrenics with all this bloodletting.) In recent times it has been demonstrated that antipsychotic drugs block a dopamine receptor (see Chapter 9). Although this biochemical reaction takes place immediately in test tubes containing brain tissue, patients who are given antipsychotic medication do not show beneficial effects for about 2 weeks. The inference, therefore, is that the drug itself and its biochemical reaction do not produce the ameliorative effect; rather it is the adaptation of the brain to the presence of the drug that is beneficial. The question then is what is this adaptation; the answer is we still don't know what to measure (but see Chapter 13)

Deciding *where* to measure something in neuroscience is complicated by the heterogeneity of nervous tissue: In general, unless one has a particular axon to grind, it is preferable to use peripheral nerve rather than the CNS. Suburban neurochemists have an easier time than their CNS counterparts, since it is not only a question of which region of the brain to use for the test preparation but which of the multitude of cell types within each area to choose. If a project involved a study of amino-acid transport in nervous tissue, for example, would one use isolated nerve-ending particles (synaptosomes), glial cells, neuronal cell bodies in culture, a myelinated axon, or a ganglion cell? Up to the present time most investigators have used cortical brain slices, but the obvious disadvantage of this preparation is that one has no idea which cellular organelle takes up the amino acid.

How to measure something is a surprisingly easy question to answer, at least if one is dealing with simple molecules. With the recent advances in microseparation techniques and in fluorometric, radiometric, and immunological assays, there is virtually nothing that cannot be measured with a high degree of specificity and sensitivity. In this regard one should be careful not to overlook the classic bioassay, which tends to be scorned by young investigators but is in fact largely responsible for striking progress in our knowledge of both the prostaglandins and the opiate receptor with its peptide agonists. The major problem is with macromolecules. How can neuronal membranes be quantified, for example, if extraneuronal constituents are an invariable contaminant and if markers to identify unequivocally a cellular constituent are often lacking? The quantitative and spatial measurement of receptors utilizing autoradiography is also a key problem (see Chapter 4).

This harangue about measurement is meant to point out that what would on the surface appear to be the simplest part of research can in fact be very difficult. It is vital that students learn not to accept data without a critical appraisal of the procedures that were employed to obtain the results.

Finally, although the theme is not explicitly dealt with in this book, students may find it educational and often entertaining to attempt to define patterns of research design in neuropharmacology as well as current trends in research areas. One common pattern is for someone to observe something in brain tissue, trace its regional distribution in the brain, and then perform a developmental study of the phenomenon in laboratory animals from prenatal through adult life. Another common pattern is for someone to develop a technique and then search (sometimes with what appears to be desperation) for projects that will utilize the technique. Yet another somewhat simplistic idea is that of attempting to relate a behavioral effect to a changing level of a single neurotransmitter, invariably the one that a team has just learned how to measure. Current trends in the neural sciences that are related to neuropharmacology include identifying subclasses of ion channels, utilizing molecular genetics to uncover new peptides, neural cartography (the mapping of transmitters and neuroactive peptides in the CNS), searching for toxins with specific effects on conduction or transmission, cloning and characterizing receptors and ion channels, and identifying trophic factors involved in synaptogenesis and neuronal regulation. It can also easily be predicted that within the next few years an intensive search will be undertaken to explain the function, as well as the second messenger systems, for the plethora of receptor subtypes. Clearly, neuropharmacologic agents will be invaluable probes in this search.

2 | Cellular Foundations of Neuropharmacology

As we begin to consider the particular problems that underlie the analysis of drug actions in the central nervous system, it may be asked, "Just what is so special about nervous tissue?" Nerve cells have two special properties that distinguish them from all other cells in the body. First, they can conduct bioelectric signals for long distances without any loss of signal strength. Second, they possess specific intercellular connections with other nerve cells and with innervated tissues such as muscles and glands. These connections determine the types of information a neuron can receive and the range of responses it can yield in return.

CYTOLOGY OF THE NERVE CELL

We do not need the high resolution of the electron microscope to identify the characteristic structural features of the nerve cell. The classic studies of the Spanish Nobel Prize–winning cytologist Santiago Ramón y Cajal demonstrated the heterogeneous size and shape of neurons as individual cells. An inescapable rule of neurocytology and neuroanatomy is that structures have several synonymous names. So, for example, we find that the body of the nerve cell is also called the soma and the perikaryon—literally, the part that surrounds the nucleus. A fundamental scheme classifies nerve cells by the number of cytoplasmic processes they possess. In the simplest case, the perikaryon has one process, called an axon; the best examples of this cell type are the sensory neurons whose perikarya occur in groups in the sensory or dorsal root ganglia. In this case, the axon conducts the signal—which was generated by the sensory receptor in the skin or other viscera—centrally through the dorsal root into the spinal cord or cranial nerve nuclei. At the next step of complexity we find neurons possessing two processes: the bipolar nerve cells. The sensory receptor nerve cells of the retina, the olfactory mucosa, and the auditory nerve are of this form, as is a class of small nerve cells of the brain known as granule cells.

All other nerve cells tend to fall into the class known as multipolar nerve cells. These cells possess only one axon or efferent-conducting process, which may be short or long, branched or straight, and which may possess a recurrent or collateral branch that feeds back onto the same type of nerve cell from which the axon arises. Their main differences relate to the extent and size of the receptive field of the neuron, termed the dendrites or dendritic tree. In silver-stained preparations for the light microscope, the branches of the dendrites look like trees in wintertime, although the branches may be long and smooth, short and complex, or bear short spines like a cactus. It is on these dendritic branches as well as on the cell body where the termination of axons from other neurons makes the specialized interneuronal communication point known as the synapse.

The Synapse

The characteristic specialized contact zone that has been presumptively identified as the site of functional interneuronal communication is the synapse. It contains special organelles. As the axon approaches the site of its termination, it exhibits structural features not found more proximally. Most striking is the occurrence of dilated regions of the axon (varicosities) within which are clustered large numbers of microvesicles (synaptic vesicles). These synaptic vesicles tend to be spherical in shape, with diameters varying between 400 and 1200 Å. Depending upon the type of fixation used, the shape and staining properties of the vesicles can be related to their neurotransmitter content. The nerve endings also exhibit mitochondria, but do not exhibit microtubules unless the varicosity is a "pre-terminal" region of an axon as it extends toward its terminal target. One or more of these varicosities may form a specialized contact with one or more dendritic branches before the ultimate termination. Such endings are known as en passant terminals. In this sense, the term "nerve terminal" or "nerve ending" connotes a functional transmitting site rather than the end of the axon.

Electron micrographs of synaptic regions in the central nervous system reveal a specialized contact zone between the axonal nerve ending and the postsynaptic structure (Fig. 2-1). Cell types arising from the embryonic ectoderm often have such specialized intercellular contact zones, which are generally presumed to maintain the structural integrity of the cells within a layer. In the nervous system, the specialized contact zone at synapses has been viewed as the site of active chemical transmission and response. This conjecture posed substantial controversy in the era before any of the molecules associated with presynaptic transmitter release or postsynaptic transmitter

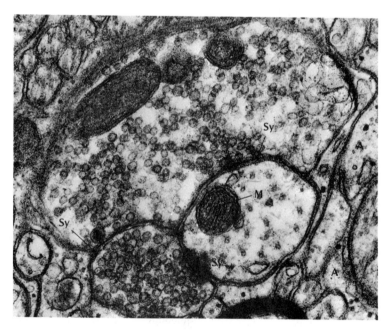

Figure 2-1. High-power view of two nerve terminals contacting a small dentritic spine. At this magnification the synaptic vesicles can be seen clearly, as can the zones of specialized contact (Sy). Mitochondria (M) are seen in the dendrite and large nerve terminal. Astrocyte processes containing glycogen (A) can be seen. Note that the larger nerve terminal makes a specialized contact on the small terminal (axo-axonic) as well as on the dendrite (axo-dendritic) × 12,000.

response had been characterized. However, the controversy now seems to have ended with the identification of many of the specific proteins that have precise functional properties in and on the surface of the synaptic vesicles and at sites along the inside of the presynaptic membrane associated with vesicle docking and release. Similarly, the direct chemical identification at discrete sites within the postsynaptic surface of the specialized contacts of receptor proteins and other proteins capable of modifying the response to neurotransmitters has given near concensus to the functional inference of a synaptic active zone. In some neurons, especially the single-process small granule cell types, the "dendrite" may also be structurally specialized to store and release transmitter.

Glia

A second element in the maintenance of the neuron's integrity depends on a type of cell known as neuroglia. There are two main types of neuroglia, together termed macroglia. The first is called the fibrous astrocyte, a descriptive term based on its starlike shape when viewed in the light microscope and on the fibrous nature of its cytoplasmic organelles, which can be seen with both light and electron microscopy. Astrocytes are found mainly in regions of axons and dendrites; they tend to surround and closely contact the adventitial surface of blood vessels. Functions such as insulation (between conducting surfaces) and organization (to surround and separate functional units of nerve endings and dendrites) have been empirically attributed to the astrocyte, mainly on the basis of its structural characteristics. However, studies of carbohydrate-metabolizing properties of astrocytes in culture have demonstrated their capacity to accumulate glucose, synthesize glycogen, and provide energy substrates to neurons. Furthermore, the link between glucose uptake into astrocytes is also activity dependent and is regulated by extracellular cations, as well as by at least some neurotransmitters. These findings have led to the suggestion that the regional localization of glucose uptake, as demonstrated in human brain by positron emission tomography (PET) and in animal brains by 2-deoxyglucose autoradiography, reflects primarily the uptake of glucose by astrocytes and not by neurons.

The second major type of neuroglia is known as the oligodendrocyte. It is called the satellite cell when it occurs close to nerve cell bodies and Schwann cell when it occurs in the peripheral nervous system. The cytoplasm of the oligodendrocyte is characterized by rough endoplasmic reticulum, but its most prominent characteristic is the enclosure of concentric layers of its own surface membrane around the axon. These concentric layers come together so closely that the oligodendrocyte cytoplasm is completely squeezed out and the original internal surfaces of the membrane become fused, presenting the ringlike appearance of the myelin sheath in cross section. Along the course of an axon, which may be many centimeters, many oligodendrocytes are required to constitute its myelin sheath. At the boundary between adjacent portions of the axon covered by separate oligodendrocytes, there is an uncovered axonal portion known as the node of Ranvier.

Many central axons and certain elements of the peripheral autonomic nervous system do not possess myelin sheaths. Even these axons, however, are not bare or exposed directly to the extracellular fluid; rather they are enclosed within single invaginations of the astrocyte surface membrane. Because of this close relationship between the conducting portions of the nerve cell, its axon, and the astrocyte, it is easy to see the origin of the proposition that the astrocyte may contribute to the nurture of the nerve cell.

There is yet a third nonneuronal cell class in the brain, termed microglia, that is gaining increasing importance as a pharmacological target. The microglia are of mesodermal origin and are related to macrophages and monocytes. Some microglia are resident within the brain, often around blood vessels. During events creating tissue necrosis, such as stroke, trauma, or infection, however, macrophages enter the brain and secrete chemical signals (cytokines, see Chapter 12) to recruit lymphocytes and leukocytes to seal off and repair the tissue damage.

Brain Permeability Barriers

While the unique cytological characteristics of neurons and glia are sufficient to establish the complex intercellular relationships of the brain, there is yet another histophysiological concept to consider. Numerous chemical substances pass from the bloodstream into the brain at rates that are far slower than for entry into all other organs in the body. There are similar slow rates of transport between the cerebrospinal fluid and the brain, although there is no good standard in other organs against which to compare this latter movement.

These permeability barriers appear to be the end result of numerous contributing factors that present diffusional obstacles to chemicals on the basis of molecular size, charge, solubility, and specific carrier systems. The difficulty has not been in establishing the existence of these barriers, but in determining their mechanisms. When the relatively small protein (mol wt = 43,000) horseradish peroxidase is injected intravenously into mice, its location within the tissue can be demonstrated histochemically with the electron microscope. As opposed to the easy transvascular movement of this substance across muscle capillaries, in the brain the peroxidase molecule hardly penetrates the continuous layer of vascular endothelial cells at all. Recent studies suggest that a very, very reduced transcytosis may occur. The endothelial cells of brain capillaries differ from those of other tissues in that the intercellular zones of membrane apposition are much more highly developed in the brain and are virtually continuous along all surfaces of these cells. Furthermore, cerebral vascular endothelial cells lack pinocytotic vesicles, considered to be the transvascular carrier systems of both large and small molecules in other tissues.

Since the enzyme marker can barely go through or between the endothelial cells, an operationally defined barrier exists. Whether the same barrier is also applicable to highly charged lipophobic small molecules cannot be determined from these observations. As neuropharmacologists, what concerns us more are the factors that retard the entrance of these smaller molecules, such as norepinephrine and serotonin, their amino-acid precursors, or drugs that affect the metabolism of these and other neurotransmitters. Charged molecules can, however, diffuse widely through the extracellular spaces of the brain when permitted entry via the cerebrospinal fluid. Astrocytes are currently thought to elicit expression in cerebral capillary endothelial cells of the proteins that constitute the blood–brain barrier. Recent research suggests that a model blood–brain barrier system may be attainable under culture conditions. So long as central astrocytes are viable, brain fragments transplanted to vascular beds, such as the anterior chamber of the eye, that would normally lack a blood–brain barrier nevertheless retain a functional barrier when revascularized. This suggests that the properties of the barrier reside not in the endothelial cells themselves, but in some functional response to the adjacent central astrocytes.

Substances that find it difficult to get into the brain, in general, also find it exceedingly difficult to leave. Thus, when monoamines are increased in concentration by the blocking of their catabolism (see Chapter 9), high levels of amine persist until the inhibiting agents are metabolized or excreted. One such excretory route is the "acid transport" system by which the choroid plexus and/or brain parenchymal cells actively secrete acid catabolites, as well as drugs such as penicillin or zidovudine, which one might like to keep in. This step can be blocked by the drug probenecid, resulting in increased brain and cerebrospinal fluid amine catabolite and drug levels.

The choroid plexus may also be an interface between the peripheral vascular system and the immune response system, where antigens for which immuno-surveillance is required can be recognized and immune responses can be mounted. Lymphocytes may normally "wander" through the brain in search of immune targets, although how they can slip through the endothelial cell barriers without reducing their normal barrier functions remains unclear.

Since the precise nature of these barriers still cannot be formulated, students would be wise to avoid the "great wall of China" concept and lean toward the possibility of a series of variously placed, progressively selective filtration sites that discriminate substances on the basis of several molecular characteristics. With lipid-soluble weak electrolytes—a characteristic of most centrally acting drugs—transport occurs by a process of passive diffusion. Thus, a drug will penetrate the endothelial cell only in the undissociated form and at a rate consonant with its lipid solubility and its pKa.

Specialized sets of neurons known as the circumventricular organs exist within discrete sites along the linings of the cerebroventricles, but are functionally on the blood side of the blood–brain barrier. Considered to be "windows" through which the normally excluded central milieu can monitor the components of the bloodstream, these neurons can communicate directly with neurons well within the enclosure of the blood–brain barrier.

BIOELECTRIC PROPERTIES OF THE NERVE CELL

Given these structural details, we can now turn to the second striking feature of nerve cells, namely, their bioelectric property. However, even for this introductory presentation, we must understand certain basic concepts of the physical phenomena of electricity in order to have a working knowledge of the bioelectric characteristics of living cells.

The initial concept to grasp is that of a difference in potential existing within a charged field, as occurs when charged particles are separated and prevented from randomly redistributing themselves. When a potential difference exists, the amount of charge per unit of time that will flow between the two sites (i.e., current flow) depends upon the resistance separating them. If the resistance tends to zero, no net current will flow since no potential difference can exist in the absence of a measurable resistance. If the resistance is extremely high, only a minimal current will flow, and that will be proportional to the electromotive force or potential difference between the two sites. The relationship between voltage, current, and resistance is Ohm's law: $V = I \cdot R$.

When we come to the measurement of the electrical properties of living cells, these basic physical laws apply, but with one exception. The pioneer electrobiologists, who did their work before the discovery and definition of the electron, developed a convention for the flow of charges based not on the electrons but on the flow of positive charges. Therefore, since in biological systems the flow of charges is not carried by electrons but by ions, the direction of flow is expressed in terms of movement of positive charges. To analyze the electrical potentials of a living system, we use small electrodes (a microprobe for detecting current flow or potential), electronic amplifiers for increasing the size of the current or potential, and oscilloscopes or polygraphs for displaying the potentials observed against a time base.

Membrane Potentials

If we take two electrodes and place them on the outside of a living cell or tissue, we will find little, if any, difference in potential. However, if we injure a cell so as to break its membrane or insert one ultrafine electrode across the otherwise intact membrane, we will find a potential difference such that the inside of the cell is 50 or more millivolts negative with respect to the extracellular electrode (Fig. 2-2). This transmembrane potential difference has been found in almost all types of living cells in which it has been sought; such a membrane is said to be electrically polarized. By passing negative ions into the cell through the microelectrode (or extracting cations), the inside can be made more negative (hyperpolarized). If positive current is applied to the inside of the cell, the transmembrane potential difference is decreased, and the potential is said to be depolarized. The potential difference across the membrane of most living cells can be accounted for by the relative distribution of the intracellular and extracellular ions.

The extracellular fluid is particularly rich in sodium (Na) and relatively low in potassium (K). Inside the cell, the cytoplasm is relatively high in K content and very low in Na. While the membrane of the cell permits potassium ions (K^+) to flow back and forth with relative freedom, it resists the movement of the sodium ions (Na^+) from the extracellular fluid to the inside of the cell. Since the K^+ can cross the membrane, they tend to flow along the concentration gradient—which is highest inside the cell. K^+ diffusion out of the cell

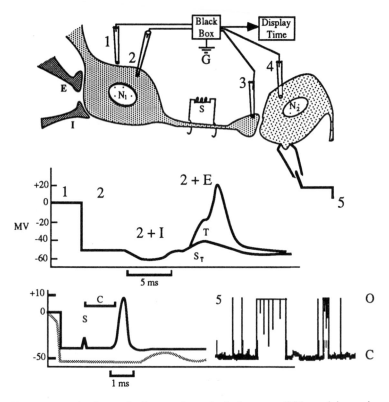

FIGURE 2-2. At the top is shown a hypothetical neuron (N_1) receiving a single excitatory pathway (E) and a single inhibitory pathway (I). A stimulating electrode (S) has been placed on the nerve cell's axon; microelectrode 1 is extracellular to nerve cell 1, while microelectrode 2 is in the cell body, and microelectrode 3 is in its nerve terminal. Microelectrode 4 is recording from within postsynaptic cell 2. The potentials and current, recorded by each of these electrodes, are being compared through a "black box" of electronics with a distant extracellular grounded electrode and displayed on an oscilloscope screen. When the cell is resting and the electrode is on the outside of the cell, no potential difference is observed (1). In the resting state, electrode 2 records a steady potential difference between inside and outside of approximately -50 mV (2). While recording from electrode 2 and stimulating the inhibitory pathway, the membrane potential is hyperpolarized during the inhibitory postsynaptic potential (2 + I). When recording from electrode 2 and stimulating the excitatory pathway, a subthreshold stimulus (S_T) produces an excitatory postsynaptic potential indicated by a brief depo-

18 |

leaves a relative negative charge behind owing to the negative charges of the macromolecular proteins. As the negative charge inside the cell begins to build up, the further diffusion of K^+ from inside to outside is retarded. Eventually an equilibrium point will be reached that is proportional to certain physical constants and to the relative concentrations of intracellular and extracellular K^+ and chloride (Cl^-) ions. These concepts of ionic diffusion potentials across semipermeable membranes apply generally, not only to nerve and muscle, but to blood, glandular, and other cells large enough to have their transmembrane potential measured.

Membrane Ion Pumps

When the nerve cell or muscle fiber can be impaled by electrodes to record transmembrane potential, the relation between the membrane potential and external K^+ concentration can be directly tested by exchanging the extracellular fluid for artificial solutions of vary-

---

larization of the resting membrane potential (2 + E). When the excitatory effects are sufficient to reach threshold (T), an action potential is generated which reverses the inside negativity to inside positivity (2 + E). On the lower scale, potentials recorded by electrodes 3 and 4 are compared on the same time-base following axonal stimulation of nerve cell 1, which is assumed to be excitatory. The point of stimulus is seen as an electrical artifact at point S. The action potential generated at the nerve terminal occurs after a finite lag period due to the conduction time (c) of the axon between the stimulating electrode and the nerve terminal. The action potential in the nerve ending does not directly influence postsynaptic cell 2 until after the transmitter has been liberated and can react with nerve cell 2's membrane, causing the excitatory postsynaptic potential indicated by the dotted line. The time between the beginning of the action potential recorded by microelectrode 3 and the excitatory postsynaptic potential recorded by electrode 4 (A) is the time required for excitation secretion coupling in the nerve terminal and the liberation of sufficient transmitter to produce effects on nerve cell 2. Electrode 5 is a patch clamp electrode attached to neuron 2; it indicates the effects of transmitters acting through receptors located elsewhere on the neuron's surface, mediating channel opening (o) and closing (c) events through intracellular second messengers.

ing K concentration. When this experiment is performed on muscle cells, we find that the membrane potential bears a linear relationship to the external K concentration at normal to high K concentrations but that it deviates from this linear relationship when the external K concentration is less than normal. To account for this discrepancy we must now reexamine an earlier statement. While the plasma membranes of nerve and muscle cells and other types of polarized cells are relatively impermeable to the flow of Na ions along the high concentration gradient from extracellular to intracellular, they are not completely impermeable. Radioisotope experiments can establish that a certain amount of Na "leaks" into the resting cell from outside. The amount of measurable Na entry into the cell occurs at a rate sufficient to double the intracellular Na concentration in approximately 1 hour if there were not some opposing process to maintain the relatively low intracellular Na concentration. The process that continuously maintains the low intracellular Na concentration is known as active Na transport or, colloquially, as the "sodium pump." This pump mechanism ejects Na from the inside of the cell against the high concentration and electrical gradients forcing it in. However, the "pump" does not handle Na exclusively but requires the presence of extracellular K^+. Thus, when a Na ion is ejected from the cell, a K ion is incorporated into the cell.

When the external K^+ concentration is near normal, the transmembrane potential, which is based mainly on K concentration differences, behaves as if there were actually more extracellular K than really exists. This is because the Na-K exchange mechanism elevates the amount of K coming into the cell. Remember that K permeability is relatively high and that K^+ will tend to diffuse out of the cell because of its concentration gradient but to diffuse into the cell because of charge attraction. Therefore, two factors operate to drive K^+ into the cell in the presence of relatively low external K^+ concentration: (1) the electrical gradient across the membrane and (2) the Na-K pump mechanism. The latter system could be considered "electrogenic" since at low external K^+ concentrations it modifies the electrical status of the muscle membrane. Other metabolic pumps operate simply to exchange cationic species across the membrane and are "nonelectrogenic." The relative "electrogenicity" of a pump may depend on the ratio of the exchange cations (i.e., 1:1 or 2:2 or 3:2). The pump is immediately dependent upon metabolic energy and can be blocked by several metabolic poisons such as dinitrophenol and the rapid-acting cardiac glycoside, ouabain. Astrocytes have such pumps too (see below).

The Uniqueness of Nerve

All that we have said regarding the transmembrane ionic distributions applies equally as well to the red blood cell or glia as to the nerve membrane. Thus, the possession of a transmembrane potential difference is not sufficient to account for the bioelectric properties of the nerve cell. The essential difference between the red blood cell and the nerve cell can be brought out by applying depolarizing currents across the membrane. When the red blood cell membrane is depolarized, the difference in potential across the cell passively follows the imposed polarization. However, when a nerve cell membrane, such as the giant axon of an invertebrate, is depolarized from a resting value of approximately −70 millivolts to approximately −10 to −15 millivolts, an explosive self-limiting process occurs by which the transmembrane potential is reduced not merely to zero but overshoots zero, so that the inside of the membrane now becomes positive with respect to the outside. This overshoot may extend for 10 to 30 millivolts in the positive direction. Because of this explosive response to an electrical depolarization, the nerve membrane is said to be electrically excitable, and the resultant voltage polarity shift is the action potential. While astrocytes can also show variable membrane potential, they do not exhibit this excitability.

Analysis of Action Potentials

In an elegant series of pioneering experiments that are now classic, Hodgkin, Huxley, and Katz were able to analyze the various ionic steps responsible for the action potential. When the cell begins to depolarize in response to stimulation, the current flow across the membrane is carried by K. As the membrane becomes more depolarized, the resistance to Na decreases (i.e., Na conductance increases), and more Na enters the cell along its electrical and concentration gradients. As Na enters, the membrane becomes more and more depolarized, which further increases the conductance to Na and thus depolarizes the membrane more and faster. Such conductance changes are termed voltage dependent. This self-perpetuating process continues, driven by the flow of Na ions moving toward their equilibrium distribution, which should be proportional to the original extracellular and intracellular concentrations of Na.

However, the peak of the action potential does not attain the equilibrium potential predicted on the basis of transmembrane Na concentrations because of a second phase of events. The voltage-dependent increase in Na^+ conductance and the consequent depolarization also activates a voltage-dependent K^+ conductance, and K flow then also increases along its concentration gradient from inside to outside the cell. This process restricts the height of the reversal potential, since it tends to maintain the inside negativity of the cell and also begins to reduce the membrane conductance to Na, thus making the action potential a self-limiting phenomenon. In most nerve axons, the action potential lasts for approximately 0.2 to 0.5 milliseconds (ms), depending on the type of fiber and the temperature in which it is measured.

Once the axon has been sufficiently depolarized to reach threshold for an action potential, the wave of activity travels at a rate proportional to the diameter of the axons (through which the bioelectric currents will flow). In large axons, the rate is further accelerated by the insulation provided by myelin sheaths, restricting the flow of transmembrane currents to the opening at the nodes of Ranvier. Therefore, instead of the action potential propagating from minutely contiguous sites of the membrane, the action potential in the myelinated axon leaps from node to node. This saltatory conduction is consequently much more rapid.

The threshold level for an all-or-none action potential is also inversely proportional to the diameter of the axon: Large myelinated axons respond to low values of imposed stimulating current, whereas fine and unmyelinated axons require much greater depolarizing currents. Local anesthetics appear to act by blocking activation of the Na conductance, preventing depolarization.

Once the threshold has been reached, a complete action potential will be developed unless it occurs too soon after a preceding action potential, during the so-called refractory phase. This phase varies for different types of excitable nerve and muscle cells and appears to be related to the activation process increasing Na conductance, a phenomenon that has a finite cycling period; that is, the membrane cannot be reactivated before a finite interval of time has occurred. K conductance increases with the action potential and lasts slightly longer than the activation of Na conductance. This results in a prolonged phase of after-hyperpolarization due to the continued redistribution of potassium from inside to outside the membrane. If the axonal membrane is artificially maintained at a transmembrane potential equal to the K^+ equilibrium potential, no after-hyperpolarization can be seen.

Ion Channels

The experiments of Hodgkin and Huxley defined the kinetics of cation movement during nerve membrane excitation without constraint on the mechanisms accounting for the movement of ions through the membrane. The discovery that cation movement can be selectively blocked by drugs and that Na^+ permeability (blocked by tetrodotoxin) can be separated from K^+ permeability (blocked by tetraethylammonium [TEA]), made more detailed analysis of the ion movement mechanisms feasible. Membrane physiologists now agree that there are several ion-specific pathways that form separate and independent "channels" for passive movement of Na^+, K^+, Ca^{2+}, and Cl^-. According to Hille, the channels are pores that open and close in an all-or-nothing fashion on time scales of 0.1 to 10 ms to provide aqueous channels through the plasma membrane that ions can traverse. In his view, the channels can be conceived as protein macromolecules within the fluid–lipid plane of the membrane; these channel macromolecules can exist in several interconvertible conformations, one of which permits ion movement while the others do not. The conformational shifts from one form to another are sensitive to the bioelectric fields operating on the membrane; by facilitating or retarding the conformational shifts, the ion channels are "gated." In this concept, Ca^{2+} acts at the membrane surface to alter permeability only by virtue of the effect its charge has on the electric fields of otherwise fixed (mainly negative) organic charges. The altered fields in turn can gate the channels because a part of the channel protein is able to sense the field and thus to modulate the conformational shifts that open or close the gate. When ions flow across the membrane, the ionic current changes membrane potential and other membrane properties.

From a variety of experimental methods, including methods that can sample single ion channels in cultured neurons and other excitable cells, a large number of specific ion channels have been described. In current terminology, one recognizes (1) "non-gated" passive ion channels, previously referred to as "leakage channels," that are continuously open; (2) voltage-gated (i.e., voltage sensitive) channels whose opening and closing are affected by the membrane potential inside of the cell; (3) chemically gated channels whose openings and closing are affected by receptors on the external plasma membrane, such as those affected by drugs and other transmitters; and (4) ion-gated channels, whose openings and closings are affected by shifts in intracellular ion concentrations. Ion-gated channels are often also sensitive to membrane potential and to external regulatory receptors, and chemically gated channels are often also voltage sensitive. These various modes of interaction provide an extremely rich spectrum of responses, thus greatly complicating what were once simple rules of excitability and ion conductance regulation. Additional forms of more complex ion flow regulation provide the means by which neurons communicate to their target cells through junctional transmission.

Junctional Transmission

While these ionic mechanisms appear to account adequately for the phenomena occurring in the propagation of an action potential down an axon, they do not per se explain what happens when the action potential reaches the nerve ending. At the nerve ending the membrane of the axon is separated from the membrane of the postjunctional nerve cell, muscle, or gland by an intercellular space of 50 to 200 Å (Fig. 2-1). When an electrode can be placed in both the terminal axon and the postsynaptic cell, depolarization of the nerve terminal does not result in a direct instantaneous shift in the transmembrane potential of the postsynaptic element, except in those cases in which the connected cells are electronically coupled. With this exception, the junctional site seldom exhibits direct electrical excitability like the axon.

Postsynaptic Potentials

With the advent of microelectrode techniques for recording the transmembrane potential of nerve cells in vivo, it was possible to determine the effects of stimulation of nerve pathways that had previously been shown to cause either excitation or inhibition of synaptic transmission. From just such studies Eccles and his colleagues observed that subthreshold excitatory stimuli would produce postsynaptic potentials with time durations of 2 to 20 ms. The excitatory postsynaptic potentials could algebraically accumulate with both the excitatory and inhibitory postsynaptic potentials. Most importantly, the duration of these postsynaptic potentials was longer than could be accounted for on the basis of electrical activity in the preterminal axon or on the electronic conductive properties of the postsynaptic membrane (Fig. 2-2). This latter observation, combined with the fact that synaptic sites are not directly electrically excitable, provides the conclusive evidence that central synaptic transmission must be chemical: The prolonged time course is compatible with a rapidly released chemical transmitter whose time course of action is terminated by local enzymes, diffusion, and reuptake by the nerve ending.

By such experiments it was possible to work out the basic ionic mechanisms for inhibitory and excitatory postsynaptic potentials. When an ideal excitatory pathway is stimulated, the presynaptic element liberates an excitatory transmitter that activates an ionic conductance of the postsynaptic membrane. This response leads to an increase in one or more transmembrane ionic conductances, depolarizing the membrane toward the Na equilibrium potential; in the resting state, as has already been discussed, the membrane resides near the K equilibrium potential. If the depolarization reaches the threshold for activating adjacent voltage-dependent conductances, an all-or-none action potential (spike) will be triggered. For many neurons the axon hillock has the lowest spike threshold. If the resultant depolarization is insufficient to reach threshold, the cell can still discharge if additional excitatory postsynaptic potentials summate to threshold.

The postsynaptic potential resulting from the stimulation of an ideal inhibitory pathway to the postsynaptic cell has been explained in terms of the fact that inhibitory transmitter selectively activates channels for Cl^- or K^+ resulting in a diffusion of ions and a hyperpolarization of the membrane. This counterbalances the excitatory postsynaptic potentials.

Because the sites of synaptic or junctional transmission are electrically inexcitable, the postsynaptic membrane potential can be maintained at various levels by applying current through intracellular electrodes and changing the intracellular concentrations of various ions. By such maneuvers, it is possible to poise the membrane at or near the so-called equilibrium potentials for each of the ionic species and to determine the ionic species whose equilibrium potential corresponds to the conductance change caused by the synaptic transmitter. This is the most molecular test for the identification of actions of a synaptic transmitter substance. (However, certain objections can be raised to this test in terms of those nerve endings making junctional contacts on distal portions of the dendritic tree. Here, the postsynaptic potentials may be incompletely transmitted to the cell body where the recording electrode is placed.)

On re-reading the above section, note the insidious use of the term "ideal." It is generally considered that classically acting neurotransmitters produce their effects on receptor-coupled ion conductances that are voltage independent, that is, the receptor will alter the coupled ion channel regardless of the membrane potential at the moment. Nevertheless, many nonclassic transmitters do seem to operate on receptors coupled to voltage-sensitive mechanisms. Transmitters whose receptors are associated with intracellular second messenger systems (such as activation of cyclic nucleotide synthesis) frequently produce these more complex forms of interaction (see Chapter 5). Similarly, many neuropeptides appear to affect certain of their target cells by modifying responses to other transmitters while not showing any direct shifts in membrane potential or conductance when tested for actions on their own. For example, the β-adrenergic actions of locus ceruleus neurons on their central targets produce excitability changes that depend upon which other afferent systems are activated synchronously. These β-adrenergic actions can enhance either excitatory afferents or inhibitory afferents, a general effect referred to as "enabling" or, more ambiguously, "modulatory." Some pharmacological actions of neuropeptides have been described as having the opposite effect, or "disenabling" (such as the effects of opioid peptides on the excitatory actions of sensory transmitters within the spinal cord (see Chapter 12). This story is more complex (surprised?) because neuropeptides coexist with amino acid and amine transmitters, and therefore their effects are more properly assessed as conceptually coordinating the responses of target cells to the array of transmitters that the presynaptic neuron can release, thereby "enabling," "disenabling," or affecting transmission in some other way not yet defined (see Iversen, Bloom).

However, two other aspects of ionic mechanisms bear some passing mention. First, despite our preoccupation with action potentials and their modification, Bullock has pointed out that the most numerous central nervous system neurons, the small single-process type of granule-like cell, may conduct its neuronal business within its restricted small spatial domain with no need ever to fire a spike. Second, those neurons that do fire spikes may sometimes do so unconventionally by using an influx of Ca^{2+} ions (a voltage-sensitive Ca conductance) rather than Na^+. This Ca spike may represent a mechanism to transmit activity from the cell body out to the dendritic system and may play a functional role in those neurons whose dendrites can also release transmitter—such as the catecholamine cell body nuclei. Thus, the simplified ideal version of ionic mechanisms may be only one of many regulatory mechanisms between connected cells. In addition, astrocytes in long-term cell cultures can exhibit ionic responses that look very much like voltage-dependent K and Ca channels of neurons. As the data accumulate, there are fewer and fewer distinctions between classic transmitters acting to alter voltage-dependent receptor mechanisms and "modulators" acting to modify the effects of the classic transmitters on their targets. The student is advised to maintain an appreciative awareness of these potentially complex interaction systems. In the following section, we examine some of the less-classic synaptic events and their advantageous properties.

During the 1980s, it became fashionable to accept transmitter response data only when they were recorded from stable neurons in vitro. In addition to slices of mammalian brains, other preferred systems included "model" neurons (i.e., neuronally derived cell lines, neuron–glia hybrid cells, and endocrine cells of normal or tumor origin previously exploited in lieu of real neurons). As a result, many experimental findings previously obtained with great difficulty from intact living brains were neglected in deference to the new "simpler" preparations. According to Llinas, in vitro electrophysiological assay systems (i.e., brain slices, cultured neurons, and patch-clamp methods of whole cell and single channel observation) provided the rigorous data needed to expand the principles of specialized mammalian receptor transduction mechanisms beyond the classic modes of excitation and inhibition and to define precisely the ion conductances through which different transmitters act. Still beyond the pale, given the frenzy to work in vitro, is the reexamination of these transmitter-related effects in the intact nervous system, where their role in specific circuitry operations and interactions could actually be evaluated under living conditions.

The transductive mechanisms now considered to be part of the normal repertoire of regulatory processes include a variety of transmitter-regulated (transmitter-dependent) and voltage-sensitive ionic conductances that were formerly regarded as "unusual" for the mammalian central nervous system despite their nearly simultaneous demonstration in invertebrate ganglia, mammalian autonomic ganglia, and selected central mammalian neurons. Most of these unusual transmitter-regulated conductance mechanisms relate to an unexpectedly large number of distinctive ionic conductances for Ca^{2+} and K^+, with more modest expansions in the channels for Cl^- or Na^+ currents.

Given the onslaught of molecular biological characterizations for ion channels (see Chapter 3), it is important for the student to recognize that within a specific functional category of ion conductance (i.e., Na^+, K^+, Ca^{2+}) there are subtypes of functional responses that are ligand specific and that may be carried out by more than one ion channel protein (ionophore) complex. These precisely defined channel proteins can now be examined in intimate molecular detail to dissect how drugs, toxins, and imposed voltages can alter the excitability of a neuron.

CALCIUM CHANNELS

Among the multiple voltage-sensitive Ca^{2+} conductances described in neurons, three are most consistent: The first is a transient low-threshold Ca^{2+} conductance (T). This Ca^{2+} conductance is inactive at resting membrane potentials but is "deinactivated" by modest hyperpolarizations, providing a feature of oscillatory behavior. It is most frequently inhibited by Cd^{2+} or Co^{2+}, and in some cases by Ni, Mg, or Mn as well, and can be activated by Ba^{2+}. The second Ca^{2+} conductance channel is a slowly inactivating, high-threshold Ca^{2+} conductance (L) seen mainly in nerve terminals. The third is a transient high-threshold Ca conductance (N) observed in the soma and dendrites of large neurons in neocortex, olfactory cortex, and hippocampal formation. The latter are blocked by Mn^{2+}, Co^{2+}, and Cd^{2+} and activated by Ba^{2+} and TEA, and these responses may be inhibited functionally by endogenous purinergic receptors. The N-type Ca^{2+} channels have also been well-studied in sympathetic neurons, where they are regulated through three separate transductive pathways, each of which may be engaged by different neurotransmitters and their specific intracellular mechanisms (see Chapter 5).

POTASSIUM CHANNELS

At least three types of K channels have been described in central neurons: (1) the "A" or "A-like" fast transient K conductances inhibited by 4-aminopyridine, Ba^{2+}, or Co^{2+}; (2) the so-called "anomalous rectifying" K conductances (see below) of which the M current (closed by cholinergic muscarinic receptors) is one example of the class; (3) the Ca-activated K conductances, blocked by Co, Mn, Cd, and some neurotransmitters (see Siggins and Gruol, 1986, and Hille, 1992). Although most data on these K^+ channel effects are pharmacological, the properties can clearly regulate cell firing and response patterns in distinct manners. By closing the M current, muscarinic receptors transduce cholinergic signals into a more effective depolarization, once a partial depolarization brings this channel into play. Somatostatin can oppose this effect, forcing the M channel to open. The latter effect may be mediated intracellularly by second messengers derived from arachidonic acid metabolism (see Chapter 5). By blocking the Ca-activated K channel of central neurons, the transmitter receptors for β-adrenergic agonists, 5-HT, histamine, and corticotropin-releasing hormone enhance the ability of responsive neurons to follow long depolarizing pulses, thereby generating longer trains of spikes per afferent impulse.

Depending on the specific cells in which they were recognized (even Ca-activated K channels exist in many glands and completely nonneural cell types) and the conditions and possible inhibitors that may have been evaluated, as many as 12 different K conductances have been proposed. For example, many K channels are linked to second messenger mediation (e.g., the channels activated by $GABA_B$ receptors and by D2 receptors on rat substantia nigra neurons). With patch-clamp analyses of single K channels in locus ceruleus neurons acutely isolated from 1- to 7-day-old rats, opioids (at μ receptors), somatostatin, and $α_2$-adrenergic agonists all seem to open a K channel that is not voltage dependent and seems to be directly regulated by receptor occupancy through a G protein (see Chapter 4) but with no known intervening second messenger.

OTHER ION-SPECIFIC CHANNELS

A "persistent" Na^+ conductance was first observed in cerebellar Purkinje neurons and later in hippocampal pyramidal neurons, as well as neurons throughout the neuraxis. This conductance provides a long-lasting but low amplitude depolarization that does not lead directly to neuronal firing but rather provides a bias from which the conventional fast Na channels can produce full spike initiation. Persistent Na channels are typically blocked by tetrodotoxin and activated by TEA.

Slow Postsynaptic Potentials

Most of the postsynaptic potentials described by Eccles and his colleagues were relatively short, usually 20 ms or less, and appeared to result from passive changes in ionic conductance. Postsynaptic potentials of slow onset and several seconds' duration have been described (Fig. 2-2), both of a hyperpolarizing nature and of a depolarizing nature. While such prolonged postsynaptic potentials could be the result of either a prolonged release of transmitter or a persistence of the transmitter at postsynaptic receptor sites, there is substantial support for the possibility that slow postsynaptic potentials could also be caused by other forms of synaptic communication. Many of these slower synaptic potentials are not accompanied by the expected increase in transmembrane ionic conductances, but are instead accompanied by increased transmembrane ionic impedance. The most simple explanations for membrane potential changes accompanied by increased membrane resistance are (1) that the effect is generated at a "presynaptic" site whereby tonically active excitatory or inhibitory synapses are inactivated by synapses on their terminals or (2) that the action of the transmitter released by a slow synapse inactivates a normal finite resting conductance, such as that of Na or K. Since most of the sites where slow synaptic potentials have been observed do not fulfill the anatomical prerequisite of axo-axonic presynaptic synapses, the second explanation may be more likely. A third explanation for such potentials has also been advanced, namely, that the transmitter activates an electrogenic pump mechanism, that is, a metabolically driven ionic "pump" that exchanges unequal numbers of similarly charged ions across the membrane. Studies on invertebrate neurons indicate that such electrogenic mechanisms should have several distinguishable properties: They should be virtually independent of the distribution of the ions to be pumped; they should not exhibit an equilibrium potential; they should be temperature dependent; and they should be sensitive to metabolic poisons and uncouplers of oxidative phosphorylation. Such data are extremely complex to analyze because several membrane properties also include temperature dependency and metabolic poison sensitivity. Thus, at the present time, a true electrogenic synapse in the mammalian central nervous system remains to be described. However, it also seems likely that transmitters such as the catecholamines can activate the synthesis of cyclic nucleotides, which in turn can activate intraneuronal protein kinases that can phosphorylate specific membrane proteins. The phosphorylation of a membrane protein could be expected to alter its ionic permeability, and perhaps such changes lie at the root of the membrane effects of several types of neurotransmitters.

Conditional Actions of Transmitters

Frequently, transmitters produce novel actions unlike those of classically conceived transmitters. These unconventional actions suggest that broader definitions are useful for conceptualizing the range of regulatory signals involved in interneuronal communication and are helpful for examination of transmitter actions. For example, when the β-adrenergic effects of locus ceruleus stimulation are examined, target cell responses no longer adhere to standard concepts of inhibition. Rather, they appear to fit better the designation of "biasing" or "enabling." The latter term indicates that the enabling transmitter (in this example, norepinephrine) can enhance or amplify the effectiveness of other transmitter actions converging on the common target neurons during the time period of the enabling circuit's activity (see Chapter 7).

   To reexplore the issue of time course on the more complex interactions, it may be useful to speak of "conditional" and "unconditional" actions. Unconditional actions are those that a given transmitter evokes by itself (i.e., in the absence of other transmitters acting on the common target cell). Conditional actions, occurring either pre- or postsynaptically, include but would not be limited to the type of enhancement that is subsumed by "enabling." In such a conditional interaction each transmitter would act at its own pre- or postsynaptic transmitter receptor and would interact on that target cell when both transmitters occupy their receptors simultaneously.

Thus, it can be seen that there are abundant circuits, abundant transmitters, and, for each of these, many classes of chemically coupled systems that can transduce the effects of active transmitter receptors. These receptors can operate either actively or passively, conditionally or unconditionally, over a wide range of time through nonspecific, dependent, or independent metabolic events. Clearly, neurons have a broad but finite and, as yet, incompletely characterized repertoire of molecular responses that messenger molecules (transmitters, hormones, and drugs) can elicit. The power of the chemical vocabulary of such components is their combinatorial capacity to act conditionally and coordinatively and to integrate the temporal and spatial domains within the nervous system.

Transmitter Secretion

We have already seen that the cellular machinery of the neuron suggests it functions as a secretory cell. The secretion of synaptic transmitters is the activity-locked expression of neuronal activity induced by the depolarization of the nerve terminal. Recently, it has been possible to separate the excitation–secretion coupling process of the presynaptic terminal into at least two distinct phases. This has been made possible through an analysis of the action of the puffer fish poison tetrodotoxin, which blocks the electrical excitation of the axon but does not block the release of transmitter substance from the depolarized nerve terminal. The best of these experiments have been performed in the giant synaptic junctions of the squid stellate ganglion, in which the nerve terminals are large enough to be impaled by recording and stimulating microelectrodes and with recording from the postsynaptic and presynaptic neurons. In this case, when tetrodotoxin blocks conduction of action potentials down the axon, electrical depolarization of the presynaptic terminal still results in the appearance of an excitatory postsynaptic potential in the ganglion neuron. Since tetrodotoxin selectively blocks the voltage-dependent Na^+ conductance, the excitation secretion must be coupled more closely to other ions. Present evidence strongly favors the view that a voltage-sensitive Ca^{2+} conductance is required for transmitter secretion. In review, we see that the spike-generating and conducting events rest on voltage-dependent ion conductance changes, while synaptic events rest on voltage-independent or voltage-sensitive conductances.

Biochemical, ultrastructural, and physiological experiments have led to the concept that transmitter molecules are stored within vesicles in the nerve terminal and that the Ca-dependent excitation–secretion coupling within the depolarized nerve terminal requires the transient exchange of vesicular contents into the synaptic cleft. It is unclear whether the vesicle simply undergoes a rapid fusion with the presynaptic specialized membrane to allow the transmitter stored in the vesicle to diffuse out or whether the process of exocytotic release simultaneously requires the insertion of the vesicle membrane into the synaptic plasma membrane, reappearing later by the reverse process, namely, endocytosis. Information on the lipid and protein components of the two types of membranes once suggested that long-term fusion–endocytosis cycles were unlikely, but more recent data are compatible with either fusion release or contact release. In noradrenergic vesicles, for example, the transmitter is stored in very high concentrations in ternary complexes involving ATP, Ca, and possibly additional lipids or lipoproteins. Unfortunately, neurochemically homogeneous vesicles from central synapses have never been completely purified, and therefore all such analyses remain somewhat open to interpretation. For other molecules under active consideration as neurotransmitters, storage within brain synaptic vesicles has been extremely difficult to document chemically. The difficulties arise from the fact that homogenization of the brain to prepare synaptosomes disrupts both structural and functional integrity, and under these conditions the failure to demonstrate that amino-acid transmitters are stored in vesicles is rationalized as uncontrollable leakage. With electron microscopy and autoradiography, however, sites accumulating transmitters for which there is a high-affinity, energy-dependent uptake process can be demonstrated. Under these conditions, authentic "synaptic terminals" are identified, but glial processes are also labeled. The vesicle story is further discussed in Chapter 8.

In some cases, release of the transmitter can be modulated "presynaptically" by the neuron's own transmitter (autoreceptors) or by the effects of transmitters released by other neurons in the vicinity of the terminal or the cell body. Autoreceptors are conceived to be receptors that are generally distributed over the surface of a neuron and are sensitive to the transmitter secreted by that neuron. In the case of the central dopamine-secreting neurons, such receptors have been related to the release of the transmitter and to its synthesis. Such effects seem to be achieved through receptor mechanisms different from those by which the same transmitter molecule acts postsynaptically.

ANALYSIS OF MEMBRANE ACTIONS OF DRUGS AND TRANSMITTERS IN VITRO

The development of methods for the nearly complete functional maintenance of central neurons in vitro for several hours, such as the tissue slice preparations, and for several days to weeks in single-cell or in explant culture systems has led to a proliferation of additional electrophysiological methods to examine transmitter and drug action (see Aston-Jones and Siggins, 1994). In slice preparations, neuronal targets can be readily localized by inspection, and intracellular electrodes can be inserted into suitably large neurons under visual control, while sources of afferent circuitry within the slice may be activated by additional stimulating electrodes. Transmitters and drugs can then be applied to the whole slice by superfusion within oxygenated buffered salt solutions or more locally by micropressure pulse or iontophoretic application methods. In many cases, excellent intracellular recordings can be obtained for long periods of time because there are no annoying respiratory, cardiac, or other movements to dislodge the electrode.

When long-term intracellular recordings can be obtained, two additional sources of information on transmitter actions can be analyzed. In *voltage-clamp* analysis, the experimenter inserts one or two electrodes into the cell and by injecting current holds the membrane potential of the neuron at a constant value. The cell is usually poised at a membrane potential more negative than resting in order to prevent spontaneous spikes. Transmitter action is monitored by the amount of current required to keep the membrane potential constant and thus measures transmembrane current flow directly. However, since the membrane potential stays constant, any of the non-linear properties of that neuron's response that could occur when sufficient depolarization has occurred will be prevented.

The clamp can also be quickly changed to a new level of membrane potential, and the neuron's responses to this shift provide the basis for a pharmacological dissection (e.g., with ion channel blockers or ion substitutions) of the degree to which the effects of a transmitter can be explained as ion dependent or voltage dependent. Many of the actions ascribed to neuropeptides and some of those ascribed to monoamines fit the concept of voltage dependent, since they are modest effects at best at resting membrane potential levels, but they emerge as more substantive effects when the responding neuron is depolarized or hyperpolarized by other convergent transmitters.

Another method, termed *noise analysis*, also examines ion channel activity more directly than the standard in vivo methods. This method assumes that ion channels are either open or closed and that they switch instantaneously, and do so independently, between the two conditions. Using intracellular electrodes, the fluctuations in membrane potential (or, if voltage clamping is used, the fluctuations in membrane current flow) are analyzed statistically to infer the conductance of individual types of channels and the mean time they are open in the absence or presence of the transmitter to be analyzed.

When the neurons considered to be the appropriate targets of a specific transmitter can be maintained in long-term tissue culture, the method of *patch-clamp analysis* can be applied, as the current superstar of membrane action analysis methods. This method offers the ability to study the behavior of single-ion channels and to do so under conditions of almost unbelievable precision. Special "fire-polished" microelectrodes are placed on the neuron's surface and a slight vacuum applied to the pipette to attain a very tight junction with the exposed surface of the neuronal membrane, thus requiring near-nude neurons for best application. The resulting cell–electrode junction will have such a high electrical resistance (gigohms!) that the patch of enclosed membrane within the microelectrode's tip will be essentially isolated from the rest of the cell. Current flowing within that patch can then be analyzed independently of the responses of the rest of the neuron. With state-of-the-art, low-noise amplifiers, current flow through individual channels can be monitored and transmitter actions evaluated in terms of open time, amplitudes (number of channels opened), and closing times.

In addition, with clever micromanipulations, the patch clamps can be done in three configurations. In the "cell-attached" mode, the pipette is sealed to the intact cell and the measurements made with no further physical disruption. However, further application of slight vacuum allows the patch of enclosed membrane to be removed from the cell, but with enclosed ion channels still viable and responsive. In the "inside-out" patch, the previously intracellular surface will be on the exterior of the sealed membrane patch, and the ion channel can be examined for regulation by simulations of changes in intracellular ions or, for example, catalysts of protein phosphorylations. It is also possible to demonstrate that, with clever handling and with further negative pressure before pulling the membrane patch off the cell, an "outside-out" patch can be obtained. Here the original patch is ruptured, the perimeter remains attached, and then the surrounding external membrane segments reseal once they are excised from the cell surface. By placing the outside surface into solutions with differing doses of transmitter, drug, or ion channel toxin, it is possible to analyze very discrete, single-channel pharmacology.

AN APPROACH TO NEUROPHARMACOLOGICAL ANALYSIS

One can now see that the business of analyzing bioelectrical potentials can be very complicated, even when restricted to changes in single neurons or to small portions of contiguous neurons. But if we restrict our examination of centrally active drugs to analyses of effects on single cells, we can ask rather precise questions. For example, does drug X act on resting membrane potential or resistance, on an electrogenic pump, or on the Na- or K-activation phase of the action potential? Or does it act by blocking or modulating the effects of junctional transmission between two specific groups of cells?

Unfortunately we have not had these precise electrophysiological tools for very long. Earlier neuropharmacologists were thus required to examine effects of drugs on populations of nerve cells. This was usually done in one of two ways. Large macroelectrodes were employed to measure the potential difference between one brain region and another. These electroencephalograms reflect mainly the moment-to-moment algebraic and spatial summations of slow synaptic potentials, and almost none of their electrical activity is due to actual action potentials generated by individual neurons (unit discharges). A second type of analysis was based on evoked changes occurring when a gross sensory stimulation (such as a flash of light or a quick sound) was delivered. Changes in recordings from cortical or subcortical sites along the sensory pathway were then sought during the action of a drug. While we can criticize the technical and interpretative shortcomings of such methods of central drug analysis, these methods were able to reflect the population response of a group of neurons to a drug, something that single unit analysis can do only after many single recordings are collated. The macroelectrode methods are receiving increased attention again, since, as noninvasive methods, they can be used to examine drug actions clinically.

Approaches

If, as modern-day neuropharmacologists, we are chiefly concerned with uncovering the mechanisms of action of drugs in the brain, there are several avenues along which we can organize our attack: We could choose to examine the way in which drugs influence the perception of sensory signals by higher integrative centers of the brain. This is compatible with a single-neuron and ionic conductance type of analysis, directed, say, at how drugs affect inhibitory postsynaptic potentials. Drugs that cause convulsions, such as strychnine, have been analyzed in this respect, but all types of inhibitory postsynaptic potentials are not affected by strychnine.

A second basic approach would be to use both macroelectrodes and microelectrodes to compare the drug responses of single units and populations of units in the same brain region. This approach is clearly limited, however, unless we understand the intimate functional relations between the multiple types of cells found even within one region of the brain.

A third approach is also possible. We could choose to separate the effects of drugs between those affecting generation of the action potential and its propagation and those acting on junctional transmission. For this type of analysis, we must identify the chemical synaptic transmitter for the junctions to be studied. Many of the interpretative problems already alluded to can be attacked through this approach. Thus, as you might expect, there is likely to be more than one type of excitatory and inhibitory transmitter substance, and a convulsant drug may affect the response to one type of inhibitory transmitter without affecting another. Moreover, a drug may have specific regional effects in the brain if it affects a unique synaptic transmitter there. In fact, by using this approach it may be possible to find drug effects not directly reflected in electrical activity at all but related more to the catabolic or anabolic systems maintaining the required functional levels of transmitter. We shall conclude this chapter by considering the techniques for identifying the synaptic transmitter for particular synaptic connections. The chapters that follow are organized to present in detail our current understanding of putative central neurotransmitter substances.

IDENTIFICATION OF SYNAPTIC TRANSMITTERS

How then, do we identify the substance released by nerve endings? The entire concept of chemical junctional transmission arose from the classic experiments of Otto Loewi, who demonstrated chemical transmission by transferring the ventricular fluid of a stimulated frog heart onto a nonstimulated frog heart, thereby showing that the effects of the nerve stimulus on the first heart were reproduced by the chemical activity of the solution flowing onto the second heart. Since the phenomenon of chemical transmission originated from studies of peripheral autonomic organs, these peripheral junctions have become convenient model systems for central neuropharmacological analysis.

Certain interdependent criteria have been developed to identify junctional transmitters. By common-sense analysis, one would suspect that the most important criterion would be that a substance suspected of being a junctional transmitter must be demonstrated to be released from the prejunctional nerve endings when the nerve fibers are selectively stimulated. This criterion was relatively easily satisfied for isolated autonomic organs in which only one or at most two nerve trunks enter the tissue and the whole system can be isolated in an organ bath. In the central nervous system, however, satisfaction of this criterion presumes (1) that the proper nerve trunk or set of nerve axons can be selectively stimulated and (2) that release of the transmitter can be detected in the amounts released by single nerve endings after one action potential. This last subcriterion is necessary since we wish to restrict our analysis to the first set of activated nerve endings and not examine the substances released by the secondary and tertiary interneurons in the chain, some of which might reside quite close to the primary endings. The biggest problem with this criterion in the brain, however, is the lack of a method for detecting release that does not in itself destroy the functional and structural integrity of the region of the brain being analyzed. Such techniques as internally perfused cannulae or surface suction cups are chemically at the same level of resolution as the evoked potential and the cortical electroencephalogram. Each of these methods records the resultant activity of thousands if not millions of nerve endings and synaptic potentials. Release has also been studied in brain slices or homogenate subfractions incubated in "physiological buffer solutions" in vitro. These techniques can demonstrate the effects of depolarizing drugs or electric fields to simulate events in the living brain.

Localization

Because it is difficult, if not impossible, to identify the substance released from single nerve endings by selective stimulation, the next-best evidence might be to prove that a suspected synaptic transmitter resides in the presynaptic terminal of our selected nerve pathway. Normally, we would expect the enzymes for synthesizing and catabolizing this substance also to be in the vicinity of this nerve ending, if not actually part of the nerve-ending cellular machinery. In the case of neurons secreting peptides or simple amino-acid transmitters, however, these metabolic requirements may need further consideration. To document the presence of neurotransmitter, several types of specific cytochemical methods for both light microscopy and electron microscopy have been developed. More commonly employed is the biochemical population approach, which analyzes the regional concentrations of suspected synaptic transmitter substances. However, presence per se indicates neither releasability nor neuroeffectiveness (e.g., acetylcholine in the nerve-free placenta or serotonin in the enterochromaffin cell). Although it has generally been conceived that a neuron makes only one transmitter and secretes that same substance everywhere synaptic release occurs, neuropeptide exceptions to this rule have become common.

Synaptic Mimicry: Drug Injections

A third criterion arising from the peripheral autonomic nervous system analysis is that the suspected exogenous substance mimics the action of the transmitter released by nerve stimulation. In most pharmacological studies of the nervous system, drugs are administered intravascularly or onto one of the external or internal surfaces of the brain. The substances could also be directly injected into a given region of the brain, although the resultant structural damage would have to be controlled and the target verified histologically.

The analysis of the effects of drugs given by each of these various gross routes of administration is quite complex.

We know that diffusional barriers selectively retard the entrance from the bloodstream of many types of molecules into the brain. These barriers have been demonstrated for most of the suspected central synaptic agents. In addition, we suspect that extracellular catabolic enzymes could destroy the transmitter as it diffuses to the postulated site of action. A further complicating aspect of these gross methods of administration is that the interval of time from the administration of the agent to the recording of the response is usually quite long (several seconds to several minutes) in comparison with the millisecond intervals required for junctional transmission. The delay in response further reduces the likelihood of detecting the primary site of action on one of a chain of neurons.

Microelectrophoresis

The student will now realize how important it is to have methods of drug administration equal in sophistication to those with which the electrical phenomena are detected. The most practical micromethod of drug administration yet devised is based upon the principle of electrophoresis. Micropipettes are constructed in which one or several barrels contain an ionized solution of the chemical substance under investigation. The substance is applied by appropriately directing the current flow. The microelectrophoretic technique, when applied with controls to rule out the effects of pH, electrical current, and diffusion of the drug to neighboring neurons, has been able to overcome the major limitations of classic neuropharmacological techniques. Frequently, a multiple-barreled electrode is constructed from which one records the spontaneous extracellular discharges of single neurons while other attached pipettes are utilized to apply drugs. One can also construct an intracellular microelectrode glued to an extracellular drug-containing multielectrode so that the transmembrane effects of these suspected transmitter agents can be compared with the effects of nerve pathway stimulation. The intracellular electrode can also be used to poise the relative polarization of the membrane and let us detect whether the applied suspected transmitter and that released by nerve stimulation cause the membrane to approach identical ionic equilibrium potentials.

Considerable experimentation with this technique has made possible certain generalizations regarding the actions of each putative neurotransmitter that has been studied. These substances are reviewed in detail in each of the chapters that follows. However, it should be borne in mind that certain substances have more or less invariable actions; for example, γ-aminobutyrate and glycine always act to inhibit, while glutamate and aspartate always excite. Insofar as we know, these actions arise from increased membrane conductances to Na, K, Ca, or Cl in every case. Other substances have many kinds of effects, depending on the nature of the cell whose receptors are being tested. Thus, acetylcholine frequently excites but can also inhibit, and the receptors for either response can be nicotinic or muscarinic. Similarly, dopamine, norepinephrine, and serotonin almost always inhibit, but they have a few excitatory actions that are probably not completely artifactual.

Pharmacology of Synaptic Effects

The fourth criterion for identification of a synaptic transmitter requires identical pharmacological effects of drugs potentiating or blocking postsynaptic responses to both the neurally released and the administered samples. Because the pharmacological effects are often not identical (most "classic" blocking agents are extrapolated to brain from effects on peripheral autonomic organs), this fourth criterion is often satisfied indirectly with a series of circumstantial pieces of data. Recently, with the advent of drugs that block the synthesis of specific transmitter agents, the pharmacology for certain families of transmitters has been improved.

The electrophysiological analysis of drug and transmitter actions in the central nervous system was traditionally accomplished in terms of single cell activity in vivo. Four types of physiological responses served as the major indices to compare exogenously applied transmitter candidates and drugs with the effects of endogenous transmitters: (1) spontaneous activity, (2) orthodromic synaptically evoked activity, (3) antidromic activity, and (4) relative responses to independently acting excitatory or inhibitory transmitter released from another experimental source, such as another barrel of a multi-barrel pipette. These techniques have been most successful when applied to large neurons, whose selected afferent pathways can be stimulated, and the transmembrane effects specifically analyzed. However, when the unit recording techniques are applied to the mammalian brain, visualization and selection of the neuron under investigation are almost impossible. Although we can generally select a relatively specific brain region in which to insert the micro-electrode assembly, we must utilize electrophysiological criteria for cell identification. We must also depend on geometrical attributes of the nerve cell to encounter its electrical activity as the electrode is navigated by blind mechanical means through the brain. Our ability to encounter cells in the brain depends partly on the size and type of the electrode assembly we use, partly on the size and spontaneous activity of the cell we are approaching, and partly on the surgical or chemical means by which we have prepared the animal (viz., presence or absence of anesthesia, homeostatic levels of blood pressure, and tissue oxygen and carbon dioxide). Even with very fine micro-electrodes, the analysis does not take place in a completely undisturbed system, since many connections must be broken by the physical maneuvering of the electrode. The leakage of cytoplasmic constituents, including potential transmitter agents and metabolic enzymes, can only be considered as part of the background artifact of the system. In the past few years, additional electrophysiological methods pertinent to transmitter action have been developed that can be applied to neuronal systems analyzed in vitro to provide additional insight into ionic channel regulation.

THE STEPS OF SYNAPTIC TRANSMISSION

Let us now conclude this chapter by briefly examining the mechanisms of presumed synaptic transmission for the mammalian central nervous system. Each step in such transmission constitutes one of the potential sites of central drug action (Fig. 2-3). A stimulus activates an all-or-none action potential in a spiking axon by depolarizing its transmembrane potential above the threshold level. The action potential propagates unattenuated to the nerve terminal where ion fluxes activate a mobilization process leading to transmitter secretion and "transmission" to the postsynaptic cell. From companion biochemical experiments (to be described in the next chapters), the transmitter substance is believed to be stored within the microvesicles or synaptic vesicles seen in nerve endings by electron microscopy. In certain types of nerve junctions, miniature postsynaptic potentials can be seen in the absence of conducted presynaptic action potentials. These miniature potentials have a quantal effect on the postsynaptic membrane in that occasional potentials are statistical multiples of the smallest measurable potentials. The biophysical quanta have been related to the synaptic vesicles, although the proof for this relationship is still circumstantial.

When the transmitter is released from its storage site by the presynaptic action potential, the effects on the postsynaptic cells cause either excitatory or inhibitory postsynaptic potentials, depending on the nature of the postsynaptic cell's receptor for the particular transmitter agent. If sufficient excitatory postsynaptic potentials summate temporally from various inputs onto the cell, the postsynaptic cell will integrate these potentials and give off its own all-or-nothing action potential, which is then transmitted to each of its own axon terminals, and the process continues.

In trying to solve the problem of interneuronal chemical communication, it may be useful, nevertheless, to maintain an open mind with regard to three dimensions by which neuronal circuits can be characterized: (1) the spatial domain (those areas of the brain or peripheral receptive fields that feed onto a given cell and those areas into which that cell sends its efferent signals); (2) the temporal domain (the time spans over which the spatial signals are active); and (3) the functional domain (the mechanism by which the secreted transmitter substance operates on the receptive cell). When the receptive cell is closely coupled in time, space, and function to the se-

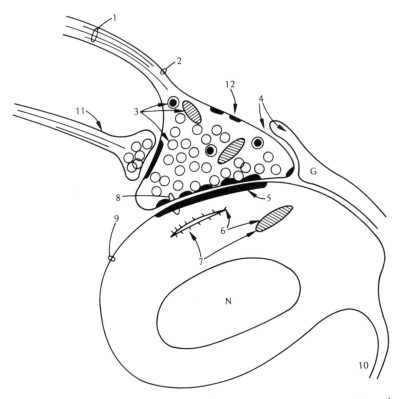

FIGURE 2-3. Twelve steps in the synaptic transmission process are indicated in this idealized synaptic connection. Step 1 is transport down the axon. Step 2 is the electrically excitable membrane of the axon. Step 3 involves the organelles and enzymes present in the nerve terminal for synthesizing, storing, and releasing the transmitter, as well as for the process of active reuptake. Step 4 includes the enzymes present in the extracellular space and within the glia for catabolizing excess transmitter released from nerve terminals. Step 5 is the postsynaptic receptor that triggers the response of the postsynaptic cell to the transmitter. Step 6 shows the organelles within the postsynaptic cells which respond to the receptor trigger. Step 7 is the interaction between genetic expression of the postsynaptic nerve cell and its influences on the cytoplasmic organelles that respond to transmitter action. Step 8 includes the possible "plastic" steps modifiable by events at the specialized synaptic contact zone. Step 9 includes the electrical portion of the nerve cell membrane that, in response to the various transmitters, is able to

creting cell, almost everyone would agree that "real" synaptic actions have occurred. When the effects are long lasting and widely distributed, however, many would prefer to call this action something else, even though the molecular agonist is stored in and released presynaptically from neurons onto the nerve cells they contact.

SELECTED REFERENCES

Adams, M. E. and G. Swanson, (1994). Neurotoxins supplement. *T.I.N.S. 17 (Suppl.), 1–31.*

Aston-Jones, G. and G. R. Siggins, (1994). Electrophysiology. In *Psychopharmacology: The Fourth Generation of Progress.* (F. E. Bloom and D. J. Kupfer, eds.), 41–64, Raven Press, New York.

Bloom, F. E. (1988). Neurotransmitters: Past, present and future directions. *FASEB J. 2, 32–41.*

Bloom, F. E. (1995). Neurohumoral transmission and the central nervous system. In *The Pharmacological Basis of Therapeutics,* 9th ed. (L. Limbird et al., eds.), (in press) Pergamon Press, New York.

Bourne, H. R., and R. Nicoll (1993). Molecular machines integrate coincident synaptic signals. *Cell, 72,* 65–75.

Bullock, T. H. (1980). Spikeless neurones: Where do we go from here? *Soc. Exp. Biol. Semin. Ser. 6,* 269–284.

Catterall, W. A. (1993). Structure and function of voltage-gated ion channels. *Trends Neurosci. 16,* 500–506.

Eccles, J. C. (1964). *The Physiology of Synapses.* Academic Press, New York.

Hille, B. (1992). *Ionic Channels of Excitable Membranes.* 2nd ed. Sinauer Associates, Sunderland, MA.

---

integrate the postsynaptic potentials and produce an action potential. Step 10 is the continuation of the information transmission by which the postsynaptic cell sends an action potential down its axon. Step 11, release of transmitter, is subjected to modification by a presynaptic (axo-axonic) synapse; in some cases an analogous control can be achieved between dendritic elements. Step 12, release of the transmitter from a nerve terminal or secreting dendritic site, may be further subject to modulation through autoreceptors that respond to the transmitter which the same secreting structure has released. Glia (G) can accumulate (4) released transmitters.

Hodgkin, A. L., and A. F. Huxley (1952). Currents carried by sodium and potassium ions through the membrane of the giant axon of *Loligo*. *J. Physiol. 116*, 449.

Iversen, L. L. (1984). The Ferrier Lecture, 1983. Amino acids and peptides: Fast and slow chemical signals in the nervous system? *Proc. R. Soc. Lond. Biol. 22*, 245–260.

Jan, L. Y. and Y. N. Jan, (1992). Structural elements involved in specific K^+ channel functions. *Annu. Rev. Physiol. 54*, 537–558.

Katz, B. (1966). *Nerve, Muscle and Synapse*. McGraw-Hill, New York.

Llinas, R. R. (1988). The intrinsic electropysiological properties of mammalian neurons: Insights into central nervous system function. *Science 242*, 1654–1660.

Loewi, O. (1921). Über humorale Übertragbarkeit der Herznervenwirkung. *Pfluegers Arch. 189*, 239.

Madison, D. V., R. C. Malenka, and R. A. Nicoll, (1991). Mechanisms underlying long-term potentiation of synaptic transmission. *Annu. Rev. Neurosci. 14*, 379–397.

Magistretti, P. J., L. Pellerin, and J.-L. Martin, (1994). Brain energy metabolism: An integrated cellular perspective. In *Psychopharmacology: The Fourth Generation of Progress*. (F. E. Bloom and D. J. Kupfer, eds.), 657–670. Raven Press, New York.

Miller, R. J. (1989). Multiple calcium channels and neuronal function. In *Molecules to Models: Advances in Neurosciences* (K. L. Kelner and D. E. Koshland, eds.), 1–17. American Association for the Advancement of Science, Washington, D.C.

Nicoll, R. A. (1988). The coupling of neurotransmitters to ion channels in the brain. *Science 241*, 545–553.

Schweitzer, P., S., Madamba, J. Champagnat, and G. R. Siggins, (1993). Somatostatin inhibition of hippocampal CA1 pyramidal neurons: Mediation by arachidonic acid and its metabolites. *J. Neurosci. 13(5)*, 2033–2049.

Siggins, G. R. and D. L. Gruol (1986). Synaptic mechanisms in the vertebrate central nervous system. In *Handbook of Physiology, Intrinsic Regulatory Systems of the Brain* vol IV. (F. E. Bloom, ed.). The American Physiological Society, Bethesda, MD.

3 | Molecular Foundations of Neuropharmacology

Complete understanding of the basis for a drug's actions on the brain requires knowledge of all the molecules involved. A drug acts selectively when it elicits responses from discrete populations of cells that possess "drug-recognizing" macromolecules. Such recognition sites are known more colloquially as receptors. Drug receptors with important regulatory actions on the nervous system are generally on the external surface of neurons if they involve sites where neurotransmitters act. Some drugs can act on the nervous system because they resemble, at the molecular level, endogenous intercellular signals like those of a neurotransmitter. However, drugs may also act by regulating intracellular enzymes critical for normal transmitter synthesis or breakdown. Receptors recognize drugs for a variety of reasons that this book explores in subsequent chapters. Once having made that recognition, the activated receptor usually interacts with other molecules to alter membrane properties or intracellular metabolism. These cellular changes in turn regulate the interactions between cells in circuits. These circuit changes regulate the performance of functional systems (like the sensory, motor, or vegetative control systems) and eventually the behavior of the whole organism.

Thus, understanding the actions of drugs on the function of the brain, whether it be on single cells or behavior, is a multilevel, multifaceted process that begins with and builds upon the concept of molecular interactions. Even beginning students of drug action on the nervous system will probably accept this statement as a reasonable hypothetical principle. In practice, however, this principle is severely compromised because most molecules in a very complex organ like the brain are unknown. Furthermore, until the recent explosion of knowledge in molecular biology, there was little hope that the necessary details would be obtained.

When Watson and Crick deduced the three-dimensional, double-helical structure of DNA in 1953, the implications for the coding and replication of genetic information were recognized but could not be experimentally tested. Almost 25 years of effort were required before the new biological technology was launched. During this interval it became clear how to combine genes and gene fragments from multicelled organisms with those of viruses, fungi, and bacteria to produce new genetic instructions and novel gene products. At last, the concept of a *gene* and its cellular product attained concrete form. Almost immediately, neuroscientists, who are always ready to exploit new technologies, began to apply these methods to the most complex cellular system known—the brain.

The power of molecular biological methods is gained from several related but independent developments: (1) the ability to *clone* genetic information (i.e., to isolate a selected segment and accurately reproduce it in large amounts); (2) the ability to determine the nucleic acid sequence of the selected gene segment (i.e., to read the complete molecular structure of a gene); and (3) the ability to practice *genetic engineering* (i.e., to perturb and control gene expression and alter the structure of gene products by chemically modifying precise sites in the molecular structure of the genes). Within a decade the possibilities for applying this basic triad of powerful tools were dramatically revealed by two additional innovative technologies: the *polymerase chain reaction* or *PCR* (by which large amounts of specific nucleic acid sequences can be produced without their prior purification, cloning, or even a complete knowledge of their sequences) and the ability to create *transgenic animals* (i.e., transfer synthetic genes into embryonic cells to make new mice, pigs, and cows to the experimentalist's specifications). Second-generation refinements of the latter two technologies have allowed PCR to serve as a means to find novel gene products within a family of genes in which some segments have been conserved (like the transmembrane domains of certain receptors and transporters). The transgenic strategy can now reproducibly provide mice that are good disease models—either by creating mice lacking a specific gene ("knockout" mutations) presumed to be essential for one or another transductive pathway or by extending the original application of transgenic mice to overexpress selected genes.

All of these developments have contributed to a very rapid advance toward a truly molecular basis for the understanding of the nervous system and the way it can be altered by drug actions. In this chapter we explore these molecular foundations.

CELLULAR VARIATION

Before we can deal effectively with the critical details of molecules that regulate the function of the nervous system and mediate the responses to drugs that act there, we must begin by briefly considering how the cells of the brain differ from the other cells of the body and what it is that allows for differences between types of cells.

Except for erythrocytes, all mammalian cells have a nucleus that separates the basic units of genetic information from the cytoplasm. The cytoplasm and its organelles allow the cell to generate energy that the cell uses to synthesize the structural and enzymatic molecules that give it and its enveloping plasma membrane the functional properties by which it contributes to the overall operation of the organism. Liver cells, kidney cells, skin cells, white blood cells, and cells of the nervous system all possess these basic similarities and these basic individual properties. In a single individual organism, all of these somatic cells (lymphocytes being the only exception), including the cells of the central and peripheral nervous system, possess entirely the same basic set of genetic information. Lymphocytes can rearrange their genes extensively to refine the antibodies they must produce to meet all challenges.

The total set of potential genetic instructions of an individual, its *genotype*, is composed of basic instructional units—the *genes*—each having a specific location on a specific chromosome. Each type of specialized cell expresses a subset of genes that encode the special structural and enzymatic proteins that endow the cell with its size, shape, location, and other functional characteristics. This set of characteristic features that are expressed by a cell is termed its *phenotype*.

Classically the "phenotype" of the cell classes in the brain (i.e., neurons or glia) has been addressed through description of their physical features (e.g., location, size, shape, connections) and their transmitter-specific neurochemistry. Understanding the genetic basis for cell typing affords the opportunity to explain how neurons of the same shape and size can have different connections or transmitters, what this diversity achieves, and why some neurons may be more vulnerable to destructive insults than others. For now, we can assume that these cellular archetypes and the almost innumerable variants within each of those two large classes are the outward reflection of their corresponding specific expression of subsets of their genes. The next section provides a brief overview of this process, with additional complexities in the expanded coverage that follows. Neuropharmacology students recently exposed to molecular biology or biochemistry courses may find the next few sections "old hat." However, we include this material so that everyone reaching the end of this chapter will have the fundamentals necessary for a full appreciation of the material to follow (see Fig. 3-1).

1. The genetic information, stored in the form of long strands of deoxyribonucleic acid (DNA) chains, is selectively expressed. That is, each specialized cell attains its specialized functional and structural status by expressing a subset of all of its genetic instructions. To express selective segments of the genome, the DNA-encoded information is converted, or *transcribed*, into a second similar molecular form as strands of ribonucleic acid (RNA) under the control of special proteins and RNA-synthesizing enzymes (polymerases) that perform the transcription steps. We will deal in slightly more detail with the actual molecular properties of DNA and RNA below.

2. The primary transcript form (the heterogeneous nuclear RNA or hNRNA) is then edited by several rapid steps and

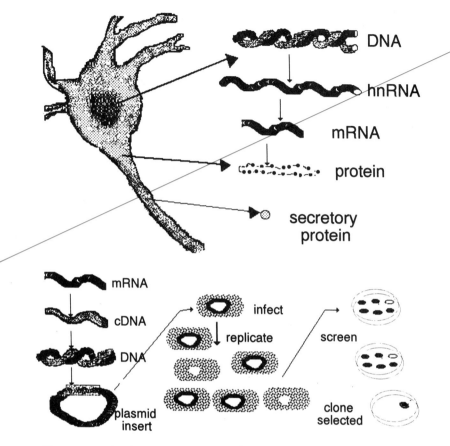

FIGURE 3-1. A schematic overview of the basic steps and cellular compartments involved in determining the specific phenotype of a neuron (above) and in cloning the mRNAs that allow the neuron to translate its genetic information into specific proteins. An mRNA is converted to a single-stranded DNA that is further converted into a double-stranded segment that is then inserted into a plasmid. Bacteria are infected with individual plasmids, and individual, plasmid-infected bacteria are grown into colonies. Replicates of the culture plate are screened by nucleic acid or antibody probes to select clones of interest.

exported from the nucleus to the cytoplasm. The edited RNA transcript, or *messenger RNA* (mRNA), is then *translated* by special cytoplasmic organelles, also made of RNA and proteins, called *ribosomes*. The translation is a chemical language shift from the nucleic acid code of the RNA into the amino acid sequence of the protein that is to be expressed.

3. In some brain cells, such as the glia, that make their proteins largely for use within the cell, the ribosomes occur free within the cytoplasm. In neurons and other dedicated secretory cells, the translated protein undergoes *post-translational processing*, in which the protein's structure may be modified to attain the folded, globular, or linear structural properties that allow it to become associated with the proper intracellular compartments (e.g., within the plasma membrane or within the cytoplasm) where it is intended to function.

4. In cells, like neurons, that transport large amounts of protein products within the cell's interior for purposes of secretion, more extensive post-translational processing occurs. In such cells, the ribosomes are physically associated with a special set of endoplasmic reticular membranes, giving the membranes a "rough" appearance, for which they are known as the "rough ER." Within the channels of this inside-the-cell network, the newly synthesized proteins are led to a set of smooth endoplasmic reticular membranes, the *Golgi apparatus*, where they are packaged into secretory organelles for transport to the secretory, or releasing, segments of the cell.

All of these organelle systems of the cell are essential for the selective transcription, translation, and packaging or compartmentalization of the specific proteins by which a given class of cells attains its specific phenotype. These structural elements of cell biology were well known to classic cytologists long before the molecular mechanisms underlying these events were understandable. Although the details of these molecular revelations are well beyond the capacities of this book, an exploration of the basic features will help the interested reader begin to comprehend the special analytic advantages that arise from the methods of molecular biology.

FUNDAMENTAL MOLECULAR INTERACTIONS

The cornerstone discovery of molecular biology was the formulation by Watson and Crick in 1953 of the *double-stranded helix* model of DNA structure. The insightful model they developed provided a coherent integration of the regular x-ray crystallographic structure of partially purified DNA with the previously known quantitative chemical data on the relative frequency within DNA of its four nucleotide bases, thus explaining why the purine–pyrimidine pairs adenine (A)–thymidine (T) and guanine (G)–cytosine (C) occur in precisely equal frequency. The Watson-Crick molecular model for DNA also accurately predicted the basic mechanism of DNA replication.

Nucleic Acid Base Pairing Complementarity

In the Watson-Crick double helix, two right-handed helical polynucleotide chains coil around the same central axis, making a complete helical turn every 10 nucleotides (Fig. 3-2) In the interior of the helix, the purine and pyrimidine bases (A with T and G with C) are paired through hydrogen bonding of their complementary structures, placing the phosphate groups around the outside of the helix. The crucial structural feature that is the focal point for gene expression is the precise molecular complementarity between the primary sequences of nucleotide bases in one strand of the DNA helix with the antiparallel sequence of the second strand. The strand that encodes the genetic information is termed the sense strand. Wherever a particular base occurs in the sense strand, there will be a complementary base, and only that base in the antisense strand, such that A always pairs with T, and vice versa, and G always pairs with C, and vice versa.

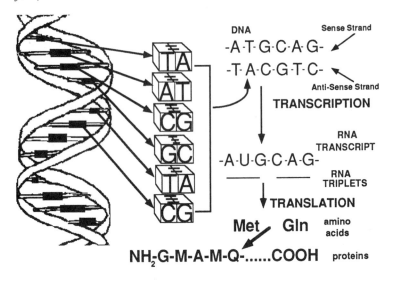

FIGURE 3-2. The arrangement of bases within the DNA double helix and the relationships between nucleic acid base pairs, RNA, and amino acids during the process of transcription and translation.

The base pair complementarity allows for duplication of the genetic information in dividing cells. This is accomplished by enzymes known as DNA polymerases that open the helix and replicate each single strand back into double strands according to the single strand's template. The double-stranded complementarity also provides a means to repair the DNA should it be damaged, since whichever single strand survives the damage can act as a template for the repair.

In a similar manner, the information-bearing, or sense, strand of the helical DNA chain is copied into a complementary single-stranded RNA during the process of transcription. RNA is thus a single-stranded complementary copy of the DNA antisense strand (so that its sequence resembles closely that of the DNA sense strand; see Fig. 3-2). RNA differs chemically with the substitutions of uridine for thymidine and ribose phosphates for deoxyribose phosphates. The process of transcription is accomplished by enzymes known as RNA polymerases. The affinity of the base pairs along sequences of a single DNA strand for their complementary base pair sequences in DNA or RNA are so precise that small segments can be used as *probes* for the detection of homologous sequences between large domains of DNA and RNA because the molecular complementarity will allow the probe to bind to its complementary structure only when there is a long sequence of consistent match. The ability of a single-stranded nucleic acid to bind, or hybridize, to its complementary sequence is an essential component of many molecular biological techniques.

The Genetic Code

To translate genetic information from sequences of RNA into linear sequences of amino acids in proteins requires a strict coding in order that the 20 or so amino acids commonly found in proteins can be specified by various combinations of the four nucleotides. Through an ingenious series of important experiments, which space does not permit us to describe here, it was demonstrated that sets of three RNA bases (triplets) provide the code words that specify the order of amino acids to be incorporated into protein (Fig. 3-2) and that other triplet sequences mark the point at which synthesis would begin or end or be modified in other ways.

Often, mRNAs encode far more protein sequence than is represented in the final form of the processed gene product. One feature that was only detected through recent cellular biological insight was the *signal peptide:* a 15- to 30-amino acid sequence at the N-terminal end of the encoded gene product in which the amino acids are highly hydrophobic. The signal peptide is a near-constant feature of proteins intended for secretion, such as neuropeptides; its function seems to be to guide the nascent protein chain through the endoplasmic reticulum membrane for subsequent packaging by the Golgi membrane apparatus. (A corollary that might be anticipated is that proteins that are not intended for secretion would lack signal peptides and thus represent a means to predict from the structure of novel proteins whether they are secretory products; regrettably, this corollary does not always hold.)

DNA Segments for Genes Are Interrupted

Almost as essential to the progress in application of this base pairing complementarity and nucleic acid to protein translation was another totally unanticipated wrinkle in the molecular basis for gene regulation in eukaryotic cells that stemmed from even earlier observations on the adenoviruses. In the late 1970s, with the analysis of the genes encoding immunoglobulins and the hemoglobins, it was recognized that the basic organization of genes in eukaryotic cells (cells with nuclei, as in all multicellular organisms) did not follow the principles that had been uncovered from the study of prokaryotes (cells, like bacteria, that lack a separable genetic compartment). Instead, higher organisms (and some viruses) have their gene segments split up into coding regions that are expressed *(exons)* separated by intervening regions *(introns)* that are not found in the mRNA.

This interrupted DNA coding structure leads to two outcomes: (1) The primary gene transcript, formally termed the heterogeneous nuclear RNA, or hnRNA, will contain extra RNA sequences, and the introns must be edited out before the mRNA can successfully direct protein synthesis by ribosomes (see Fig. 3-3). The editing process opens up the hnRNA, removes introns, and resplices the cut ends. (2) In some cells, including neurons, the composition of the transcribed mRNA can also be edited by splicing out certain exon segments. The editing–splicing process (or "alternative splicing") provides a means by which a gene containing several exons can give rise to several different gene product proteins with some shared protein domains and some unique domains. When gene products share similar nucleotide and protein sequences, we often speak of them as a "structural family."

This editing and splicing may seem to be an unnecessarily complicated route to follow for a process that is intended to translate important genes with great fidelity into equally important enzymatic

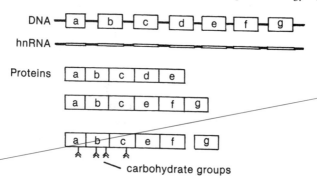

FIGURE 3-3. The relationship between DNA base sequences in introns and exons (a–g), the resulting primary RNA transcript, the subsequently edited forms of mRNA (two different forms of hypothetical editing and splicing are depicted), and the resulting proteins, which can then undergo post-translational processing to yield small peptides or to add carbohydrates or other chemical modifications to the molecule.

and structural proteins. However, the reader should be cautioned that such biological complexity almost always implies important and unanticipated regulatory control and enrichment. In the case of gene regulation and expression, the added complexities offer the means by which new life forms can evolve.

Nucleic Acid Sequence Determinations

The final fundamental procedural development that accelerated discoveries made with molecular biological methods was the body of techniques to determine the sequence of DNA molecules, even those that are several thousand base pairs in length. The methods for doing this sequencing are clearly outside the province of this book. Two different approaches to sequence determination (that of Maxam and Gilbert and that of Sanger), each based on some very clever chemistry, had profound enough importance to merit Nobel recognition. From these DNA sequences, it is possible to deduce the nucleotide sequences of the RNA and thereby the amino-acid sequence of the protein product. The structures of the DNA and of the gene product can then be analyzed, often via computer, to compare their sequences with data banks of previously characterized proteins or nucleic acids. In addition to structural comparison of gene or RNA sequences, additional important clues to the functional features of the encoded molecule may be inferred from the domains of hydrophobic, hydrophilic, or other consensus structural sites for possible post-translational modification. The science of nucleotide sequencing is now so advanced that the procedure can be automated, employing robotic devices for the chemical reactions and computerized readers to determine the sequences. Even though this procedure has an accuracy of greater than 95%, humans are still required to evaluate the results for consistency.

Once-Over Quickly Cloning

When mRNAs are converted into DNA by the enzyme *reverse transcriptase*, the copied double-stranded DNA form, or *cDNA*, can be incorporated (inserted) into specific sites within an infectious vector, or plasmid. The sites for insertion are selected by identifying DNA sequences that can be "cut" by the actions of *restriction endonucleases*, enzymes from purified bacterial sources that cleave DNA sequences at specific palindromic repeated sequence sites. By using plasmids of known DNA sequence that can be tailored to include the proper restriction cleavage sites to allow for the DNA insertion, the same enzyme can then later be used to cleave out the insert. The restriction sites for insertion are typically chosen within plasmid genes that code for some discernible functional property (such as antibiotic resistance). Thus, interruption of its coding and expression (when insertion has been successful) leads to the loss of that functional property and results in a means to identify which plasmids have successful inserts. Each plasmid can generally only incorporate one cDNA insert, and, with a great excess of host bacteria, each insert-bearing plasmid will infect only a single host (almost always *Escherichia coli*).

By growing these infected bacteria in such a way that each individual bacterium gives rise to a *colony* of identical bacteria bearing identical replicates of the plasmid and insert, the DNA has been *cloned.* The cDNA can then be recovered from the plasmid through another exposure to the restriction enzyme selected for the original opening of the plasmid insertion site. Thus, in a relatively few steps, one can start with a mixture of mRNAs in widely differing proportions from common to very rare and purify them individually, as well as prepare virtually unlimited pure samples of the DNA insert.

The Polymerase Chain Reaction

In early 1988, the final touches were put onto a startling new technology for preparing large amounts of specific and rare nucleic acid sequences through amplification in vitro without the necessity of first purifying the desired product through cloning. The changes in molecular biological research wrought by this new technology were immediate and dramatic, and the chemistry has already been rewarded with a Nobel Prize.

As originally conceived by scientists at the biotech company Cetus, the goal was to develop a quick diagnostic tool for the genetic disorder sickle cell hemoglobinopathy. Using previously selected restriction enzymes, human DNA was sliced into small sections, one of which contained the complete hemoglobin gene sequence. When the DNA is heated to 95° C, the double-stranded DNA "melts" and the strands separate. Two small DNA sequences were synthesized (oligonucleotides) to be complementary to opposite strands of the hemoglobin gene separated by the part of the gene structure that was known to be altered in patients with the sickle cell mutation. These oligonucleotide "primers" will hybridize to their complementary sequences on the single-stranded DNA. In the presence of large amounts of a purified DNA polymerase and large amounts of all the deoxynucleotides, the primers will be extended to the end of the single strand. In the original experiments, this span covered only a distance of 20–30 bases; subsequent modifications now permit the amplification of nucleotide sequences that are several hundred bases long.

By then separating through heat denaturation the dual helical strands of the newly synthesized material, cooling the reaction mixture, and adding fresh DNA polymerase and nucleotides, the cycle of *denature, anneal,* and *extend* reactions, each lasting only a few seconds, can be rapidly repeated. As long as the polymerase and the nucleotide substrates remain in excess, the extended sequences of the first reaction each serve as templates for opposite strand synthesis in subsequent cycles, thereby providing a geometric rate of amplification. By identifying a DNA polymerase from a bacterial strain that survived in the heat of a Yellowstone geyser (*Thermus aquaticus,* or *Taq* for short), it was possible to develop a kit that would allow large amounts of polymerase to survive the heat denaturation step. Thus, with the *Taq* polymerase and large starting amounts of deoxynucleotides, a series of rapid cycles allow for the exponential in vitro amplification of the desired gene segment.

If the sequences selected are known to be generally constant in the genome of the species to be studied, the same primers can be used to select, amplify, and then evaluate the same intervening gene segment from many individuals. If the amplified segment is simply evaluated for its length, it is possible to determine whether a given individual has had a major structural gene mutation, such as a deletion or insertion. However, many other clever applications of the technology have pushed its advantages even more broadly than initially imagined. By first using reverse transcriptase to make double-stranded copy DNA (cDNA), one can also apply PCR to mRNAs and even make the assessment quantitative. By modifying the ends of the probes to be used, it is possible to incorporate special synthetic sequences that will later make it easier to clone or sequence the amplified segments or to re-insert modified versions to assess the functional importance of a specific sequence domain. By using special nucleotides that will form complementary base pairs even if the sequences do not match precisely, it is possible to amplify homologous sequences from different but related genes (such as the transmembrane domains for the seven-transmembrane domain receptors like the noradrenergic, cholinergic, or serotonergic receptors).

This highly superficial survey should indicate the facility of cloning DNA segments (taken directly from genomic digests or from mRNA copies) and then sequencing the cloned segments and deducing the structure of their product. However, we have glossed over the one potentially sour note in this rhapsody of high-tech molecular music making: How do you identify which clone is carrying the insert that encodes the gene product you want to identify? Suppose you do not even know what it is you want to look for and clone? As it turns out, through more cleverness and a few good breaks from Mother Nature, biologists have found multiple methods to do these feats of molecular magic, although some of the screening methods have given new meaning to the term "a search for needles in haystacks."

MOLECULAR STRATEGIES IN NEUROPHARMACOLOGY

The immediate applications of molecular biological strategies within neuropharmacology are shortcuts in molecular isolation and sequence determination—for instance, to uncover new peptides or to provide more complete understanding of enzymes, receptors, channels, or other integral proteins of the cell. A rather likely premise holds that the phenotype of a cell within the nervous system depends on the structural, metabolic, and regulatory proteins by which it establishes its recognizable structural and functional properties. If this is valid, then complex, multifaceted neurons will probably rely upon hundreds, if not thousands, of special-purpose proteins, many of which may exist in rather limited amounts. Purifying such rare proteins by the methods that existed before molecular cloning, especially in the absence of a functional assay to guide the purification process, is an overwhelming task requiring exceptional patience, resources, and a very large supply of the proper starting tissue material. For some of the very rare hypothalamic hypophysiotrophic releasing factors (see Chapter 12), hundreds of thousands of hypothalami were required, as well as the development of unique purification schemes for each subsequent factor to be pursued.

Converting the quest for the structure of specific proteins into a molecular biological quest for the mRNA or gene segment that encodes this protein greatly facilitates the experimental analysis, simply because of the powers of cloning, complementarity, and rapid sequencing. Several methods have been developed that increase the chances of finding whatever the researcher is seeking. These methods depend in part on the nature of the cDNA being pursued and how the investigator is able to probe either for the insert or for the translation of the fusion gene product in a cell system capable of processing the primary translation product into a structural form that will resemble its natural configuration and sometimes even its natural function.

In addition to their capacity to accelerate the discovery of new molecules participating in the nervous system's response to disease or to self-administered drugs, molecular biological strategies can also be used to determine how critical a particular gene product may be in mediating a cellular event with behavioral importance—for example, the role of a specific transmitter's receptor subtype in a cellular event (like long-term potentiation) that may in turn underlie behavioral phenomena such as memory formation or recall (see Chapter 12). These functional probes can be achieved through advanced methods for transgenic animal construction (see below) in which specific mutations are engineered into any targetted gene of an experimental animal's genome, leading eventually to the production of homozygous animals lacking that gene completely.

Less demanding, and less vulnerable to potential confounding roles of the targetted gene in the developing nervous system, are more short-term manipulations such as the intracerebral or intraventricular injection of special nucleotide constructs to deliver *ribozymes* (RNA constructions that can target and degrade specific mRNAs) or *antisense nucleotides* (which can bind to mRNAs, delay their translation, and enhance their degradation). In addition to preparing novel mutant animals by transgenic technologies, it is also possible to insert the special constructs into "null cells" (cells that normally do not express the transduction system under investigation) and, when they express a new receptor in their membrane, to probe these receptors for more conventional pharmacological characterization of agonists and antagonists. One might also develop novel drugs that may be unique to such receptors (see Chapter 13).

GENERAL STRATEGIES FOR CLONE SCREENING AND SELECTION

Enrichment by Tissue Selection and Preparation

The basic starting point is to select the brain cells or regions that are presumed to express the molecule to be studied and then to enrich sources of mRNA to favor the detection of the one being pursued. Cell lines and even tumors that produce large amounts of a hormone (such as pheochromocytomas or VIP-omas) or bear large numbers of the desired receptors or channels (like electroplax or striated muscle) have proved to be excellent starting materials. Once the cell source is selected, the desired mRNAs can be further enriched (e.g., by sucrose gradient centrifugation or electrophoresis) provided that some characteristics of the mRNA being sought are known.

Recognizing the Wanted Clone

A general strategy for detecting desired colonies of cloned bacteria is known as *colony hybridization*. In this approach, the bacteria are grown on a special culture plate from which their colonies can be copied (transferred as a group by lightly pressing them to another supporting surface, called "replica plating," and thereby sampling and preserving the spatial identity of all the colonies on the plate). The bacteria on the replicate supports are treated to expose their DNA and are screened with radioactive nucleic acid probes. Colonies that hybridize with the probe can then be identified by autoradiography. Alternatively, if the plasmid-carrying inserts were tailored to allow for expression of the protein encoded by the transferred genetic material, it might also be possible to identify the desired clones by immunological reactivity. When a reactive colony has been identified, its original bacterial colony is recovered from the original culture plate, and the living bacteria are then grown in large amounts to provide the starting material for DNA sequence analysis.

Building Your Own DNA

If partial protein (or peptide) sequences are known, it is possible to make predictions of what the mRNA sequence should be (by back-translating the genetic code for amino acids) and including enough alternatives to overcome the ambiguous cases where a specific amino acid may be encoded by several variant triplets. From this predicted RNA structure, it is then possible to design and synthesize the hypothetical cDNA. This approach has been used, for example, to create hypothetical cDNAs for the hypophysiotrophic hormones whose amino-acid sequences had been accurately determined "the old-fashioned way," by earning it one amino acid at a time from highly purified brain extracts. The main reason to do this with a biologically active protein or peptide whose structure is already known would be to get at the complete structure of its prohormone or to obtain its complete genomic structure and analyze its regulatory control and expression mechanisms. As will be seen in Chapter 12, when this is done, more often than not, the prohormone of the known peptide is found to encode more than one active product. However, because of the redundancy of triplet RNA codons for some amino acids (there are six different codes for leucine alone) it is generally difficult to acquire a functional full-length mRNA by predictive synthesis.

An alternative approach is to synthesize a shorter complementary single-stranded DNA (a so-called "oligodeoxynucleotide probe") and use it as a *probe* to screen libraries of clones. Such libraries may be prepared from mRNA extracts or from special digests of the whole genome. In the former case, the starting material would be brain, while in the latter case, in theory, any somatic cells could be used to prepare the library. With the availability of automated "gene machines" it is now possible to synthesize a proper probe or two overnight and use them to screen an awaiting genomic or cDNA library, thereby determining the complete coding sequences for a partially purified protein within a few weeks.

On such a screening expedition, likely candidate colonies can be cross-screened by a second synthetic oligodeoxynucleotide probe, based upon another separate domain of the full protein. Clones positive for both probes would then have to contain the gene sequence that encodes the two sequences against which the probes were made as well as the sequence between them. This strategy has been used with many neuropeptide mRNAs (see Chapter 12).

When You Haven't Got a Clue

It is also possible to penetrate the large treasure trove of cellular proteins that have not yet been identified by conventional strategies. Given the length of the mammalian genome, and the relatively short list of identified specific molecules in cells of all classes, we must conclude that there is an awful lot left to be identified and few clues as to what it is we don't know about. The methods of molecular biology can help here too.

On the other hand, there would appear to be a rather select group of "ancient conserved sequences," numbering probably less than 1000, that nature has found to be important enough to keep relatively unmodified throughout thousands of years of evolution. This core must then be significantly enriched by the other 99% of the genome. Interestingly for neuropharmacologists, most neurotransmitter receptor molecules are not among these ancient conserved sequences, although at least one 5-HT receptor and some of the transporters are.

In nonneural tissues it has been possible to identify the unique proteins expressed in a male versus a female liver, or those that are unique to thymus-derived versus bone marrow-derived lymphocytes, by exploiting the fact that a very high proportion of the proteins expressed in these pairs of tissues are, for the most part, very similar. Depending on how the pairs of tissues or cell types are defined and how similar or dissimilar they actually turn out to be, this method can be made highly sensitive and can reveal unique differences in cell-specific gene expression.

The discovery strategy can also be broadened to look for large sets of tissue-specific genes. For example, Sutcliffe, Milner, and Bloom have employed molecular cloning methods to ask what proteins are generally made by brain that are not found in other major tissues and to determine the degree to which neurons differ in specific proteins underlying their phenotypes that at present are only defined empirically. In their general approach, a cDNA library was prepared from mRNAs extracted from whole rat brain, and the individual cDNAs were characterized for their ability to hybridize to mRNAs extracted from brain, liver, or kidney. Those expressed in brain, but not detectable in extracts of liver or kidney, were found to represent well over half of the total brain mRNA population (approximately 30,000 out of an estimated total of 50,000), suggesting that much of the genome contains information pertinent to the generation of neuronal function. Individual brain-specific clones are then analyzed further by determining their nucleotide sequence and deducing the amino acid sequence of its encoded protein. Proteins that are unique to the data base of known sequences can then be identified further by raising antisera against synthetic peptides that mimic selected regions of the deduced protein structure. One such clone appears to resemble a possible precursor for neuropeptides.

Yet another strategic approach involves injecting mRNAs from enriched or prepared sources into frog oocytes where complex eukaryotic genes may be expressed more efficiently. Because the oocytes are relatively large, their expression of a novel functional protein can be assessed physiologically or biochemically to identify species of mRNA for isolation and cloning. By injecting groups of mRNAs and evaluating the oocytes for the response one seeks, it is possible through trial and error to identify the mRNA for a specific functional protein, like an ion channel or a cell surface receptor.

An All-Points Search

Would-be molecular neuropharmacologists attending a seminar involving these approaches may find initial exposure to the jargon confusing without some orientation. Orientation is an appropriate term, for such presentations are sprinkled with what may sound like references to compass points. When DNA is fractionated by restriction enzymes, and the resulting fragments are separated by gel electrophoresis, it is possible to transfer, or "blot," the resulting fragments, separated mainly on the basis of their lengths, from the acrylamide gel to a nitrocellulose or nylon support and there to analyze them for the ability to hybridize with cDNA or RNA probes. This method is referred to as a "Southern" blot analysis, named for the scientist (Dr. E. M. Southern) who started the evolution of this method. Later, a similar approach was devised in which the starting material was RNA. Here, the separated RNAs were blotted for probing with radioactive, single-stranded cDNA probes. Because the starting material is, speaking in terms of nucleic acid, the opposite of the Southern, this RNA blot is referred to as a "Northern blot."

More recently, immunological methods have been used to probe protein extracts that were separated by acrylamide gel electrophoresis and then "blotted" (by an electrical transfer) for identification by peroxidase or radioisotope-labeled antibodies to specific protein antigens. The result is termed a "Western blot." If RNA or protein samples are simply dried directly onto nitrocellulose for probing analysis without first separating them for size, the resulting blots are termed "slot blots" or "dot blots." These blots are useful when screening a large number of clone extracts for insert or expressed products quickly. As of this writing, no method has yet earned the accolade of being an "Eastern blot."

Given the resourcefulness with which molecular biologic methods are being applied to all aspects of specialized cell biology, including the actions of drugs on the nervous system, one senses that we are about to experience a logarithmic increase in the number of specific molecules that will be fully characterized and that will enable pharmacological engineers to shape drug molecules precisely to fit the pocket of receptors or enzymes for the ultimate in specificity. Any biological phenomenon in the brain that is mediated by proteins, which seems to exclude hardly any brain function, is therefore amenable to the ultimate in molecular analysis. The range of such specific molecules would then include (but is definitely not limited to) the enzymes needed for transmitter synthesis, storage, release, and catabolism, the receptors and related macromolecules needed for response, and response mediation to broader scale events such as establishing the shape and orientation of a cell by the proper anchoring of its microtubules and filaments to the plasma membrane and the repair and maintainence of synaptic connectivity.

BEYOND THE CLONES

Although we may rightly marvel at the advances that have been achieved through the use of molecular biological techniques, all that we have in essence discussed so far in this chapter is a set of methods that provides a novel, powerful, and accurate way to identify, isolate, and characterize the amino acid sequences of a host of intracellular proteins and their possible subsequent metabolic products. While this is unquestionably a very major advance in the research armamentarium, it still does not really begin to deal with a wide range of other important questions that are also approachable through the molecular tools that recombinant methodologies provide.

For example, once a cDNA has been proven to represent the mRNA for a specific molecule, the deduced protein's sequence can be inspected to infer potential functional properties. Thus, the acetylcholine receptor molecule and the myelin proteolipid protein exhibit several stretches of 20–24 hydrophobic amino acids in a row, which are strong presumptive evidence of membrane-crossing domains and thus suggestive of plasma membrane constitutive proteins. In any case, when the protein structure can be deduced, the entire molecule or selected fragments of it can be synthesized and used to raise antisera. These antisera can then be used to develop radioimmunoassays for the protein. The antisera can also be used for immunocytochemical analysis of the nervous system to determine which cell and which compartments of those cells exhibit the protein that has been identified. The synthetic fragments can be used to determine whether the protein's domains may be substrates for post-translational modification, being processed by further proteolytic cleavage or by structural modification with glycosylation, phosphorylation, sulfation, or acylation. Subcellular fractions and ultrastructural cytochemistry may suggest organelle specialization or cell surface associations.

The products of cDNA cloning can also be taken back to the genome, to probe the regions around the location of the exons to search for their molecular mechanisms for control of expression. Once the location of the surrounding elements in the genome have been located, the cDNA probes can be used to determine the degree to which the mRNA or the underlying gene exons have been conserved across eukaryotic species and to determine the position on the chromosomes to which the gene can be mapped. The chromosomal location of many proteins has been determined in this or a similar manner. However, the human genome, and that of most mammals, is estimated to be on the order of 3×10^9 base pairs long, of which fewer than 1000 genes have been mapped, most of them being on the relatively small X chromosome. What this means is that there are vast expanses of the genome with no known markers of any kind. For students of genetic diseases of the nervous system, the situation is even worse, because much less than 1% of the genome is associated with known nervous system markers even though a significant fraction of the genome may be expressed selectively in the brain.

These considerations of length and complexity of neuronal gene expression suggest that efforts to link specific patterns of DNA polymorphism may be a critical future development. The genetic linkages are defined on the basis of Southern blot analysis of family members whose genomic DNA has been treated with various different restriction enzymes to provide *restriction fragment length polymorphisms*. Because there can be considerable individual variation in nucleotide sequences without disturbing the function of the encoded proteins, the degree to which the patterns of restriction endonuclease-digested fragments differ reflects this individual variation. The ability to link fragments of DNA with inheritance of genetic disorders and to specific markers within the digested fragments helps to establish approximate locations of genetic mutations, such as the localization of the gene for Huntington's disease to human chromosome 4.

Although new modifications of these basic strategies are reported continuously, and although there is a steady growth of molecular information on the elements of neuronal function, most of the details that follow in this book deal with those still relatively rare molecules (e.g., the major known neurotransmitters, their synthetic and catabolic enzymes, their receptors and response mediators), whose nature is already partly established. While this list may well be lengthened ever more rapidly by the shortcuts made possible by molecular biological methods, the student should recognize that identifying new molecules per se is merely a first step toward an important pharmacological end but is nowhere near the true strength of what molecular biology may have to offer our field.

We have already spoken of some important tools as though they were only to be used for molecular identification, and it would be somewhat misleading to let those impressions stand without a pedagogical challenge. The awareness of how to "tailor" purified mRNAs can be taken far beyond simply sticking them into a plasmid: More complex tailoring, leading to real "designer genes," can be used to prepare synthetic genes that carry with them the basis for being incorporated and being expressed by novel cellular hosts, bacterial or mammalian. Directly injecting mRNAs into frog oocytes and allowing them to be translated and incorporated into the oocyte's cellular machinery is useful for detecting specific mRNAs for specific functional proteins, but again there is more: The cell bearing a known receptor whose complete amino-acid sequence is known could be used as an ideal template for drug design by combining x-ray crystallographic information on the gene product protein, amino-acid sequences, and functional responses in a responsive cell. Such methods have already been employed in part to derive potent enzyme inhibitors for membrane lipolytic enzymes. Furthermore, once that mRNA has been identified, synthetic versions of it that are lacking certain nucleotide segments can be evaluated to determine the functional consequences of tinkering around with the molecule in specifically designated sites. This strategy would be capable of indicating the active sites of a receptor, or its membrane anchoring, or its sites of interactions with the ion channels or enzymes that the receptor can regulate. Such mutagenesis can also be achieved by other treatments for the same objective.

The use of viable gene expression systems can go considerably beyond the frog oocyte, however. Effective use has already been made of cell lines, most of them derived from spontaneous or chemical- or virus-induced tumors of the central or peripheral nervous systems or from fused representatives of such originally tumorous cells. The ideal cell lines characteristically differ, depending on the situations in which they will be used—such as an abundant receptor (to look for genes that transduce it when activated), or no receptors and abundant second messenger systems, so that a gene encoding a receptor can be screened for its natural ligand or for its mechanisms of transduction.

Even more promising in its implications for neuropharmacology is the microinjection of a segment of cloned DNA into the pronucleus of a single cell zygote (a fertilized ovum) to create a *transgenic mouse*. After injecting a large number of fertilized ova (a success rate of 10% is considered good), one transfers the eggs to the uterine cavity of a foster mother that has been pseudo-bred with a sterile male. When the pups are born, they are evaluated through skin fibroblast cultures for incorporation of the injected DNA and, if possible, for expression of the gene product (e.g., overproduction of a circulating hormone). After puberty, their sperm can also be evaluated for integration of the foreign DNA sample into germ cells; these are best evaluated by their ability to transmit the integrated gene to the mouse's own progeny. If the integration of the foreign DNA is successful, and if the resultant gene structure (at the site of integration) does not itself produce mutational consequences that may be troublesome, the founder mice give rise to lines of mice bearing the transgene. By preparing gene constructs that induce the overproduction of growth hormone, it has been possible to use the technology to make not only supermice, but super pigs and super goats as well.

Transgenic mice are now being used for a variety of experimental purposes. They are used to express the products of a virus such as the hepatitis B or the human immunodeficiency virus, and they determine which, if any, of the virus gene products may be directly cytotoxic, immunogenic, or immunocytotoxic. Other partially tested strategies include removing a natural gene to determine the effects on tissue or organ function, as, for example, producing an artificial demyelination disorder by eliminating the gene for myelin basic protein. The reverse application of transgenic technology was actually done first, namely, preparing a transgene for expression of myelin basic protein and using that to "cure" the gene mutation of the myelin-deficient mouse, which is ordinarily unable to make the protein.

Among the more experimental applications of transgenic technology is the preparation of a gene construct in which the regulatory domains of a known gene (say for a neuropeptide) are coupled with the expression of a novel "reporter" gene (such as an enzyme that is normally absent in the mouse); under these conditions, one can evaluate whole animal treatments (or whole cell treatments if cell lines are transfected instead of embryonic cells) that "turn on the reporter gene," thus providing inferential evidence of what controls (i.e., elevated cyclic AMP) the activation of the natural gene. Finally, applications rumored to be in progress include the use of a retrovirus to carry a gene segment to be transfected into cells beyond the single cell embryo level; this is especially useful in systems where partial development can reveal the effects of the added gene. A third creative but not yet well-developed application involves the transfection of embryonic "stem" cells, undifferentiated blast-like cells taken from the blastocyst level of embryonic development. When these transfected stem cells are then returned to blastocyst embryos, they spread throughout the developing organism, and if the researcher is fortunate, the foreign gene will appear in the germ cell lines as well. Should gene markers ever be identified that predict the development of single gene-dependent neuropsychiatric diseases (such as Huntington's, familial Alzheimer's, or scrapie), it should be possible to prepare true animal models of the diseases and their treatment.

THE MOLECULAR MOTIFS OF TRANSMITTER RELEASE AND RESPONSE

We conclude this foray into molecular neuropharmacology by quickly considering the consistent structural patterns or "molecular motifs" that have been identified in selected categories of molecules that are essential for synaptic transmission and thus critical for the content of subsequent chapters. This survey will also illuminate some of the clever ways that molecular biology has moved beyond "mere" discovery to an illumination of the molecular mechanisms of synaptic transmission.

Ion Channels

The three basic channels underlying neuronal excitability were cloned by starting with known short segments of the proteins obtained from highly purified Na channel preparations. Oligonucleotides encoding these peptide segments and antibodies raised against them were used to probe a cDNA library of electric organ mRNA and led to the sequence determination for what turned out to be the α-subunit of the Na channel (there is also a simpler β-subunit). Modeled from the distribution of hydrophobicity plots for the deduced amino-acid sequences, the hypothetical α-subunit structure contained four repeats of highly similar six transmembrane α-helical domains, in which the fifth and sixth transmembrane domains are separated by extracellular loops, while the N-terminal, the C-terminal, and the cytoplamsic loops linking the sixth transmembrane domain to the first transmembrane domain of the adjacent repeat are all intracellular cytoplasmic loops.

These sequence data were then used to derive the rat brain Na channels, which also turned out to have a similar structure, at least in the transmembrane domains though not the connecting loops. These modeled conformations have been refined by a variety of functional analyses, including expression in oocytes to confirm their functional properties (including sensitivity to channel blocking toxins) and to determine pore- and voltage-sensitive domains by mutation, deletion, and transposition of sequences.

Subsequent work, again starting with highly purified protein sources and cloning from the first few short segments of amino acid sequences wrung from them, revealed that the hetrero-oligomeric L-type Ca channel protein (at least five subunits) also contains an α_1-subunit with a similar structural motif and has significant sequence similarity to that of the Na channel, also showing four homologous repeats of a presumptive six-transmembrane domain structure. Its specific functional domains for channel function and subunit sensitivity have also been defined by expression analysis of the natural and modified forms in oocytes. Presumably these features will soon be known for the modulation of these channels through convergent intracellular effects of ligand-gated receptors.

Potassium Channels

Definition of the structures of K channels began with the cloning of the *Shaker* gene of *Drosophila*, which then provided the structural tools to characterize the multiple forms of voltage-sensitive and then ligand-gated channel proteins carrying K currents in mammalian neurons. Interestingly, the basic voltage-dependent K channel is a presumptive six-transmembrane structure, almost identical to the conformation of one of the repeat segments in the Na and Ca α-subunits. In this K channel, however, there is a similar special connecting peptide between the fifth and sixth presumptive transmembrane domains, known to its afficianados as "H5." This segment has been found (through analysis of mutated versions) to perform the voltage-sensing function. Similarly, the cytoplasmic loop between transmembrane domains 4 and 5 provides the means to inactivate the channel from the inside of the cell. When the ligand-gated form of K channel, termed the inward rectifier, was eventually cloned, it showed an even simpler structure, resembling only the two transmembrane domains with an intervening "H5"-like sequence. The brain has more than a dozen mRNAs encoding K channels, and many show a high degree of neuronal selectivity in their expression, with the proteins often highly compartmentalized to perikarya or dendrites.

Neurotransmitter Receptors

Here the starting point was protein sequences from highly purified preparations of electroplax or muscle receptors (for acetylcholine) or brain (for GABA and glycine) or blood cells (for norepinephrine). Once those structures were in hand, however, multiple additional members of their receptor superfamilies were snared by probing cDNA libraries with RNA probes designed to accept loose (or low stringency) matches with similar but not identical nucleotide sequences

As noted in greater detail in Chapter 5, neurotransmitter receptor cloning studies have so far split the spoils into two large groups: the ionophore receptors in which the receptor is a multimeric ion channel composed of four or five subunits, each exhibiting a similar presumptive four transmembrane domain structure, and the metabotropic or "G-protein-coupled" receptors, each of which is a monomer with seven conceptual transmembrane domains. Within the ionophore receptors, the protein subunits for a given neurotransmitter (acetylcholine nicotinic receptors, GABA and glycine receptors, most of the glutamate receptors, and one of the 5-HT receptors), there is enough similarity to allow for recognition of further receptor subtyping of different assembled multimers structurally. Different combinations of GABA subunits expressed in neurons with a high degree of heterogeneity, for example, can confer cell-type-specific sensitivity to ethanol or benzodiazepines (see Chapter 7).

The second large category of neurotransmitter receptors comprises the guanine nucleotide binding protein coupled receptors (for cholinergic muscarinic receptors; all catecholaminergic, histaminergic, and neuropeptide receptors; all the other 5-HT receptors; and the metabotropic glutamate receptor). All of these exhibit a presumptive seven-transmembrane domain configuration first recognized in rhodopsin and then found in the first of the cloned adrenergic receptors. Through expression of mutated mRNA constructs in null cells, it has been possible to determine quite definitively which segments of which transmembrane domains are responsible for ligand recognition, and which segments of which cytoplasmic loops are responsible for interactions with the G-proteins.

Neurotransmitter Transporters

We now turn our attention to a family of molecules that are essential to the process of transmitter conservation-after-release. This process that we used to call simply "reuptake" has been upgraded with the term "transporter" following the cloning. In a relatively brief burst of discoveries in the early 1990s, the transporters for GABA, glycine, all the monoamines, as well as proline and betaine were cloned and found to express a very similar structural motif consisting of 12 transmembrane domains with substantial conservation across the transmembrane domain sequences, allowing for the information from the initial cloning of the GABA transporter to be extended to the others. In all of these molecules, concentration of the specific small molecule from the extracellular space back into the interior of the neuron, generally the presynaptic terminals, seems to be driven by the co-transport of Na^+ and Cl^- ions, and the molecules bear a strong resemblance to the previously known glucose transporters and to adenylyl cyclase for unclear reasons.

A second variety of transporter has also been defined that concentrates cytoplasmic neurotransmitter into the synaptic vesicles, driven by either a pH or an ionic gradient. While these transporters also exhibit apparent 12-transmembrane domain structures, the actual sequences distinguish them from those that operate in the presynaptic plasma membrane.

Synaptic Vesicle Proteins

In an era of neuropharmacology not too long ago, a minor parlor game was made of the ability to classify synaptic vesicles as to their transmitter content on the basis of their size, shape, and cytochemistry. The advent of molecular biology has now radically transformed our views of vesicles from mere morphological description to arguably the most completely detailed nervous system organelle in terms of specifiable inventory of proteins and interactions during transmitter secretion. Again the initial studies started with highly purified brain synaptic vesicles and a search for the proteins they contain, and in a marvelous display of Mother Nature's generosity managed to reveal the molecular machinery for secretion across wide realms of the phylogenetic tree. Given the estimates for the molecular mass of the typical 500 Å, and the roughly 600 kD size of the vesicle transporter, there could be room for about a dozen more molecules of approximately 50 kD. The inventory to date reveals the following three groups:

1. Vesicle-bound proteins: *synapsins* (which are substrates for cyclic AMP dependent protein kinase and can regulate transmitter release); *synaptotagmin* (possibly a Ca-sensor protein) and *synaptobrevin* (also known as VAMP for vesicle-associated membrane protein, and a substrate for proteolysis by tetanus toxin and by some of the botulinum toxins) both of which seem to be essential for docking on the presynaptic plasma membrane; and *synaptophysin*, a vesicle protein of as yet unknown function (but antibodies to it have served as a useful means to quantify synapses). Each of these vesicle proteins has several isoforms such that, it is believed, one of each class will be present on every synaptic vesicle.

2. Synaptic membrane proteins: the synaptic soup thickens with two other proteins, *SNAP 25*, a synapse associate protein of 25 kD, and *syntaxin*, which are both substrates for other botulinum toxin proteases, and thus also essential players in the docking and/or fusion steps to release transmitter.

3. A series of soluble proteins called NSF (an *N*-ethylmaleimide–sensitive factor that alters the capacity to transport vesicles across the cisternae of the Golgi complex in a clever in vitro assay) and its three soluble NSF-associated proteins (called α-, β-, and γ-SNAPs, of which the β-form is quite enriched in brain, while the others are more pro-

nounced in nonneuronal secretory cells, and which bear a strong resemblance to secretory proteins recognized in yeasts for their role in constitutive secretion).

In the case of the vesicle-related secretion proteins, cloning provided the detailed sequences and the related isoforms, the capacity to make small mutations and deletions to define the docking and interaction domains, and the ability to reassemble the entire array in null cell assays and to establish the potential roles for each protein. There are a few more paralytic bacterial toxins whose substrates in secretion are still to be accounted for, so we are not likely to be at the end of this very interesting search. Furthermore, we still do not know exactly how the vesicles are re-formed within the synaptic terminal following the docking, fusion, and release steps or when it is time to ship the whole vesicle back up to the perikaryon for an overhaul.

SELECTED REFERENCES

Barondes, S. H. (1993) *Molecules and Mental Illness*. W. H. Freeman, New York.

Cohen, S. A., A. Chang, S. Boyer, and R. Helling (1973). Construction of biologically functional bacterial plasmids in vitro. *Proc. Natl. Acad. Sci. U.S.A. 70*, 3240–3244.

Crick, F. H. C., L. Barnett, S. Brenner, and R. J. Watts-Tobin (1961). General nature of the genetic code for proteins. *Nature 192*, 1227–1232.

Gilbert, W. (1985). Genes-in-pieces revisited. *Science 228*, 823.

Gusella, J. F., R. E. Tanzi, M. A. Anderson, W. Hobbs, K. Gibbons, R. Raschtchian, T. C. Gilliam, M. R. Wallace, N. S. Wexler, and P. M. Conneally (1984). DNA markers for nervous system diseases. *Science 225*, 1320–1325.

Magistretti, P. J., L. Pellerin, and J.-L. Martin (1994). Brain energy metabolism. An integrated cellular perspective. In *Psychopharmacology: The Fourth Generation of Press* (F. E. Bloom and D. J. Kupfer, eds.), pp. 657–670, Raven Press, New York.

Maxam, A. M. and W. Gilbert (1977). A new method of sequencing DNA. *Proc. Natl. Acad. Sci. U.S.A. 74*, 5463–5467.

Milner, R. J. and J. G. Sutcliffe (1988). Molecular biological strategies applied to the nervous system. *Disc. Neurosci. 5(2),* 11–63.

Morrison, J. H. (1993). Differential vulnerability, connectivity, and cell typology. *Neurobiol. Aging 14,* 51–54.

Ramirez-Solis, R., A. C. Davis, and A. Bradley (1993). Gene targetting in embyonic stem cells. *Methods. Enzymol.* 225:890–900.

Ruddle, F. H. (1981). A new era in mammalian gene mapping. Somatic cell genetics and recombinant DNA. *Nature 294,* 115–120.

Sanger, F., and A. R. Coulson (1975). A rapid method for determining sequences in DNA by primed synthesis with DNA polymerase. *J. Mol. Biol. 94,* 444–448.

Schwarz, T. L. (1994). Genetic analysis of neurotransmitter release at the synapse. *Curr. Opin. Neurobiol. 4,* 633–639.

Southern, E. M. (1975). Detection of specific sequences among DNA fragments separated by gel electrophoresis. *J. Mol. Biol. 98,* 503–517.

Sutcliffe, J. G., R. J. Milner, T. M. Shinnick, and F. E. Bloom (1983). Identifying the protein products of brain-specific genes with antibodies to chemically synthesized peptides. *Cell 33,* 671–682.

Tjian, R. (1995). Molecular machines that control genes. *Sci. Am.* 54–61.

Usui, H., J. D. Falk, A. Dopazo, L. de Lecea, M. G. Erlander, and J. G. Sutcliffe (1994). Isolation of clones of rat striatum-specific mRNAs by directional tag PCR subtraction. *J. Neurosci.* 14:4915–4926.

Weintraub, H. M. (1990). Antisense RNA and DNA. *Sci. Am.* 40–46.

Watson, J. D., M. Gilman, J. Witkowski, and M. Zoller (1992). *Recombinant DNA,* 2nd ed. W. H. Freeman, New York.

Watson, J. D., N. H. Hopkins, J. W. Roberts, J. A. Steitz, and A. M. Weiner (1987). *Molecular Biology of the Gene,* 4th ed. Benjamin-Cummings, New York.

Watson, J. D. and F. H. C. Crick (1953). Molecular structure of nucleic acids: A structure for deoxyribose nucleic acid. *Nature 171,* 737–738.

4 | Receptors

The concept that most drugs, hormones, and neurotransmitters produce their biological effects by interacting with receptor substances in cells was introduced by Langley in 1905. It was based on his observations of the extraordinary potency and specificity with which some drugs mimicked a biological response (agonists) while others prevented it (antagonists). Later, Hill, Gaddum, and Clark independently described the quantitative characteristics of competitive antagonism between agonists and antagonists in combining with specific receptors in intact preparations. This receptor concept has been substantiated in the past several years by the actual isolation of macromolecular substances that fit all the criteria of being receptors. To date, although they have not all been cloned, receptors have been identified for all the proven neurotransmitters as well as for histamine, opioid peptides, neurotensin, VIP, bradykinin, CCK, somatostatin, Substance P, insulin, angiotensin II, gonadotropin, glucagon, prolactin, and TSH. In addition, as noted in Table 4-1, multiple receptors have been shown to exist for all the biogenic amines, ACh, GABA, histamine, opiates, and the amino-acid transmitters. If receptors for all agents, e.g., hormones, trophic factors, odorants, peptides, in addition to the neurotransmitters, were counted, a total of 1000 would not be surprising. According to a literature survey, multiple receptors appear to metastasize at an uncontrollable rate, but this should be viewed skeptically until a physiological response to the ligand has been shown or a specific gene has been cloned and expressed. Some of the receptor subtypes have only been identified by binding techniques (see below) that can lead to erroneous conclusions. All the receptors for neurotransmitters and peptide hormones that have been studied, regardless of whether they have been isolated, are localized on the surface of the cell; among other receptors, only those for steroid and thyroid hormones are intracellular.

In earlier days, after the discovery that the action of physostig-

TABLE 4-1. Neurotransmitter Receptors

Adrenergic

α_{1A}, α_{1B}, α_{1C}, α_{1D}

α_{2A}, α_{2B}, α_{2C}, α_{2D}

β_1, β_2, β_3

Dopaminergic

D_1, D_2, D_3, D_4, D_5

GABAergic

$GABA_A$, $GABA_{B1a}$, $GABA_{B1\delta}$, $GABA_{B2}$, $GABA_C$

Glutaminergic

NMDA, AMPA kainate, $mGluR_1$, $mGluR_2$, $mGluR_3$, $mGluR_4$,
$mGluR_5$, $mGluR_6$, $mGluR_7$

Histaminergic

H_1, H_2, H_3

Cholinergic:

Muscarinic: M_1, M_2, M_3, M_4, M_5

Nicotinic: muscle, neuronal (α-bungarotoxin insensitive),
neuronal (α-bungarotoxin sensitive)

Opioid

μ, δ_1, δ_2, κ

Serotonergic

$5\text{-}HT_{1A}$, $5\text{-}HT_{1B}$, $5\text{-}Ht_{1D}$, $5\text{-}HT_{1E}$, $5\text{-}HT_{1F}$, $5\text{-}HT_{2A}$, $5\text{-}HT_{2B}$,
$5\text{-}HT_{2C}$, $5\text{-}HT_3$, $5\text{-}HT_4$, $5\text{-}HT_5$, $5\text{-}Ht_6$, $5\text{-}HT_7$

Glyinergic

Glycine

SOURCE: *RBI Handbook of Receptor Classification*, 1994.

mine (eserine) resulted from its anticholinesterase activity, it was assumed that most drugs probably acted by inhibiting an enzyme. However, it now appears that with few exceptions the mechanism of action of neuroactive drugs usually stems from their effect on specific receptors. Predictably, the current search for receptors is among the most intensively investigated areas in the neurosciences. This interest is not purely academic. The recent identification of adrenergic, dopaminergic, muscarinic, serotinergic, and histaminergic receptor subtypes has led to the synthesis of highly selective drugs that are considerably more specific than their prototypes that were developed after general screening for activity. And with advances in gene cloning and expression, more and more receptor subtypes are being identified, each presumably having its own function. What this indicates is that future drugs can be designed to fit a single receptor subtype, thus precluding what are now called the "side effects" of a nonspecific drug.

DEFINITION

In this rapidly developing field, considerable confusion has arisen as to what functional characteristics are required of an isolated, ligand-binding molecule to qualify as a receptor. This confusion, a semantic problem, developed after the successful isolation of macromolecules that exhibit selective binding properties (see below), which made it mandatory to determine whether this material comprised both the binding element and the element that initiated a biological response, or merely the former. Some investigators use the term "receptor" only when both the binding and signal generation occur; they use the term "acceptor" if no biological response has been demonstrated. Others are content to ignore the bifunctional aspect and use "receptor" without specifications. In this chapter, we will define a *receptor* as the binding or recognition component and refer to the element involved in the biological response as the *effector*, without specifying whether the receptor and effector reside in the same or separate units. The criteria for receptors will be dealt with shortly; the biological response that is generated by the effector obviously has a wider range of complexity, from a simple one-step coupling to an unknown number of steps (Fig. 4-1).

ASSAYS

Basically, there are two ways to study the interaction of neurotransmitters, hormones, or drugs with cells. The first procedure (and

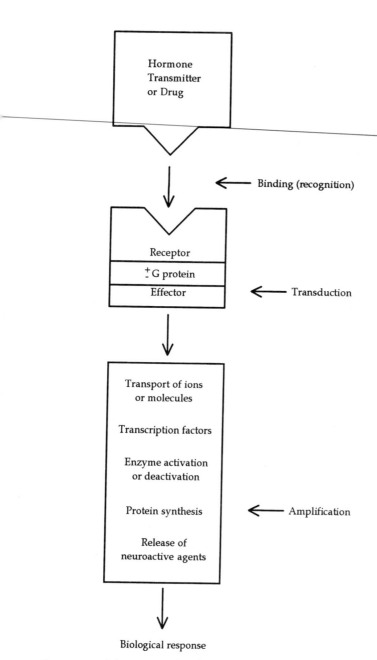

FIGURE 4-1. Schematic model of ligand–receptor interaction.

until recently the only one) is to determine the biological response of an intact isolated organ, such as the guinea pig ileum, to applied agonists or antagonists. The disadvantage of this procedure is that one is obviously enmeshed in a cascade of events beginning with transport, distribution, and metabolism of the agent before it even interacts with a receptor, ranging through an unknown multiplicity of steps before the final biological response of the tissue is measured. Thus, although studies with agonists may be interpretable, it is not difficult to envision problems when antagonists are employed, since these compounds may be competing at a different level than receptor binding. Despite the not unusual problem of a nonlinear relationship between receptor occupancy and biological response, this approach has yielded a considerable amount of information.

The second approach to studying receptors is by measuring ligand binding to a homogenate or slice preparation. This technique became feasible with the development of ligands of a high specific radioactivity and a high affinity for the receptor. Here the direct method is to incubate labeled agonist or antagonist with the receptor preparation and then separate the receptor–ligand complex from free ligand by centrifugation, filtration, or precipitation. The indirect technique is to use equilibrium dialysis where the receptor–ligand complex is determined by subtracting the ligand concentration in the bath from that in the dialysis sac.

Recently, a third, electrophysiological approach to identifying receptors has emerged. Intracellular stimulation and recording via microelectrodes inserted into a brain slice or neurons in culture, combined with the application of receptor agonists and antagonists, can functionally identify receptor subtypes.

Though not always appreciated, the advantage of using isolated tissues is that both the efficacy and the functional activity of an agonist is assessed, in contrast to binding procedures using broken cell preparations where only affinity or a biochemical sequella of binding can be appraised. Ideally, both techniques should be used, but, alas, this is rarely done. It should be noted in passing that efficacy and affinity are independent. To date it appears that the drugs that exhibit a high affinity but low efficacy have a more efficient coupling to the effector than the reverse situation; therefore, these are the most potent and selective agents. For detailed discussion of the above, the review by Kenakin, et al. is recommended.

IDENTIFICATION

In the midst of an intensive drive to isolate and characterize receptors, some zealous investigators have lost sight of basic tenets that must be satisfied before it is certain that a receptor has indeed been isolated. Thus, on occasion, enzymes, transport proteins, and merely extraneous lipoproteins or proteolipids that exhibit binding properties have been mistakenly identified as receptors. Authentic receptors should have the following properties:

1. *Saturability.* The great majority of receptors are on the surface of a cell. Since there are a finite number of receptors per cell, it follows that a dose–response curve for the binding of a ligand should reveal saturability. In general, specific receptor binding is characterized by a high affinity and low capacity, whereas nonspecific binding usually exhibits high capacity and low-affinity binding that is virtually nonsaturable.

2. *Specificity.* This is one of the most difficult and important criteria to fulfill because of the enormous mass of nonspecific binding sites compared with receptor sites in tissue. For this reason, in binding assays it is necessary to explore the displacement of the labeled ligand with a series of agonists and antagonists that represent both the same and different chemical structures and pharmacological properties as the binding ligand. One should also be aware of the avidity with which inert surfaces bind ligands. For example, Substance P binds tenaciously to glass, and insulin can bind to talcum powder in the nanomolar range. With agents that exist as optical isomers, it is of obvious importance to show that the binding of the ligand is stereospecific. Even here problems arise. With opiates it is the levorotatory

enantiomorph that exerts the dominant pharmacological effect. Snyder, for example, has found glass fiber filters that selectively bind the levorotatory isomer. Specificity obviously means that one should find receptors only in cells known to respond to the particular transmitter or hormone under examination. Furthermore, it is a truism that a correlation should be evident between the binding affinity of a series of ligands and the biological response produced by this series. This correlation, the *sine qua non* for receptor identification, is unfortunately a criterion that is not often investigated.

3. *Reversibility.* Since transmitters, hormones, and most drugs act in a reversible manner, it follows that the binding of these agents to receptor should be reversible. It is also to be expected that the ligand of a reversible receptor should be not only dissociable but recoverable in its natural (i.e., nonmetabolized) form. This last dictum distinguishes receptor–agonist interactions but not receptor–antagonist binding from enzyme–substrate reactions.

4. *Restoration of function on reconstitution.* Following the isolation and identification of the components of the receptor system, to put Humpty Dumpty back together again is the goal of all receptorologists. Although a difficult task, this has been accomplished with the nicotinic ACh receptor and the peripheral β-adrenergic receptor.

5. *Molecular neurobiology.* The ultimate identification is to isolate the gene for a receptor, express it, and demonstrate the exact similarity of the cloned receptor to the natural one.

It is important to recognize that the quantitative and spatial measurement of receptors utilizing autoradiography is also a key problem. Where labeled ligands are employed to map receptors in brain via light microscopy, a mismatch is often encountered. Reasons offered for this problem are (1) except for autoreceptors, neurotransmitters and receptors are located in different neurons; (2) in addition to the synapse, receptors and transmitters are found throughout the neuron and in glial cells; (3) ligands may label only a subunit of a receptor or only one state of the receptor; (4) autoradiography is subject to quenching. With immunohistochemical peptide mapping, a possible problem is the recognition by the antibody of a prohormone or, alternatively, a fragment of a peptide hormone in addition to the well-recognized problem of crossreactivity of the antibody with a physiologically different peptide. Finally, it should be recognized that all drugs do not necessarily act directly on a receptor. They could bind to a site that is adjacent to a receptor and thus influence the activity of the receptor.

KINETICS AND THEORIES OF DRUG ACTION

From the law of mass action, the binding of a ligand (L) to its receptor (R) leads to the equation

$$L + R \xleftarrow[k_2]{k_1} LR$$

$$\text{thus } K_D = \frac{k_2}{k_1} = \frac{[L][R]}{[LR]}$$

where k_1 = association rate constant

k_2 = dissociation rate constant

[L] = concentration of free ligand

[R] = concentration of free receptors

[LR] = concentration of occupied binding sites

K_d = dissociation equilibrium constant

Since the total number of receptors = $[R_t]$ = [R] + [LR]

$$[R] = [R_t] - [LR]$$

$$\text{therefore } K_d = \frac{[L]([R_t] - [LR])}{[LR]}$$

$$\text{and } [LR] = \frac{[L][R_t]}{[L] + K_d}$$

Since the fraction of receptors occupied (r) $= \dfrac{\text{bound}}{\text{total}} = \dfrac{[LR]}{[R_t]}$

therefore $r = \dfrac{[L]}{[L] + K_d}$

If experiments are performed in which the receptor concentration is kept constant and the ligand concentration is varied, then a plot of r versus [L] will produce a rectangular hyperbole, the usual Langmuir adsorption isotherm. Here r approaches the saturation value of one. If r is plotted against log [L], a sigmoid curve will result; log [L] at half saturation will give log K_d on the horizontal axis (Fig. 4-2). This equation can be rearranged as follows:

$$K_d r + r[L] = [L]$$

$$\frac{r}{[L]} = \frac{-r}{K_d} + \frac{I}{K_d}$$

Now if r/[L] is plotted against r, a straight line will result (assuming only one set of binding sites) with two intercepts, the one on the x axis giving the number of binding sites per molecule and the y intercept yielding I/K_d. This type of plot is the Scatchard plot (more

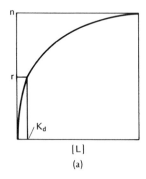

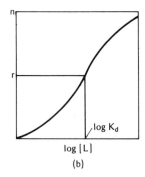

(a)

(b)

FIGURE 4-2. Ligand–receptor interactions plotted in two ways where [L] is the concentration of ligand (drug, hormone, or neurotransmitter) and r is the biological response, proportional to the moles ligand bound per mole of protein. The number of binding sites per molecule of protein is designated by n.

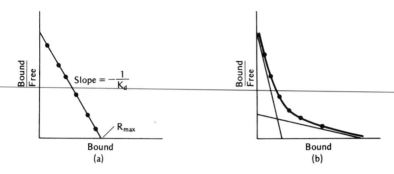

FIGURE 4-3. Scatchard plots of the binding of a ligand to a receptor. In (a) only one binding affinity occurs, but in (b) both a high- and low-affinity binding sites are suggested.

correctly, Rosenthal plot) widely used in studying receptor–ligand interactions (Fig. 4-3). Among the pitfalls that are encountered in a Scatchard analysis is the problem that the system is not in true equilibrium.

Another useful representation that can be derived from the general equation is the Hill plot (Fig. 4-4).

If $\log E/(E_{max} - E)$ is plotted versus $\log [L]$ when E is the effect produced and E_{max} is the maximum effect, then the slope, indicative

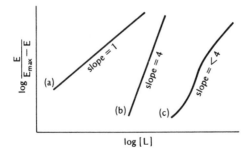

FIGURE 4-4. Hill plots for receptor–ligand binding: (a) noncooperative binding; (b) idealized plot of cooperative binding with four sites; (c) a typical Hill plot of multiple binding sites but less than four. (Modified after Van Holde, *Physical Biochemistry*, Englewood Cliffs, N.J.: Prentice-Hall, Inc., 971, p. 63)

of the nature of the binding, gives a Hill coefficient of unity in cases where E/E_{max} is proportional to the fraction of total number of sites occupied (r). In many situations, the slope turns out to be a noninteger number different from unity. This finding indicates that cooperativity may be involved in the binding of the ligand to the receptor. Cooperativity is the phenomenon whereby the ligand binding at one site influences, either positively or negatively, the binding of the ligand at sites on other subunits of the oligomeric protein. This idea, originally suggested in 1965 by Monod, Wyman, and Changeux to explain allosteric enzyme properties, currently offers the most attractive hypothesis for studying reactions of receptors with hormones, transmitters, or drugs. We will utilize this hypothesis later in an attempt to explain drug action, including the problem of efficacy (intrinsic activity), spare receptors, and desensitization.

Another useful analysis when competitive inhibitors of the receptor binding ligand are studied is the equation derived by Cheng and Prusoff in their kinetic analysis of enzyme inhibitors:

$$K_i = \frac{I_{50}}{1 + (L)/K_d}$$

where K_i is the equilibrium dissociation constant of the competitive inhibitor, and I_{50} is the concentration of the inhibitor producing 50% inhibition at the concentration of the labeled ligand that is used in the study.

Clark produced the first model of drug–receptor action, known as the "occupation theory," in which the response of a drug was held to be directly proportional to the fraction of receptors that were occupied by the drug. Here, as mentioned earlier, one should find the usual Langmuir absorption isotherm. But instead of the expected rectangular hyperbole when drug concentration was plotted against the drug bound, in most cases a sigmoid curve resulted. A second problem with the occupancy theory is that in many instances only a small fraction of the total receptors available are occupied, and yet a maximum response is obtained. These additional receptors, which may represent as much as 95–99% of the total, are referred to as "spare receptors." The occupancy theory is further complicated when the activity of a series of related agents is explored and the biological response varies from maximum to zero, even though the agents all occupy the receptor. In other words, these agents could be full agonists, partial agonists that gave less than a maximal response, and antagonists whose occupancy produced no response. This phenomenon is referred to as "efficacy" or "intrinsic activity" of a drug and is obviously not directly related to the binding affinity of the drug. A fourth characteristic of drug–receptor interaction that is sometimes observed is "desensitization," which is defined as the lack or decline of a response to a constant stimulus.

These problems—the sigmoidal dose–response curve, some anomalous effects with antagonists that spare receptors might account for, efficacy, and desensitization—can all be comfortably fitted into a two-state model of a receptor analogous to the allosteric model of Monod, Wyman, and Changeux, proposed independently by Changeux and by Karlin in 1967.

According to the two-state model, receptors exist in an active (R) and an inactive (T) state, and each is capable of combining with the drug (A):

$$A + R \overset{K_{AR}}{\leftharpoonup} AR$$
$$\downarrow \qquad \qquad \downarrow$$
$$A + T \overset{K_{AT}}{\leftharpoonup} AT$$

Here an agonist prefers the R (active configuration of the receptor), and the efficacy (i.e., intrinsic action) of the drug will be determined by the ratio of its affinity for the two states. In contrast, competitive antagonists prefer the T form of the receptor and will shift the equilibrium to AT. The sigmoid relationship between the fraction of receptors activated and the drug concentration (i.e., cooperativity) can be explained by this model with its equilibrium between R and T, if one designates T as a subunit of R that binds A and thereby influences the further binding of A to R. Cooperativity can also explain anomalous effects of antagonists whenever the effect of an antagonist persists even in the presence of a high concentration of agonist. Here it could be postulated that by tightly binding to one conformational state of the receptor, the antagonist inhibits the binding of the agonist. One might even use this two-state model to account for desensitization where T would be a receptor that has been desensitized perhaps by a local change in the ionic environment. The ionic environment, in fact, may be one of the factors that dictate the two conformational states of a receptor. Other possibilities include polymerization (clustering) of the subunits, depolymerization, or phosphorylation. It should be emphasized that this model is conjectural, subject to modifications as the need arises.

When one considers the finite number of receptors per cell and the fact that virtually all receptors are found on the plasma membrane, it is not surprising that progress in this area has been slow and laborious. It has been calculated that, assuming one binding site and an average molecular weight of 200,000 for a receptor, complete purification of a receptor protein would require about a 25,000-fold enrichment. The extraordinary density of ACh receptors in electric tissue has made this preparation so popular a choice that considerable information is now available on this cholinergic receptor. Two snake neurotoxins, *Naja siamensis* and α-bungarotoxin, which specifically bind nicotinic cholinergic receptors, have been the key agents that have helped in isolating this receptor.

Also to be considered is the relationship between receptors and effectors as they interact in the fluid mosaic membrane. As envisaged by Singer and Nicolson, membranes are composed of a fluid lipid bilayer that contains globular protein. Some of these proteins extend through the membrane, and others are partially embedded in or on the surface. The hydrophilic ends of the protein protrude from the membrane, while the hydrophobic ends are localized in the lipid bilayer. Some of the proteins are immobilized, but others, floating in an oily sea of lipids, are capable of free movement. In this concept of membrane fluidity, the receptor protein would be on the surface of the membrane and the effector within the membrane. Although the ratio of receptors to effectors may in some cases be unity (thus explaining instances in which the occupancy theory is satisfied), it may also be greater than one. Consequently, in a situation where multiple hormones activate a response (e.g., there is only one form of adenylyl cyclase in a fat cell, but it may be activated by epinephrine, glucagon, ACTH, or histamine), it would be concluded that an excess of receptors over effectors is present. This model would also explain spare receptors and is exemplified by the fact that only 3% of insulin receptors need to be occupied in order to catalyze glucose oxidation in adipocytes. With the possibility of easy lateral movement of effector in the membrane, it is also understandable why one receptor may activate several types of effector. Membrane fluidity will account for the observation that hormones can influence the state of aggregation of the receptor, thus giving rise to either positive or negative cooperativity as determined in kinetic studies of binding. Although direct evidence of the interaction and migration of receptors and effectors is difficult to obtain (but see the review article by Poo), current information is easily accommodated by the fluid mosaic membrane model.

A recently emerging view of the structure and activity of neurotransmitter receptors, as reviewed by Strange, is that they can be divided into two classes, fast and slow. Class I (fast) receptors, currently referred to as receptor ionophores or ionotropic receptors, are directly linked to an ion channel and mediate millisecond responses when activated by a transmitter. Structurally, this ligand-gated ion channel family appears to contain four hydropathic, membrane-spanning segments. Examples of ionotropic receptors are the nicotinic ACh receptor and the receptors for $GABA_A$, glutamate, aspartate, glycine, and one subtype for 5-HT. These ion channels are regulated by phosphorylation. In contrast, class II (seconds to minutes) receptors, currently denoted as G-protein-coupled receptors, mediate slower responses that are generally modulatory, either dampening or enhancing the signal that acts on ionotropic receptors. These receptors are coupled to G proteins (discussed below), which in turn may be directly coupled to ion channels or linked to second messenger systems. Structurally, all known G-protein-coupled receptors contain seven hydrophobic transmembrane domains that are linked by hydrophilic groups. Examples of these receptors are adrenergic, muscarinic cholinergic, dopaminergic, glutamate metabotropic, serotonergic, and opiate. It is likely that most neu-

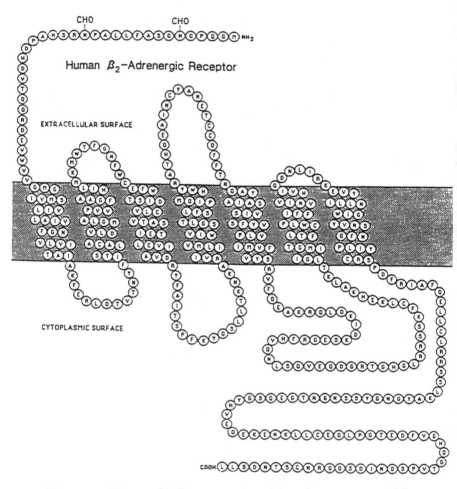

FIGURE 4-5. Topographical representation of the primary sequence of the human β₂-adrenergic receptor. The receptor protein is illustrated as possessing seven hydrophobic regions, each capable of spanning the plasma membrane, thus creating extracellular as well as an intracellular loops as well as an extracellular terminus and a cytoplasmic C-terminal region. (From Lefkowitz et al., 1989).

ropeptide receptors are also of the slow, G-protein-coupled type. It is also likely that presynaptic receptors that, when activated, alter the amount of transmitter released, are of the same group. The structure of these receptors as exemplified by the β_2-adrenergic receptor is illustrated in Fig. 4–5.

These transducing proteins are a family of guanine nucleotide-binding proteins with a heterotrimer structure consisting of α-, β-, and γ-subunits. The G protein signal transduction cycle is shown in Figure 4–6. At last count, 21 different G protein α-subunits and at

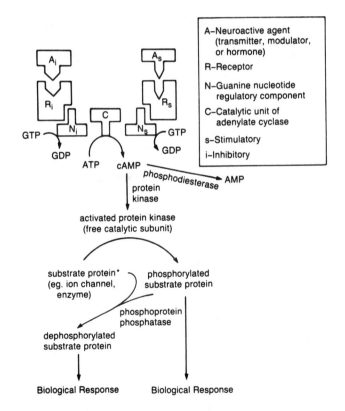

FIGURE 4-6. Components of a receptor-activated, cyclic nucleotide-linked system. (After Lefkowitz et al., 1984).

least 10 β- and γ-subunits had been identified. These proteins can be roughly classified into four groups, G_s, G_i, G_q, and G_{12}. Activation of the G_s subunit family increases adenylate cyclase activity, opens Ca^{2+} channels, and inhibits Na^+ channels. The G_i subunit family opens K^+ channels, closes Ca^{2+} channels, inhibits adenylate cyclase, and promotes cGMP phosphodiesterase and probably phospholipase A_2. Increased phospholipase C is the effector for G_q, while with G_{12}, a ubiquitous subunit, there is yet no known function (a dysfunctional family?). Also recently recognized are G-protein-coupled receptor kinases that phosphorylate a G-protein-activated receptor and terminate signaling. To add to this bewildering complexity in neuronal signaling, it should be noted that in addition to the fact that a single receptor can activate multiple G proteins that may or may not interact, cloning studies have revealed multiple isoforms of adenylate cyclase, phospholipase C, phospholipase A_2, and calcium and potassium channels. Furthermore, when one realizes that, in turn, second messenger systems promote protein phosphorylation, transcription factors, secretion, membrane depolarization, metabolic activity, and cytoskeleton reorganization, it is understandable why puzzled neuroscientists longingly think of turning to a simpler occupation like growing tomato plants in their garden.

An observation that has helped assign G proteins a role in signal transduction is that bacterial toxins catalyze the NAD-dependent ADP ribosylation of the α-subunit of many of the G proteins and inhibit their activity. Cholera toxin ribosylates G_s and G_t, whereas G_i and G_0 are substrates for pertussis toxin. Some G proteins (unclassified) are resistant to both toxins.

Although classified as modulatory rather than neurotransmitter receptors, there are two other classes of receptors that deserve attention. These are the purinoreceptors and steroid receptors.

Purinoreceptors have been classified as P_1 and P_2, with adenosine as the endogenous ligand for P_1 and ATP and other purine and pyrimidine nucleotides activating P_2 receptors. Adenosine is primarily derived from the hydrolysis of ATP by ectonucleotidases. The P_1 family are G-protein-coupled receptors, subdivided into A_1, A_{1a}, A_{2b}, A_3, and A_4 subtypes; all except the A_4 receptors have been cloned.

Interest in the P_1 receptor stems from the fact that adenosine (which is found at virtually every synapse that has been examined) exhibits an extraordinary constellation of activities. Adenosine (and its agonists) inhibits the evoked release of neurotransmitters, hyperpolarizes postsynaptic membranes, decreases locomotor activity, possesses sedative and anticonvulsant activities, increases cerebral blood flow, reduces neurodegeneration that occurs with stroke, evokes premature arousal in hibernating ground squirrels, and at high doses promotes catalepsy. Except for the A_1 and A_4 subclasses, the P_1 family of purinoreceptors is blocked by xanthines, with the classic antagonists being caffeine and theophylline.

The P_2 receptor family with ATP as the major agonist can be divided into two major classes designated P_{2x} and P_{2y}, which have a variety of subclasses. The former is coupled to ligand-gated ion channels and the latter is a G-protein-linked family with subclasses that utilize UTP or ADP as the primary agonist. ATP is considered to be a cotransmitter in the peripheral nonadrenergic, noncholinergic nervous system where it is released with norepinephrine or acetylcholine. ATP also has been shown to mediate fast synaptic transmission in mammalian neurons. Subclasses of the P_2 receptor are activated by UTP and ADP. The lack of selective antagonists to the P_2 receptor has hampered understanding of the function of P_2 subclasses.

Receptors for steroid hormones have long been known to be intracellular since the hormones act on nuclear DNA to alter gene expression. This action is referred to as a genomic effect and usually takes hours to days to be observed. These genomic actions include the induction of neurotransmitter enzymes, receptors, and dendritic spines. Recently, however, evidence has accumulated that steroid receptors are also found on membranes. The hormones' nongenomic effects, occurring in seconds to minutes, on the $GABA_A$ receptor to modulate chloride flux (see Chapter 7), modulate oxytocin receptors, and alter LHRH and dopamine release. The major players in this activity are progesterone and its metabolites, estrogens, and adrenal steroids.

Continued administration of agonists can cause many receptors to desensitize and to down-regulate. Desensitization, occurring on a time scale of minutes, is reflected by a decreased response of the cell and is often related to receptor phosphorylation. Down-regulation of receptors is observed on a time scale of hours after prolonged agonist exposure when receptors are internalized and degraded. Conversely, and predictably, the continued administration of receptor antagonists generally causes an up-regulation of receptors.

Many neuroactive agents act on receptors that are coupled to the adenylyl cyclase system (see Fig. 4-6). The components of this complex are the receptor, the catalytic portion of adenylyl cyclase that converts ATP to cAMP, and two G proteins referred to as N_s and N_i that are coupled to the catalytic unit of the enzyme. When the receptor is stimulated (e.g., a β_2-adrenergic receptor), the coupling protein is N_s. Conversely, when the receptor is inhibited (e.g., an α_2-adrenergic receptor), N_i is the coupling protein. Examples of adrenergic receptors whose activity is linked to the adenylate cyclase complex are the receptors α_2, β_1, and β_2 (but not α_1, which may be coupled to phosphatidyl inositide hydrolysis).

Finally, although not legitimately classified as receptors, another class of proteins that deserves attention is the transport proteins. With the exception of acetylcholine, the action of all the neurotransmitters that are released is terminated primarily by reuptake into their presynaptic terminals (amino-acid transmitters can also be taken up by glial cells). Acetylcholine is hydrolyzed by acetylcholinesterase, and it is choline that is recaptured by the cholinergic terminal. Neurotransmitter transport proteins are relatively specific for each transmitter, are sodium dependent, exhibit high-affinity kinetics, and are dependent on the membrane potential of the neuronal terminal. The GABA and glutamate transporters have the ability to function in reverse, transporting the transmitters out of the cell. Most of the transporters have now been cloned; their structure suggests an 11 to 13 membrane spanning domain. As discussed in the chapters on each neurotransmitter, these reuptake systems are the basis of the mechanism of action of many neuropharmacological agents, particularly the antidepressant drugs and some drugs of abuse such as cocaine.

It should be noted that the transporters operate at the plasma membrane of terminals and are different from intracellular vesicular transporters that accumulate transmitters into synaptic vesicles. Vesicular transporters, some of which have been cloned, operate via a vacuolar type H^+-pumping ATPase. This proton pump generates an H^+ electrochemical gradient whereby the efflux of H^+ is coupled to the reuptake of the transmitter into the vesicles.

Because of space limitations, this chapter has not covered such topics as the factors regulating synthesis and degradation of receptors, complex kinetics, isolation techniques, or the chemical compositions of these macromolecules. Students who are interested in these subjects may consult a number of recent books and reviews listed in the selected references.

SELECTED REFERENCES

Amara, S. and M. J. Kuhar (1993). Neurotransmitter transporters: Recent progress. *Annu. Rev. Neurosci. 16*, 73.

Ariens, E. J. and A. J. Beld (1977). The receptor concept in evolution. *Biochem. Pharmacol. 26*, 913.

Bonner, T. I. (1989). The molecular basis of muscarinic receptor diversity. *Trends Neurosci. 12*, 148.

Changeux, J. P., J. Thiery, Y. Tung, and C. Kittel (1976). On the cooperativity of biological membranes. *Proc. Natl. Acad. Sci. U.S.A. 57*, 335.

Cheng, Y.-C. and W. H. Prusoff (1973). Relationship between the inhibition constant (K_i) and the concentration of inhibitor which causes 50 percent inhibition (I_{50}) of an enzymatic reaction. *Biochem. Pharmacol. 22*, 3099.

Clapham, D. E. (1994). Direct G protein activation of ion channels. *Annu. Rev. Neurosci. 17*, 441.

Clark, A. J. (1937). General pharmacology. In *Heffter's Hanbuch d-exp. Pharmacol. Erg. Band 4*. Springer, Berlin.

Goyal, R. K. (1989). Muscarinic receptor subtypes. *N. Engl. J. Med. 321*, 1022.

Hepler, J. R. and A. G. Gilman (1992). G proteins. *TIBS 17*, 383.

Hille, B. (1992). G protein-coupled mechanisms and nervous signaling. *Neuron 9*, 187.

Karlin, A. (1967). On the application of a plausible model of allosteric proteins to the receptor for acetylcholine. *J. Theor. Biol. 16*, 306.

Kenakin, T. P., R. A. Bond and T. I. Bonner (1992). Definition of pharmacological receptors. *Pharmacol. Res. 44*, 351.

Laduron, P. M. (1984). Criteria for receptor sites in binding studies. *Biochem. Pharmacol. 33*, 833.

Lefkowitz, R. J., B. K. Kobilka, and M. G. Caron (1989). The new biology of drug receptors. *Biochem. Pharmacol. 38*, 2941.

Lefkowitz, R. J., M. G. Caron, and G. L. Stiles (1984). Mechanisms of membrane-receptor regulation. *N. Engl. J. Med. 310*, 1570.

Lester, H. H., S. Mager, M. W. Quick, and J. L. Corey (1994). Permeation properties of neurotransmitter transporters. *Annu. Rev. Pharmacol. Toxicol. 34*, 219.

McEwen, B. S. (1991). Non-genomic and genomic effects of steroids on neural activities. *TIPS 12*, 141.

Neubig, R. R. (1994). Membrane organization in G-protein mechanisms. *FASEB J. 8*, 939.

Poo, M. (1985). Mobility and localization of proteins in excitable membranes. *Annu. Rev. Neurosci. 8*, 369.

Rakic, P., P. S. Goldman-Rakic, and D. Gallager (1988). Quantitative autoradiography of major neurotransmitters in monkey striate and extrastriate cortex. *J. Neurosci. 8*, 3670.

Singer, S. J. and G. L. Nicolson (1972). The fluid mosaic model of the structure of cell membranes. *Science 175*, 720.

Strader, C. D., I. S. Sigal, and R. F. Dixon (1989). Structural basis of B-adrenergic receptor function. *FASEB J. 3*, 1825.

Strader, C. D., T. M. Fong, M. R. Tota, D. Underwood, and R.A.F. Dixon (1994). Structure and function of G protein-coupled receptors. *Annu. Rev. Biochem. 63*, 101.

Strange, P. G. (1988). The structure and mechanism of neurotransmitter receptors. *Biochem J. 249*, 309.

Williams, M. (1995). Purinoceptors in central nervous system function: Targets for therapeutic intervention. Psychopharmacology: The fourth generation of progress. (F. E. Bloom and D. J. Kupfer, eds.), p. 643 Raven Press, New York.

5 | Modulation of Synaptic Transmission

Contrary to what one might assume, the more we learn about intercellular communication in the nervous system, the more complicated the situation appears. Up to about the mid-1970s, synaptologists smugly focused on a straight point-to-point transmission where a presynaptically released transmitter impinged on a postsynaptic cell. Gradually, situations emerged where previously identified neurotransmitters were observed as not acting in this fashion but rather "modulating" transmission. These departures from the previous norm and the continuing discovery of peptides and small molecules such as adenosine, which exhibited neuroactivity but which did not appear to be transmitters in the classic sense, supported the broader view of modulation of synaptic transmission. In retrospect, it is a concept that should have been apparent early on, since it imparts to neural circuitry an extraordinary degree of flexibility that is necessary in considering mechanisms to account for behavioral changes. It should also be noted that modulation is not a neurophysiological property that is seen only in higher forms: In Cole Porter's words, "Birds do it, bees do it, even educated fleas do it."

DEFINITIONS

Primarily via the activation of a receptor, synaptic transmission may be modulated either presynaptically or postsynaptically. With presynaptic modulation, regardless of the mechanism, the ultimate effect is a change in the amount of transmitter that is released. With postsynaptic modulation, the ultimate effect is a change in the firing pattern of the postsynaptic neuron or in the activity of a postsynaptic tissue (e.g., blood vessel, gland, muscle).

Because some confusion has arisen as to the correct nomenclature of pre- and postsynaptic receptors, an explanation is in order. It should be recognized that what may be classified as a presynaptic receptor on neuron B may be a postsynaptic receptor of neuron A, which is making an axoaxonic contact with neuron B (see Fig. 2-3). Thus, depending on which neuron one is investigating, the receptor will be denoted as either pre- or postsynaptic.

Procedures that have been used to fix activity at the presynaptic receptor level in a terminal include (1) the use of synaptosomes; (2) the addition of tetrodotoxin to the preparation to block action potentials in neighboring interneurons; (3) the use of neurons from the peripheral nervous system in cell culture where no postsynaptic cells exist; or (4) where feasible, either chemical destruction of terminals or lesioning of the neuron and then demonstration by ligand binding of the loss of the receptor. To complete the nomenclature on presynaptic receptors, an *autoreceptor* is located on the terminal or somatic–dendritic area of a neuron that is activated by the transmitter(s) released from that neuron. A *heteroreceptor* is a presynaptic receptor that is activated by a modulating agent that originates from a different neuron or cell. As discussed in Chapter 1, modulators differ from transmitters in that they have no intrinsic activity but modulate ongoing neural activity. However, it should be noted that a transmitter may modulate at a concentration that is subthreshold for intrinsic activity. As we will now detail, there exists an extraordinary number of possibilities for altering the point-to-point synaptic transmission that was mentioned above.

Presynaptic modulation can be affected by:

1. Receptor activation of a presynaptic neuron causing
   a. A change in the firing frequency in the presynaptic neuron: This is probably the most common type of modulation, particularly in the central nervous system (CNS), where it can be assumed that the firing rate of virtually every neuron is governed by inputs on dendrites, soma,

or axons. The firing rate determines the frequency of impulse conduction, hence the spread of action potentials into terminals or varicosities and the amount of transmitter that is liberated.

b. A change in the transport or reuptake of a transmitter or precursor, or in the synthesis, storage, release, or catabolism of a transmitter: All of these possibilities will result in a change in the concentration of a transmitter at the terminal. In practice, it has been shown that presynaptic activation of biogenic amine neurons promotes the phosphorylation of both tyrosine and tryptophan hydroxylase, which increases the synthesis of norepinephrine, dopamine, and serotonin. Curiously, phosphorylation of the pyruvate dehydrogenase complex causes a decrease in enzyme activity and in theory would decrease the levels of ACh and the amino acid transmitters. To date, however, modulation of this enzyme activity by presynaptic receptor activation has not been observed.

c. An effect on ion conductances at the terminal: The three ions and their respective channels that one might focus on would be K^+, Ca^{2+}, or Cl^-. Endogenous neuroactive agents or drugs could alter transmitter release by opening or blocking the channels or changing the kinetics of channel open time via the possible mediation of protein phosphorylation or other second messengers.

2. A direct effect of neuropharmacological agents on some element of the release process: This could be an effect of the modulating agent on vesicular apposition to a terminal, fusion, or fission if an exocytotic mechanism is operating, or, in a nonexocytotic release process, an effect possibly on Na^+, K^+-ATPase.

Postsynaptic modulation can be affected by:

1. A long-term change in the number of receptors: This is a situation that is not observed under normal physiological conditions. It is, however, commonly noted pharmacologi-

cally where the administration of a receptor agonist for a period of time will result in the down-regulation of the receptor, and, conversely, treatment with a receptor antagonist increases receptor density.

2. A change in the affinity of a ligand for a receptor: The now classic example is the salivary nerve of the cat where both ACh and VIP are co-localized. When VIP is released on electrical stimulation, it increases the affinity of ACh for the muscarinic receptor on the salivary gland up to 10,000-fold, with a consequent increase in salivation.

3. An effect on ionic conductances: As discussed in 1a above, this is frequency modulation. It is a postsynaptic effect on the first neuron in a relay, but it would be classified as a presynaptic effect on the second neuron.

References to all the neurons and all the modulating agents that have been investigated are given in the reviews by Starke et al. and Chesselet and in a book by Levitan and Kaczmarek. These reviews can be summarized by stating that virtually every neuronal pathway is modulated and virtually every endogenous neuroactive agent has been shown in one preparation or another to be capable of affecting synaptic transmission. All this information is descriptive. The question of the second messenger systems that may be involved in regulating synaptic activity is addressed in the following section. Figure 5-1 depicts a major pathway for modulation of synaptic transmission. Rapidly accumulating evidence suggests that in most cases of receptor-activated *inhibitory* presynaptic modulation, the ultimate effect is to open K channels. This hyperolarizes terminals; less Ca^{2+} enters; and, as a consequence, less transmitter is released. Another major presynaptic mechanism is the inhibition of a calcium channel that would also decrease the evoked release of a transmitter. With the less common receptor-activated *excitatory* modulation, closing of a K channel has been implicated. A well-documented postsynaptic modulatory mechanism is that of a transmitter that inhibits a voltage-dependent K current, causing a subsequent depolarizing stimulus to produce an enhanced response.

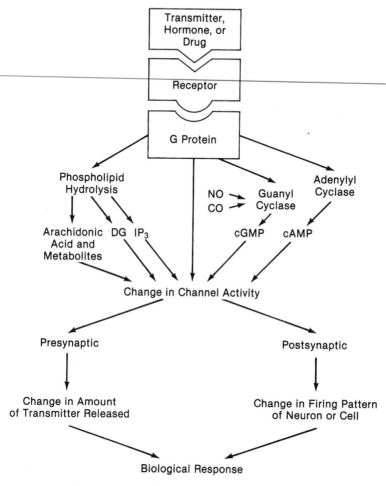

FIGURE 5-1. Major pathways for modulation of synaptic transmission. DG, diacylglycerol; IP$_3$ inositoltrisphosphate; NO, nitric oxide; CO, carbon monoxide.

SECOND MESSENGERS

Three major biochemical cascades and two new gaseous messengers, nitric oxide and carbon monoxide, have been described:

1. *Protein phosphorylation.* Following the identification of cyclic nucleotides by Sutherland and associates, and the implication that

they act as a second messenger system preceded by an initial nerve impulse or hormonal signal, E. Krebs and colleagues demonstrated a multistep sequence of events that linked cyclic AMP generation in muscle to regulation of carbohydrate metabolism. Since that time, the cyclic nucleotides have been shown to regulate an enormous diversity of processes ranging from axoplasmic transport to neuronal differentiation and including transmitter synthesis and release and the generation of postsynaptic potentials. All of these effects are attributable to the cAMP- or cGMP-activating protein kinases and thus protein phosphorylation. Second messenger activity, achieved via protein phosphorylation, can be mediated by cAMP- and cGMP-dependent protein kinases as well as by calcium-calmodulin-dependent protein kinase and calcium–phosphatidylserine-dependent protein kinase. The phosphoryl acceptor is the hydroxyl group of serine, threonine, or tyrosine. All protein kinases can themselves be autophosphorylated, a process that usually increases kinase activity. Although in most instances biological activity results from kinase-activated phosphorylation of a substrate protein, in some cases it is phosphatase-activated dephosphorylation of a phosphorylated protein that produces the biological response. Currently, eight phosphoprotein phosphatases have been identified (protein phosphatase-2B is also referred to as calcineurin). These phosphatases fall into two broad classes, phosphoserine/phosphothreonine phosphatases and phosphotyrosine phosphatases. Yet another level of regulation is suggested by the finding that protein phosphatase activity can be inhibited by other proteins, the most interesting of which is DARPP-32, found in D_1-dopaminoceptive neurons. DARPP-32, a dopamine- and cAMP-regulated phosphoprotein, by acting as a protein phosphatase inhibitor when it is phosphorylated, can regulate postsynaptic effects of dopamine in dopaminoceptive cells. Phosphorylated DARPP-32 is inactivated by protein phosphatase-2B. It should be noted that in some instances Ca acts as a second messenger without the participation of protein phosphorylation. The diversity of signals that are coupled to protein phosphorylation is depicted in Fig. 5-2. For a molecular illustration of receptor coupling, the reader is referred to Fig. 4-5.

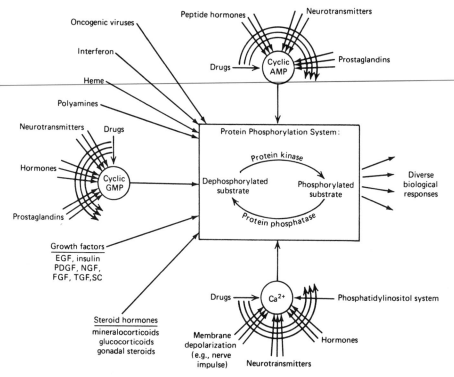

FIGURE 5-2. Schematic diagram of the role played by protein phosphorylation in mediating some of the biological effects of a variety of regulatory agents. Many of these agents regulate protein phosphorylation through altering intracellular levels of a second messenger, cAMP, cGMP, or Ca^{2+}. Other agents appear to regulate protein phosphorylation through mechanisms that do not involve these second messengers. Most drugs regulate protein phosphorylation by affecting the ability of first messengers to alter second-messenger levels *(curved arrows)*. A small number of drugs (e.g., phosphodiesterase inhibitors, Ca^{2+} channel blockers) regulate protein phosphorylation by directly altering second messenger levels *(straight arrows):* EGF, epidermal growth factor; PDGF, platelet-derived growth factor; NGF, nerve growth factor; FGF, fibroblast growth factor; TGF, transforming growth factor; SC, somatomedin C. (From Nestler and Greengard, 1989)

Despite the vast number of systems in which protein phosphorylation is implicated, there are currently only about a dozen cases where direct evidence links this process to modulation of synaptic transmission, either pre- or postsynaptically. It is not yet known if the proteins that make up the channels are phosphorylated or if the phosphorylated proteins are morphologically associated with the channels. At any rate, with over 4 dozen proteins in brain that are known to be phosphorylated (Table 5-1), the story is far from com-

TABLE 5-1. Classes of Neuronal Proteins Regulated by Phosphorylation

Enzymes involved in neurotransmitter biosynthesis
    Tyrosine hydroxylase
    Tryptophan hydroxylase
    Pyruvate dehydrogenase
Neurotransmitter receptors
    Nicotinic acetylcholine receptor
    Muscarinic cholinergic receptor
    β-Adrenergic receptor
    α_2-Adrenergic receptor
    $GABA_A$, receptor
Ion channels
    Voltage-dependent Na^+, K^+ and Ca^{2+} channels
    Ligand-gated channels
    Ca^{2+}-dependent K^+ channels
Enzymes and other proteins involved in the regulation of second messenger levels
    G proteins
    Phospholipases
    Adenylate cyclase
    Guanylate cyclase
    Phosphodiesterase
    IP_3 receptor
Protein kinases
    Autophosphorylated protein kinases

TABLE 5-1. Continued

Protein kinases phosphorylated by other protein kinases
Protein phosphatase inhibitors
  DARPP-32
  Cyclosporin A
  Okadaic acid
  Microcystin-LR
Cytoskeletal proteins involved in neuronal growth, shape, and
  motility
  Actin
  Tubulin
  Neurofilaments (and other intermediate filament proteins)
  Myosin
  Microtubule-associated proteins
Synaptic vesicle proteins involved in neurotransmitter release
  Synapsins I and II
  Clathrin
  Synaptophysin
Transcription factors
  Cyclic AMP response element binding (CREB) proteins
  Immediate early gene products (Fos, Jun, and Zif)
  Steroid and thyroid hormone receptors
Other proteins involved in DNA transcription or mRNA
  translation
  RNA polymerase
  Topoisomerase
  Histones and nonhistone nuclear proteins
  Ribosomal protein S6
  eIf (eukaryotic initiation factor)
  eEF (eukaryotic elongation factor)
  Other ribosomal proteins
Miscellaneous
  Myelin basic protein
  Rhodopsin
  Neural cell adhesion molecules

SOURCE: Modified from Hyman and Nestler (1993).

plete. Questions that remain to be answered include the regulation of the enormous number of steps in the cascade and the substrate specificity of the phosphodiesterases, protein kinases, and phospho-protein phosphatases.

2. *Phosphoinositide hydrolysis.* In 1953 Hokin and Hokin showed that the incorporation of inorganic phosphate (Pi) into phosphatidyl inositol (PI) and phosphatidic acid (PA) in pancreas slices was stimulated by ACh and ultimately resulted in the release of amylase. This receptor-activated hydrolysis of phosphoinositides is referred to as the phosphatidyl inositol effect. For nearly 30 years after the Hokin report, the literature on this effect was replete with the traditional scientific jargon "it is tempting to speculate that . . .," with no one having solid evidence as to whether the phosphoinositides or the in-ositol phosphates were the message and, if so, exactly what was the medium for the exchange. That this situation has now dramatically improved is shown in Fig. 5-3.

The signals that initiate this transduction process in neuronal systems include ACh, norepinephrine, serotonin, histamine, glutamate bradykinins, Substance P, vasopressin, TRH, neurotensin, VIP, NGF, and angiotensin acting on brain, sympathetic ganglia, salivary glands, iris smooth muscle, adrenal cortex, and neuronal tumor cells. Specific receptors that have been implicated are muscarinic cholinergic receptors, α_1-adrenergic receptors, the H_1-histaminergic receptor, Substance P, and the V_1-vasopressin receptor. In each case, Ca appears to be the intracellular second messenger that activates phosphoinositide hydrolysis. Like the specific GTP binding proteins that link receptors to the adenylate cyclase system discussed earlier, evidence is accumulating to suggest that a specific GTP binding protein is also coupled to the phosphodiesterase that catalyzes the hydrolysis of phosphatidylinositol, 4,5-bisphosphate.

As noted in Fig. 5-3, the key participants in the transduction process are the receptor, the G protein, and a phosphodiesterase, a specific phospholipase C yielding two separate second messengers— the water-soluble $InsP_3$ and the lipid-soluble diacylglycerol. The former mobilizes calcium (released in a wave form, i.e., oscillatory), which can act through calmodulin to phosphorylate specific proteins, and the latter, by activating C-kinase, a calcium–phosphatidylserine-dependent family of protein kinases, also phosphorylates specific proteins. The $InsP_3$ receptor, associated with the smooth endoplasmic recticulum, is a tetramer of identical subunits that is regulated by ATP and cAMP. Since diacylglycerol can activate guanylate cyclase to produce cGMP, an inference of a cGMP-dependent activity (protein kinase or otherwise) must be considered. With an assumed ambidexterity of neuronal cells, these two arms could function singly, cooperatively, or antagonistically, depending on the situation, thus providing subtle variations on the modulatory mechanism. In addition, as noted in Fig. 5-3, calcium may produce a physiological response directly without invoking an activation of calmodulin; so yet another control is indicated. On the subject of control, the activities of the various kinases, esterases, and phospholipases in the PI cycle would be expected to be vital control points. For example, five isoforms of phospholipase C have now been identified, one of which is stimulated when certain growth factors activate tyrosine kinase receptors. Some forms are enriched in specific brain areas. Free inositol in the brain must be derived from glycolysis since plasma inositol cannot pass the blood–brain barrier to any significant degree. Glycolysis, therefore, would be another regulatory factor in the response mechanism.

Finally, although the origin of the mobilized calcium is now clear (it is released from endoplasmic reticulum and not mitochondria), some controversy still exists as to whether phosphoinositide hydrolysis releases only internal calcium or whether external calcium influx is also invoked. Current evidence suggests that for neuronal modulation both or either may be involved, depending on the preparation. The same answer can also be given to the other controversy on whether the PI effect is presynaptic or postsynaptic. A complication in the PI cycle that has recently surfaced is the finding that the inositol trisphosphate that is produced is not always or only $Ins(1,4,5)P_3$ but may be $Ins(1,3,4)P_3$ as well as $InsP_4$, $InsP_5$, and $InsP_6$. The physiological role of these compounds is currently under investigation.

3. *Eicosanoids (arachidonic acid metabolites)*. Arachidonic acid, synthesized from dietary linoleic acid, derived on demand by either a

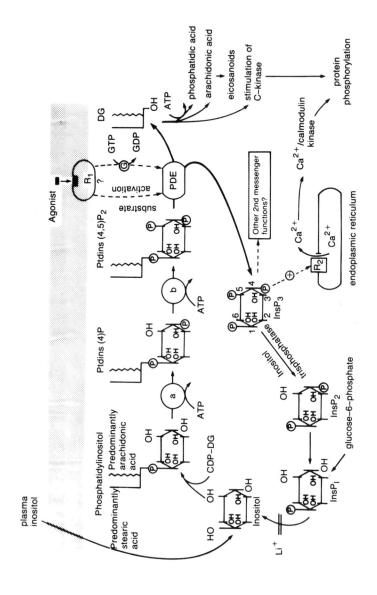

FIGURE 5-3. Receptor-activated phosphoinositide metabolism. The binding of an agonist to a receptor on the plasma membrane stimulates the hydrolysis of phosphatidylinositol, 4,5-bis-phosphate [PtdIns (4,5)P_2] by a phosphodiesterase (PDE, phosphoinositidase, phospholipase C), a specific phospholipase whose activity is controlled by a guanine nucleotide regulatory protein to form inositol 1,4,5-trisphosphate (InsP_3) and diacylglycerol (DG). The former binds to a receptor (R_2) on the endoplasmic reticulum to release calcium, which can directly produce a biological response or can activate calmodulin kinase to promote protein phosphorylation and a subsequent biological response. In some cells (e.g., mouse atria, neuroblastoma, and glioma hybrid NG108-15), receptor-activated production of InsP_3 requires extracellular Ca^{2+}.

The latter parallel arm of the cycle, diacylglycerol, can also promote a biological response via the production of prostaglandins, thromboxanes, and leukotrienes from released arachidonic acid (arachidonic acid has also been reported to stimulate guanylate cyclase to generate cGMP) or via a stimulation of protein kinase C (C-kinase) and subsequent protein phosphorylation. Diacylglycerol has also been reported to promote fusion of synaptic vesicles to terminal membrane. Diacylglycerol can also be generated from phosphatidylcholine via phospholipase D followed by phosphatidic acid phosphatase.

The phosphoinositides are synthesized from inositol with cytidine diphosphate:diacylglycerol (CDP:DG) as intermediary and the stepwise phosphorylation by kinases (a and b). As shown in the figure, lithium blocks the cycle by inhibiting inositol-1-phosphatase. Although the antimanic activity of lithium has been ascribed to this inhibitory effect, the evidence is not compelling. (Modified from Berridge and Irvine, 1984)

G-protein-regulated phospholipase A$_2$ or diglyceride lipase activation (Fig. 5-4), yields a bewildering array of bioactive metabolites as shown in Fig. 5-5. The three major groups are prostaglandins, thromboxanes, and leukotrienes.

Everybody knows that the eicosanoids, particularly the prostaglandin series, play an important modulatory role in nervous tissue, but it is difficult to write a lucid account of specifically how and where they act. This situation is primarily due to the fact that they are not stored in tissue but are synthesized on demand, particularly in pathophysiological conditions; they act briefly (some with a half-life of seconds) and at extremely low concentrations (10^{-10} M). Although indomethacin is a good inhibitor of cyclooxygenase, blocking the conversion of arachidonic acid to prostaglandins, there are few specific inhibitors available to block lipoxygenase and epoxygenase. Thus, although it had been postulated that the E series of prostaglandins modulates noradrenergic release, blocks the convulsant activity of pentylenetetrazol, strychnine, and picrotoxin (possibly by increasing the level of GABA in the brain), and increases the level of cAMP in cortical and hypothalamic slices, these effects were noted in vitro with the addition of substantial amounts of the prostaglandins. There was very little evidence in intact animals to support these neuronal findings, and since we all believe in *In Vivo Veritas*, the physiological relevance of the effect was in doubt.

Recently, however, direct evidence has implicated the eicosanoids as second messengers. The cascade begins with the binding of a neuroactive agent to its receptor. Then, according to findings from the Axelrod laboratory, the receptor is coupled to G proteins, which may either activate or inhibit phospholipase A$_2$, although this has not been conclusively established for all neural tissues. The activated enzyme promotes the release of arachidonic acid, which will then act intracellularly as a second messenger. Arachidonic acid and its metabolites can also leave the cell to act extracellularly as first messengers on neighboring cells.

To continue this saga, the electrophysiological activity of the eicosanoids has been carefully elucidated in the laboratories of Schwartz and Siegelbaum. In several identified cells of the *Aplysia*

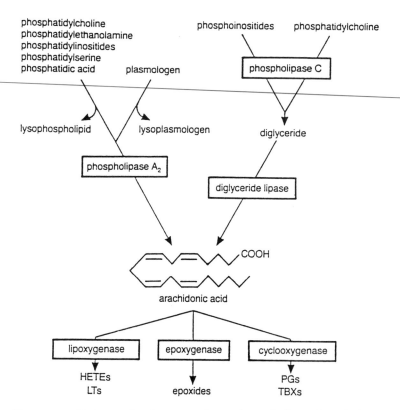

FIGURE 5-4. Pathways for the generation and metabolism of arachidonic acid. Arachidonate can arise directly from phospholipids through the action of phospholipase A_2 or prior action of phospholipase C, followed by the action of diglyceride lipase. Alternatively, the diglyceride may be phosphorylated to phosphatidic acid by action of diglyceride kinase, and arachidonate then can be released through the action of phospholipase A_1. The released arachidonate may then be metabolized by lipoxygenase, cyclooxygenase, or epoxygenase enzymes to form leukotrienes (LTs), hydroxyeicosatetraenoic acids (HETEs), prostaglandins (PGs), thromboxanes (TXs), and epoxides. (From Axelrod et al., 1988)

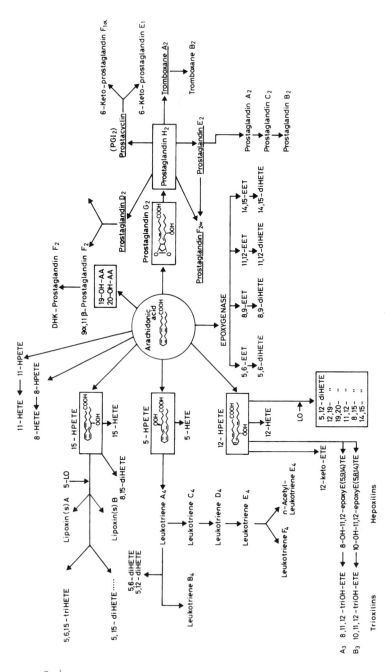

FIGURE 5-5. Metabolism of amino acids by cyclooxygenase (CO), lipoxygenases (LO), epoxygenase, and corresponding end products (prostanoids). (From Schaad et al., 1991)

abdominal ganglion, the mechanism of action of either histamine or the peptide FMRFamide, agents that cause presynaptic inhibition, is mediated by the lipoxygenase metabolite 12-HPETE. Furthermore, in L-14, a follower cell, 8-HE$_p$ETE, a metabolite of 12-HPETE, appears to be responsible for the late hyperpolarization in this cell. In cell-free patches of sensory neurons where no G proteins or protein-phosphorylation activities would be present, 12-HPETE mimics the action of FMRFamide in opening S-K$^+$ channels. In the rat hippocampus, Bliss and colleagues have found that a lipoxygenase inhibitor blocks long-term potentiation. Eicosanoids have also been shown to mediate the somatostatin-induced opening of an m channel in hippocampal pyramidal cells and the release of VIP in mouse cerebral cortical slices. It is thus becoming clear, despite enormous technological difficulties in assaying eicosanoids, that these agents are major messengers.

Before we leave the arachidonic acid story, an exciting new finding should be mentioned. Arachidonylethanolamine, referred to as anandamide, has been isolated from porcine brain and shown to be the endogenous ligand that binds to cannabinoid receptors. In behavioral tests, anandamide has exhibited cannabinoid receptor agonist activity in parallel with known psychotropic cannabinoids. Anandamide is synthesized either from arachidonic acid and ethanolamine or following hydrolytic cleavage of membrane N-arachidonoyl phosphatidylethanolamine (mechanism still unclear), and is hydrolyzed by nonspecific amidases. For a detailed description of the role of eicosanoids in synaptic transmission, the review by Piomelli is recommended.

Nitric Oxide (NO)

In 1980, Furchgott and Zawadzki observed that stimulation of the endothelium released a factor that relaxed blood vessels. This factor, referred to as the endothelium-derived relaxing factor (EDRF), was subsequently identified by the Moncada laboratory as nitric oxide (NO). Acting as a second messenger, NO is now known to be involved in an incredible number of systems. Aside from its role in the nervous system, NO is a mediator in the cardiovascular, renal, pulmonary, endocrine, and immune systems.

Nitric oxide is synthesized from arginine via the enzyme nitric oxide synthase, an FAD and FMN enzyme, requiring molecular O_2 and with NADPH as coenzyme and tetrahydrobiopterin as cofactor. The neuronal and endothelial enzyme that is constitutively expressed is activated by Ca^{2+} and calmodulin, whereas the macrophage enzyme that is inducible by cytokines is not. The synthetic reaction sequence in brain is shown in Fig. 5-6.

NO synthase is inhibited by a variety of arginine analogs, and this finding has proved to be invaluable in delineating the functions of NO in vivo. The constituitive neuronal nitric oxide synthase is regulated by phosphorylation. A number of protein kinases can phosphorylate the enzyme, and this process decreases enzyme activity. Nitric oxide itself is destroyed by reacting with hemoglobin and other iron-containing compounds. Interestingly, vasodilators such

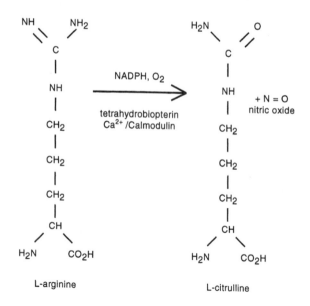

FIGURE 5-6. Synthesis of nitric oxide.

as nitroprusside and nitroglycerin produce NO, and this is considered to be their mechanism of action.

NADPH diaphorase immunoreactivity is co-localized in most but not all cells with NO synthase activity, and this has been used to map the distribution of neurons that release NO. Currently, three isoforms of NO synthase have been reported that arise from distinct genes and share approximately 50% amino acid identity. An unusual feature of this enzyme is that it is a member of the family of cytochrome P450 proteins, proteins that catalyze hydroxylation of a variety of metabolites as well as drugs.

So much for the background: Now to the subject of neuronal effects of NO. First, it should be noted that NO is not present in synaptic vesicles, is not released from terminals by exocytosis, and is not stored in any reservoir. Rather, it is released on stimulation and diffuses from a neuronal (and glial) population to act on enzymes and other elements. The first and primary action of this gaseous molecule is to complex the iron of guanylyl cyclase to stimulate the enzyme and increase the concentration of cGMP. Another target, aside from guanylyl cylase, is cytosolic ADP-ribosyl transferase. The consequence of this latter activity is unknown. The rise in cGMP will activate cGMP-dependent protein kinases that catalyze the phosphorylation of substrate proteins, largely unidentified, which then give rise to myriad effects. The first indication of a role of NO in neuronal systems came from the Garthwaite laboratory when investigators found that in cerebellar slices excitatory amino acids led to NO release, as did stimulation of nonadrenergic, noncholinergic peripheral neurons. Subsequent research has provided evidence that nitric oxide synthase inhibitors block NMDA receptor activation, preventing the well-documented NMDA-induced neurotoxicity. But the situation is confusing because prolonged glutamate stimulation produces high NO levels that can kill neuronal cells. In addition to this activity and its relaxation effect in stimulating peripheral nonadrenergic noncholinergic neurons, NO is thought to play a role in both long-term potentiation (LTP) and long-term depression (LTD). These functions concerning synaptic plasticity are complex and remain to be resolved. Part of the problem may be related to the recent finding that there are two redox forms of NO, NO· and the oxidized form NO⁺, which nitrosylates sulfhydryl groups. Clearly, the role of NO in both physiological and patho-physiological processes is still evolving.

Carbon Monoxide

Another gaseous molecule has recently surfaced as a neuromodulating second messenger that also activates guanylyl cyclase. This is carbon monoxide (CO). CO is generated via heme oxygenase that catalyzes the conversion of heme to biliverdin with the liberation of CO. Two heme oxygenases, expressed by separate genes, have been described. Heme oxygenase-1 is inducible, localized mainly in spleen and liver with a small (but inducible) concentration in glial cells and in a few neurons. In contrast, heme oxygenase-2 is constitutive, not inducible, and is found in high concentrations in brain, particularly cerebellum, olfactory bulb, and hippocampus. It is not present in glial cells under normal conditions but can be induced there.

As noted earlier, CO, like NO, raises the level of cGMP. While the localization of NO synthase does not always correlate with the localization of guanylyl cyclase, the localization of heme oxygenase and guanylyl cyclase is virtually identical and also coincides with the presence of cytochrome P450 reductase, a necessary electron donor for heme oxygenase as well as NO synthase. Recent but still preliminary evidence suggests that CO may be more important in regulating neuronal cGMP levels than NO.

Two other possible second messengers that should be noted are protein carboxyl methylation and phospholipid methylation. Both processes involve S-adenosylmethionine as the methyl donor. Although it has been shown that protein carboxyl methylation inhibits calmodulin-linked enzymes and that phospholipid methylation alters terminal membrane viscosity, these activities currently are not seen as playing an important role in modulation.

Among endogenous modulators, the neuroactive peptides are by far the most mysterious with respect to their widespread activity. These agents can be released presynaptically, postsynaptically in response to presynaptic receptor activation, from glial cells, or from blood vessels. Another mysterious modulator is adenosine, which is found at virtually every synapse that has been examined. Electrophysiologically it tends to inhibit the evoked release of transmitters, but it also acts postsynaptically, exhibiting a variety of behavioral effects ranging from evoking premature arousal in hibernating ground squirrels to anticonvulsant activity, increasing cerebral blood flow, and interacting with the benzodiazepine receptor.

Reflecting on the mechanisms available to nervous tissue to modulate synaptic transmission, one cannot fail to be overwhelmed by the almost infinite possibilities that provide the fine tuning for behavioral changes. Is this apparent redundancy of modulatory mechanisms in reality a reflection of the need for compensatory adjustments that provide survival value? It may well turn out that the key player in this scenario is calcium. Since Ca is released as a wave, its concentration as well as its translocation at discrete sites might dictate the modulatory effect of the attachment of a ligand to its receptor. It may also be the key to nonsynaptic cellular effects that involve second messenger systems such as cell movement or proliferation, gene expression, and metabolism.

Somewhere over the rainbow, one hopes that one day somebody will reveal the grand design in neuronal communication that leads to behavioral changes ranging from the subtle to the dramatic.

SELECTED REFERENCES

Axelrod, J., R. M. Burch, and C. L. Jelsema (1988). Receptor-mediated activation of phospholipase A_2 via GTP-binding proteins: Arachidonic acid and its metabolites as second messengers. *Trends Neurosci.* 11, 117.

Berridge, M. J., and R. F. Irvine (1984). Inositol triphosphate, a novel second messenger in cellular signal transduction. *Nature* 312, 315.

Berridge, M. J. (1993). Inositol trisphosphate and calcium signaling. *Nature* 361, 315.

Buttner, N., S. A. Siegelbaum, and A. Volterra (1989). Direct modulation of *Aplysia* S-K^+ channels by a 12-lipoxygenase metabolite of arachidonic acid. *Nature* 342, 553.

Chesselet, M.-F. (1994). Presynaptic regulation of neurotransmitter release in the brain. *Neuroscience 12*, 347.

Clapham, D. E., and E. J. Neer (1993). New roles for G-protein βγ dimers in transmembrane signaling. *Nature 365*, 403.

Dawson, T. M., and S. H. Snyder (1994). Gases as biological messengers: Nitric oxide and carbon monoxide. *J. Neurosci. 14*, 5147.

Dennis, E. A., S. G. Rhee, M. M. Billah, and Y. A. Hannun (1991). Role of phospholipases in generating lipid second messengers in signal transduction. *FASEB J. 5*, 2068.

Devane, W. A. and J. Axelrod (1994). Enzymatic synthesis of anandamide, an endogenous ligand for the cannabanoid receptor, by brain membranes. *Proc. Natl. Acad. Sci. U.S.A. 91*, 6698.

Di Marzo, V., A. Fontana, H. Cadas, S. Schinelli, G. Cimino, J-.C. Schwartz, and D. Piomelli (1994). Formation and inactivation of endogenous cannabinoid anandamide in central neurons. *Nature 372*, 687.

Fisher, S. K., A. M. Heacock, and B. W. Agranoff (1992). Inositol lipids and signal transduction in the nervous system: An update. *J. Neurochem. 58*, 18.

Furchgott R F, & J. V. Zawadzki (1980) The obligatory role of endothelial cells in the relaxation of arterial smooth muscle by acetylcholine. *Nature 288*, 373.

Garthwaite, J. (1991) Glutamate, nitric oxide, and cell-cell signaling in the nervous system. *Trends Neuro Sci. 14*, 60.

Halushka, P. V., D. E. Mais, P. R. Mayeux, and T. A. Morinelli (1989). Thomboxane, prostaglandin, and leukotriene receptors. *Annu. Rev. Pharmacol. Toxicol. 29*, 213.

Hokin, L. E. (1985). Receptors and phosphoinositide-generated second messengers. *Annu. Rev. Biochem. 54*, 205.

Hyman, S. E., and E. J. Nestler (1993). *The Molecular Foundations of Psychiatry*. American Psychiatric Press, Washington, D.C.

Kerwin, J. R., Jr., and M. Heller (1994). The arginine-nitric oxide pathway: A target for new drugs. *Med. Res. Rev. 14*, 23.

Kruszka, K. K. and R. W. Gross (1994). The ATP- and CoA-independent synthesis of arachidonoylethanolamide. *J. Biol. Chem. 269*, 14345.

Levitan, I. B. (1994). Modulation of ion channels by protein phosphorylation and dephosphorylation. *Annu. Rev. Physiol. 56*, 193.

Levitan, I. B., and L. K. Kaczmarek (1991). *The Neuron: Cell and Molecular Biology*. Oxford University Press, New York.

Maines, M. D., J. A. Mark, and J. F. Ewing (1993). Heme oxygenase, a likely regulator of cGMP production in the brain: Induction in vivo of HO-1 compensates for depression in NO synthase activity. *Mol. Cell. Neurosci. 4*, 398.

Mechoulam, R., L. Hanus, and B. R. Martin (1994). Search for endogenous ligands of the cannabinoid receptor. *Biochem. Pharmacol. 48*, 1537.

Moncada, S., and A. Higgs (1993). The L-arginine–nitric oxide pathway. *N. Engl. J. Med. 329*, 2002.

Nestler, E. J., and P. Greengard (1989). Protein phosphorylation and the regulation of neuronal function. In *Basic Neurochemistry: Molecular, Cellular and Medical Aspects* (G. J. Siegel, et al. eds.), 4th ed, Raven Press, New York.

Nicoll, R. A. (1988). The coupling of neurotransmitter receptors to ion channels in the brain. *Science 241*, 545.

Piomelli, D. (1994). Eicosanoids in synaptic transmission. *Crit. Rev. Neurobiol. 8*, 65.

Piomelli, D., E. Shapiro, R. Zipkin, J. H. Schwartz, and S. J. Feinmark (1989). Formation and action of 8-hydroxy-ii, 12-epoxy-5,9,14-icosatrienoic acid in *Aplysia*: A possible second messenger. *Proc. Natl. Acad. Sci. U.S.A. 86*, 1721.

Schaad, N. C., P. J. Magistretti, and M. Schorderet (1991). Prostanoids and their role in cell–cell interactions in the central nervous system. *Neurochem. Int. 18*, 303.

Schimizu, T., and L. S. Wolfe (1990). Arachidonic cascade and signal transduction. *J. Neurochem. 55*, 1.

Schuman, E. M., and D. V. Madison (1994). Nitric oxide and synaptic function. *Annu. Rev. Neurosci. 17*, 153.

Sessa, W. C. (1994). The nitric oxide synthase family of proteins. *J. Vasc. Res. 31*, 131.

Starke, K., M. Gothert, and H. Kilbinger (1989). Modulation of neurotransmitter release by presynaptic autoreceptors. *Physiol. Rev. 69*, 864.

Verma, A., D. J. Hirsch, C. E. Glatt, G. V. Ronnett, and S. H. Snyder (1993). Carbon monoxide: A putative neural messenger. *Science 259*, 381.

Vogel, Z., J. Barg, R. Levy, D. Saya, E. Heldman, and R. Mechoulam (1993). Anandamide, a brain endogenous compound, interacts specifically with cannabinoid receptors and inhibits adenylate cyclase. *J. Neurochem. 61*, 352.

Walaas, S. I., and P. Greengard (1991). Protein phosphorylation and neuronal function. *Pharmacol. Rev. 43*, 299.

Williams, J. H. and T. V. Bliss (1989). An invitro study of the effect of lipoxygenase and cyclooxygenase inhibitors of arachidonic acid on the induction and maintenance of long-term potentiation in the hippocampus. *Neurosci. Lett. 107*, 301.

Zhang, J. and S. H. Snyder (1995). Nitric oxide in the nervous system. *Annu. Rev. Pharmacol. Toxicol. 35*, 213.

6 | Amino Acid Transmitters

Over the years several amino acids have gained recognition as major neurotransmitter candidates in the mammalian CNS. Since these substances are also involved in intermediary metabolism, it has been difficult to fulfill all the required criteria that would give these substances legitimate status as neurotransmitters in the mammalian CNS. On the basis of neurophysiological studies, amino acids have been separated into two general classes: excitatory amino acids (glutamic acid, aspartic acid, cysteic acid, and homocysteic acid), which depolarize neurons in the mammalian CNS; and inhibitory amino acids (GABA, glycine, taurine, and β-alanine), which hyperpolarize mammalian neurons. Strictly from a quantitative standpoint, the amino acids are probably the major transmitters in the mammalian CNS, while the better-known transmitters discussed in other chapters (acetylcholine, norepinephrine, dopamine, histamine, and 5-hydroxy-tryptamine) probably account for transmission at only a small percentage of central synaptic sites.

GABA

Synthesized in 1883, γ-aminobutyric acid (GABA) was known for many years as a product of microbial and plant metabolism. Not until 1950, however, did investigators identify GABA as a normal constituent of the mammalian central nervous system and find that no other mammalian tissue, with the exception of the retina, contains more than a mere trace of this material. Obviously, it was thought that a substance with such an unusual distribution must have some characteristic and specific physiological effects that would make it important for the function of the CNS.

More than 45 years later we still have no conclusive proof as to the precise role this compound plays in the mammalian central nervous system. However, much evidence has accumulated supporting the hypothesis that the major share of GABA found in the brain functions as an inhibitory transmitter. The knowledge that GABA probably functions as an inhibitory transmitter in the brain has spurred a prodigious research effort to implicate GABA in the etiology of a host of neurological and psychiatric disorders. Although the present evidence is not overwhelming, GABA has been implicated, both directly and indirectly, in the pathogenesis of Huntington's disease, parkinsonism, epilepsy, schizophrenia, tardive dyskinesias, and senile dementia, as well as several other behavioral disorders.

Distribution

In mammals, GABA is found in high concentrations in the brain and spinal cord, but it is absent or present only in trace amounts in peripheral nerve tissue such as sciatic nerve, splenic nerve, and sympathetic ganglia or in any other peripheral tissue such as liver, spleen, or heart. These findings give some idea of the uniqueness of the occurrence of GABA in the mammalian central nervous system. Like the monoamines, GABA also appears to have a discrete distribution within the CNS. However, unlike the monoamines, the concentration of GABA found in the CNS is in the order of μmoles/gm rather than nmoles/gm. It is interesting that the brain also contains large amounts of glutamic acid (8–13 μmoles/gm), which is the main precursor of GABA and is itself a neurotransmitter candidate. In the rat, the corpora quadrigemina and the diencephalic regions contain the highest levels of GABA, while much lower concentrations are found in whole cerebral hemispheres, the pons, and medulla; white matter contains relatively low concentrations of GABA. It should be noted that endogenous levels of GABA increase rapidly postmortem: a 30%–45% increase in GABA occurs within 2 minutes after death in the rat if the tissue is not instantly frozen in situ. The origin of this sudden increase is uncertain, although it is believed to result in part from a transient activation of glutamic acid decarboxylase (GAD).

Progressive increases in GABA levels and in glutamic acid decarboxylase activity appear to occur in various regions of the brain during development. The high levels of GABA found in the various regions of the brain of the rhesus monkey appear to correlate well with the activity of glutamate decarboxylase, the enzyme responsible for the conversion of L-glutamate to GABA. This is not the case for the degradative enzyme GABA-transaminase, since the globus pallidus and the substantia nigra, which have the highest concentration of GABA, have a relatively low transaminase activity. Yet there does not appear to be a consistent inverse relationship between GABA concentration and transminase activity. Thus, some areas of the brain, such as the dentate nucleus and the inferior colliculus, which have relatively high concentrations of GABA, also have large amounts of transaminase activity.

Since GABA does not easily penetrate the blood–brain barrier, it is difficult if not impossible to increase the brain concentrations of GABA by peripheral administration unless one alters the blood–brain barrier. Some investigators have tried to circumvent this problem by the administration of GABA-lactam (2-pyrrolidinone) to animals in the hope that this less polar and more lipid soluble compound would penetrate more easily into the brain and be hydrolyzed to yield GABA. Although the idea seems plausible, it does not succeed: The GABA-lactam that reaches the CNS is not hydrolyzed to any extent. A more recent successful attempt has been the development of progabide (see Fig. 6-6). This agent does penetrate the brain and is subsequently metabolized into GABA.

Metabolism

Three primary enzymes are involved in the metabolism of GABA before its entry into the Krebs cycle. The relative activity of enzymes involved in the degradation of GABA suggests that, as with monoamines, they play only a minor role in the termination of the action of any neurally released GABA.

Figure 6-1 outlines the metabolism of GABA and its relationship to the Krebs cycle and carbohydrate metabolism. As mentioned previously, GABA is formed by the α-decarboxylation of L-glutamic

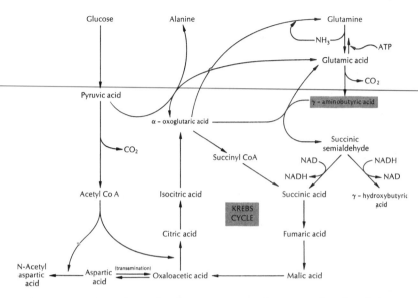

FIGURE 6-1. Interrelationship between γ-aminobutyric acid and carbohydrate metabolism.

acid, a reaction catalyzed by glutamic acid decarboxylase, an enzyme that occurs uniquely in the mammalian central nervous system and retinal tissue. The precursor of GABA, L-glutamic acid, can be formed from α-oxoglutarate by transamination or reaction with ammonia. GABA is intimately related to the oxidative metabolism of carbohydrates in the CNS by means of a "shunt" involving its production from glutamate, its transamination with α-oxoglutarate by GABA——oxoglutarate transaminase (GABA-T), yielding succinic semialdehyde and regenerating glutamate, and its entry into the Krebs cycle as succinic acid via the oxidation of succinic semialdehyde by succinic semialdehyde dehydrogenase. In essence then, the "shunt" bypasses the normal oxidative metabolism involving the enzymes α-oxoglutarate dehydrogenase and succinyl thiokinase.

From a metabolic standpoint, the significance of the "shunt" is unknown; energetically, at least, it is less efficient than direct oxidation through the Krebs cycle (3 ATP equivalents versus 3 ATP + 1

GTP for the Krebs cycle). Experimentally, it has been quite difficult to make an adequate assessment of the quantitative significance of the "shunt" in the oxidation of α-oxoglutarate in vivo. Studies with [^{14}C]-labeled glucose both in vivo and in vitro have indicated that the carbon chain of GABA can be derived from glucose. Therefore, in some cases the incorporation of radioactivity into the "shunt" metabolites following the injection of uniformly labeled [^{14}C]-glucose has been used to assess the functional aspects of this "shunt." However, these experiments are somewhat inconclusive since they do not necessarily indicate the rate of flux through this pathway. Some relatively indirect experiments have given results suggesting that about 10%–40% of the total brain metabolism may funnel through this "shunt."

Glutamic Acid Decarboxylase

GAD is the enzyme responsible for the conversion of L-glutamic acid to GABA. No convincing evidence has yet been gathered to indicate that significant amounts of GABA are formed by reactions other than the decarboxylation of L-glutamic acid. No reversal of this reaction has been demonstrable either in vivo or in vitro. In mammalian organisms this relatively specific decarboxylase is found primarily in the CNS, where it occurs in higher concentrations in the gray matter. In general, the localization of this enzyme in mammalian brain correlates quite well with the GABA content. The brain enzyme has been purified to homogeneity and its properties studied in detail. It has a pH optimum of about 6.5 and requires pyridoxal phosphate, a form of vitamin B_6 as a coenzyme. This purified enzyme is inhibited by structural analogs of glutamate, carbonyl (pyridoxal phosphate)–trapping agents, sulfhydryl reagents, thiol compounds, and anions such as chloride. Hydrazides and other carbonyl-trapping agents react with the aldehyde groups of pyridoxal phosphate and decrease its availability to act as a coenzyme for GAD. Treatment of animals with these agents gives rise to seizures that are believed to result from decreases in the amount of releasable

GABA at inhibitory synapses in the CNS. Pyridoxine-deficient animals also show a decrease in the saturation of brain GAD with coenzyme and a decrease in GAD activity without any alteration in the amount of enzyme protein. Feeding of pyridoxine to deficient animals rapidly restores brain GAD to normal. However it is produced, pyridoxine deficiency causes an increased susceptibility to seizures that can be almost immediately reversed by parenteral administration of pyridoxine. Since the observed effects are rapid, there appears to be a very rapid conversion of pyridoxine to pyridoxal phosphate in the CNS, association of this coenzyme with the apoenzyme of GAD, and normalization of the synthesis of GABA. A dramatic example is in infants with vitamin B_6-deficiency-induced seizures. Seizures can be abolished completely and quickly by an intramuscular injection of pyridoxine.

Although the properties of GAD have been actively studied for 30 years, little was known until recently about the molecular mechanisms responsible for regulation of its in vivo activity in the CNS. Studies directed at the mechanism involved in the postmortem increase in GABA levels have been partly responsible for disclosing some interesting properties of mammalian GAD, which may provide some clues to its in vivo regulation. Since GAD appears to be only partially saturated by its cofactor pyridoxal-5′-phosphate in the intact brain, it is apparent that conditions that alter the degree of saturation of this enzyme in vivo will influence its activity. Although most recent data suggest that pyridoxal phosphate is tightly bound to GAD, glutamate and adenine nucleotides can promote dissociation and block association of this important cofactor, respectively. Thus, it is conceivable that alterations in the availability of one or both of these endogenous substances could influence GAD activity and GABA formation in vivo. The possibility that the association and dissociation of pyridoxal phosphate from GAD may be important for the in vivo regulation of GABA synthesis is supported by the observation that GAD is no more than about 35 percent saturated with pyridoxal phosphate in vivo, despite the fact that the levels of this cofactor in the brain appear to be sufficiently high to fully activate this enzyme. The degree of saturation of GAD with pyridoxal phosphate in vivo may be determined in part by a balance between the rate of dissociation of pyridoxal phosphate from the enzyme brought about by glutamate and the rate of association of the cofactor with the apoenzyme that is inhibited by nucleotides. This possibility is consistent with the observation that a postmortem activation of GAD occurs in brain only when postmortem changes in ATP are allowed to take place.

Tobin and coworkers have demonstrated that brain contains two forms of GAD that differ in molecular size, amino acid sequence, antigenicity, cellular and subcellular location, and interaction with the GAD cofactor pyridoxal phosphate (PLP). These forms, GAD_{65} and GAD_{67}, derive from two genes. GABA synthesis by GAD is regulated primarily by its interaction with the cofactor PLP. Of the total GAD in rat brain, about 50% is apo-GAD, that is, it is not bound to PLP and it is enzymatically inactive. The smaller GAD form, GAD_{65}, makes up most of the apo-GAD reservoir in the brain whereas the larger form of GAD, GAD_{67}, is nearly saturated with PLP. Under depolarizing conditions, PLP associates with apo-GAD to form active holo-GAD (PLP-bound), suggesting that increased neuronal activity causes greater GABA synthesis by activating GAD. The actual levels of GAD thus appear to depend on the interaction of GAD_{65} and PLP. It is noteworthy that GAD is also expressed outside the CNS and that both GAD isozymes are found in the β cells of the pancreas where GABA is thought to play a role in pancreatic function. Studies of experimental diabetes induced with streptozotocin have shown that GAD and insulin appear to coexist in the β cell. In patients with insulin-dependent diabetes it has been shown that antibodies to the 64 kD form of GAD (which appears related to GAD_{65}) occur in almost all subjects. These antibodies appear to precede the clinical onset of the disease. It is conceivable that autoantibodies to GAD may actually underlie the development of insulin-dependent (type I) diabetes as well as the relatively rare neurological disorder known as *stiff-man syndrome*. Short of actual involvement in the triggering of the immune response responsible for destruction of pancreatic β cells, GAD antibodies might serve as a useful marker for early diagnosis of type I diabetes.

GABA-Transaminase (GABA-T)

GABA-T, unlike the decarboxylase, has a wide tissue distribution. Therefore, although GABA cannot be formed to any extent outside the CNS, exogenous GABA can be rapidly metabolized by both central and peripheral tissue. However, since endogenous GABA is present only in nanomolar amounts in cerebrospinal fluid, it is unlikely that a significant amount of endogenous GABA leaves the brain intact. The brain transaminase has a pH optimum of 8.2 and also requires pyridoxal phosphate. It appears that the coenzyme is more tightly bound to this enzyme than it is to GAD. The brain ratio of GABA-T/GAD activity is almost always greater than one. Sulfhydryl reagents tend to decrease GABA-T activity, suggesting that this enzyme requires the integrity of one or more sulfhydryl groups for optimal activity. The transamination of GABA catalyzed by GABA-T is a reversible reaction; so if a metabolic source of succinic semialdehyde were made available it would be theoretically possible to form GABA by the reversal of this reaction. However, as indicated below, this does not appear to be the case in vivo under normal or experimental conditions investigated so far.

Recent studies with more sophisticated cell fractionation techniques and electron microscopic monitoring of the fractions obtained have borne out the original claims that both GAD and GABA-T are particulate to some extent. GAD was found associated with the synaptosome fraction, whereas the GABA-T was largely associated with mitochondria. Further studies on the mitochondrial distribution of GABA-T have suggested that the mitochondria released from synaptosomes have less activity than the crude unpurified mitochondrial fraction, and it has been inferred that the mitochondria within nerve endings have little GABA-T activity. This finding has led to the speculation that GABA is metabolized largely at extraneuronal intercellular sites or in the postsynaptic neurons.

Gabaculine is the most potent GABA-T inhibitor currently available. Similar to γ-acetylenic and γ-vinyl GABA, this agent is a catalytic inhibitor of GABA-T and, unfortunately, will also inhibit GAD. However, gabaculine has a fair degree of specificity since it is about 1000-fold less effective as a GAD inhibitor than as a GABA-T inhibitor.

Succinic Semialdehyde Dehydrogenase

Brain succinic semialdehyde dehydrogenase (SSADH) has a high substrate specificity and can be distinguished from the nonspecific aldehyde dehydrogenase found in the brain. The enzyme purified from human brain has a pH optimum of about 9.2 and a Michaelis constant (K_m for succinic semialdehyde of 5.3×10^{-6}) that is quite low. SSADH from a rat brain has a similarly low K_m for succinic semialdehyde of 7.8×10^{-5} and for NAD of 5×10^{-5}. The high activity of this enzyme and the low Michaelis constant, which allow the enzyme to function effectively at low substrate concentrations, probably account for the fact that succinic semialdehyde (SSA) has not even been detected as an endogenous metabolite in neural tissue, despite the active metabolism of GABA in vivo. In contrast to SSADH isolated from bacterial sources where NADP is several times more active as a cofactor than NAD, the enzyme from monkey and human brain demonstrates a specificity for NAD as a cofactor. The regional distribution of this enzyme has been studied in human brain and has been found to parallel the distribution of GABA-T activity, although the dehydrogenase is about 1.5 times as active. The greatest activity was found in the hypothalamus, basal ganglia, cortical gray matter, and mesencephalic tegmentum. SSADH also has a marked heat activation at 38°C, and Pitts suggested that its activity in vivo might be regulated by temperature, such that fever might result in an increased flux through the GABA shunt. This appears unlikely, however, since this step is usually not considered to be rate limiting in the conversion of GABA to succinic acid, and thus small changes in its activity would not be reflected in an overall change in GABA metabolism. A sensitive and specific assay for SSADH is based on the fluorescence of NADH formed in the conversion of succinic semialdehyde to succinic acid. This method is sensitive enough to assay samples as small as 0.05 μg of freeze-dried brain tissue.

Since GABA's rise to popularity, the literature has been inundated with reports purporting to demonstrate that many pharmacological and physiological effects can be ascribed to and correlated well with changes in the levels of this substance in the brain. Since both GAD and GABA-T are dependent on the coenzyme pyridoxal phosphate, it is not surprising that pharmacological agents or pathological conditions affecting this coenzyme can cause alterations in the GABA content of the brain. Epileptiform seizure can be produced by a lack of this coenzyme or by its inactivation. Conditions of this sort also lead to a reduction in GABA levels, since GAD appears to be preferentially inhibited over the transaminase, presumably due to the fact that GAD has a lower affinity for the coenzyme than does GABA-T. A diet deficient in vitamin B_6 in infants can lead to seizures that respond successfully to treatment consisting of addition of pyridoxine to the diet. However, it must be remembered that many other enzymes including some of those involved in the biosynthesis of other bioactive substances, are also pyridoxal-dependent enzymes. A number of observations, in fact, indicate that there is no simple correlation between GABA content and convulsive activity. Administration of a variety of hydrazides to animals has uniformly resulted in the production of repetitive seizures following a rather prolonged latent period. The finding that the hydrazide-induced seizures could be prevented by parenteral administration of various forms of vitamin B_6 led to the suggestion that some enzyme system requiring pyridoxal phosphate as a coenzyme was being inhibited and that the decrease in the activity of this enzyme was somehow related to the production of the seizures observed. At this time attention focused on GABA and GAD because of their unique occurrence in the CNS and because GAD had been shown to be inhibited by carbonyl-trapping agents in vitro. The hydrazide-induced seizures were accompanied by substantial decreases in the levels of GABA and reductions in GAD activity in various areas of the brain studied. (This decrease in GABA produced by thiosemicarbazide is now believed to be due at least in part to a decrease in the rapid postmortem increase in GABA mentioned previously.) The direct demonstration of the reversal of the action of the convulsant hydrazides by GABA itself proved to be difficult because of the lack of ability of GABA to pass the blood–brain barrier in adult mammalian organisms. However, a preferential inhibition of GABA-T could be achieved in vivo with carbonyl reagents such as hydroxylamine (NH_2OH) or amino-oxyacetic acid, which resulted in an increase in GABA levels (up to 500% of control) in the CNS. Although these agents caused a decrease in the susceptibility to the seizures induced by some agents (such as metrazole), they did not exert any protective effect against the hydrazide-induced convulsions, even though they prevented the depletion of GABA and in some cases even increased the GABA levels above the controls. In fact, administration of very high doses of only amino-oxyacetic acid, instead of producing the normally observed sedation, caused some seizure activity in spite of the extremely high brain levels of GABA.

An interesting finding with amino-oxyacetic acid is that administration of this compound to a strain of genetically spastic mice in a single dose of 5–15 mg/kg results in a marked improvement of their symptomatology for 12–24 hr. This improvement is associated with an increase in GABA levels, but the GABA level increases with a similar time course and to the same extent as in normal control mice. All studies to date indicate that the principal genetic defect in these mice is not in the operation of the GABA system. However, the drug-induced increase in GABA may serve to quell an excess of or imbalance in excitatory input in some unknown area of the CNS.

Hydrazinopropionic acid has been described as a potent inhibitor of GABA-T. It has been suggested that this compound inhibits GABA-T because of its close similarity to GABA with respect to structural configuration, molecular size, and molecular charge distribution. Its inhibitory action cannot be reversed by the addition of pyridoxal phosphate. Hydrazinopropionic acid is about 1000 times more potent than amino-oxyacetic acid in inhibiting mouse brain transaminase.

Alternate Metabolic Pathways

In addition to undergoing transamination and subsequently entering the Krebs cycle, GABA can apparently undergo various other transformations in the CNS, forming a number of compounds whose importance, and in some cases natural occurrence, has not been conclusively established. Figure 6-2 depicts a variety of derivatives for which GABA may serve as a precursor. Perhaps the simplest of these metabolic conversions is the reduction of succinic semialdehyde (a product of GABA transamination) to γ-hydroxybutyrate (GHB). The transformation of GABA to GHB has been demonstrated in rat brain both in vivo and in vitro. Recent studies have demonstrated that GHB administered in physiologically relevant concentrations can also be converted to GABA by transamination.

GHB aciduria, a rare inborn error in the metabolism of GABA, has been reported in children and appears to be the result of a deficiency of succinic semialdehyde dehydrogenase, the enzyme that oxidizes succinic semialdehyde to succinic acid (see Fig. 6-1). GHB levels are increased in urine, plasma, and cerebrospinal fluid, but it is unclear whether the main clinical symptoms of motor and mental retardation, muscular hypotonia, and ataxia are all related to the elevated levels of GHB.

GABA Receptors

The term "GABA receptor" usually refers to a GABA recognition site on pre- and postsynaptic membranes that, when coupled with GABA or an appropriate agonist, causes a shift in membrane permeability to inorganic ions, primarily chloride. This change in chloride permeability results in hyperpolarization of the receptive neuron in the case of postsynaptic inhibition or depolarization in the case of pre-synaptic inhibition. However, it is clear that GABA can attach to a vast number of sites in CNS tissue, some of which may be physiologically relevant but not appropriately referred to as GABA receptors. These sites include the high-affinity GABA transport site and enzymes involved in GABA metabolism such as GAD and

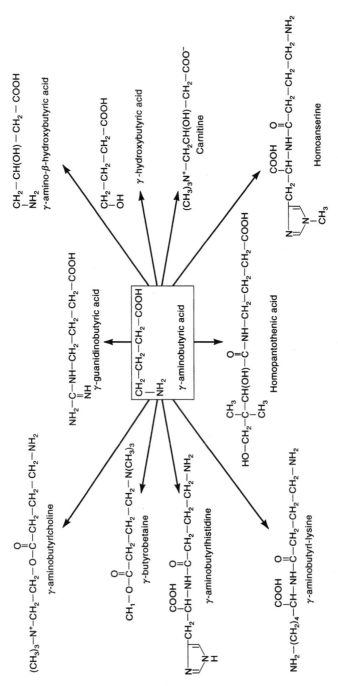

FIGURE 6-2. Possible alternate metabolic pathways for γ-aminobutyric acid.

GABA-T. In vertebrate species, GABA receptors are found primarily in nerve cell membranes and are sufficiently wide-spread that most neurons in the CNS possess them. However, GABA receptors are not exclusively associated with neurons. They are also expressed by astrocytes, where they appear to be involved in the regulation of chloride channels. Interestingly, GABA receptors are also being found outside the CNS on neurons of the autonomic nervous system.

In vertebrates there are two major types of GABA receptors—the $GABA_A$ receptor and the $GABA_B$ receptor. GABA receptors have been subdivided into these two groups based on pharmacological evidence. However, the separation also extends to second messenger mechanisms and to differences in the location of these receptor subtypes in the mammalian central nervous system. For example, radioautographic studies have revealed an abundance of $GABA_A$ sites in the granular cell layers of the cerebellum where no $GABA_B$ sites are present. Conversely, the concentration of $GABA_B$ sites in the interpeduncular nucleus is far greater than the concentration of $GABA_A$ sites. Both receptor subtypes have been shown to have pre- and postsynaptic locations and are thought to participate independently in synaptic transmission.

$GABA_A$ Receptor

The $GABA_A$ receptor is by far the most prevalent of the two known GABA receptors in mammalian CNS and has been extensively studied and characterized. Like the nicotinic acetylcholine receptor nAChR, the $GABA_A$ receptor contains an integral transmembrane ion channel that is gated by the binding of two agonist molecules. However, unlike the nAChR, the receptor-associated channel predominantly conducts chloride ions. Since the Cl equilibrium potential is near the resting potential in most neurons, increasing chloride permeability decreases the depolarizing effects of an excitatory input, thereby depressing excitability.

In the mid-1970s, the history of the GABA ($GABA_A$) receptor became intertwined with the history of the benzodiazepine receptor.

Although the anxiolytic properties of the benzodiazepines were discovered in the 1950s, the mechanism of action of benzodiazepines was not understood for many years thereafter. In the late 1970s, it became clear that there were specific binding sites in the brain for the benzodiazepines and they became known as the benzodiazepine receptors. Concrete evidence of a biochemical relationship between the GABA$_A$ and benzodiazepine receptors was found by Tallman and Gallagher. These workers clearly showed that GABA enhances the binding of benzodiazepines, but the structural details of this linkage would not be characterized for another decade.

In the late 1980s, a large research group led by Eric Barnard reported the sequencing and functional expression of the genes encoding two subunits of the GABA$_A$ receptor and showed that these two subunits could form a functional chloride channel. Two years later, a group led by Peter Seeberg identified an additional substituent of the receptor and was able to form benzodiazepine-sensitive chloride channels, thus proving that the GABA and benzodiazepine receptors are part of the same complex. Activation of the GABA$_A$ receptor by GABA agonists results in the opening of the chloride channel. The ensuing influx of chloride anions inhibits the firing of the neurons by causing hyperpolarization. Benzodiazepines increase the frequency of channel opening without appreciably altering the channel conductance or duration of opening. Barbiturates, in contrast, slightly decrease the opening frequency and prolong the duration of opening. The GABA$_A$ receptor can be functionally altered by a variety of compounds that bind to the receptor at several different sites, including the GABA site, the picrotoxin/barbiturate site, the benzodiazepine site, and the steroid site (see Fig. 6-3). The activation and modulation of chloride flux by GABA and benzodiazepines are achieved by dynamic alterations in protein configuration. Benzodiazepines and related ligands bind to a unique site on the GABA$_A$ receptor complex that does not overlap the sites for GABA or picrotoxin. Drugs that bind to the benzodiazepine site and enhance the electrophysiological effects of GABA are called benzodiazepine agonists; compounds that bind to the benzodiazepine site and decrease the effects of GABA are called inverse agonists. Subtle modifications

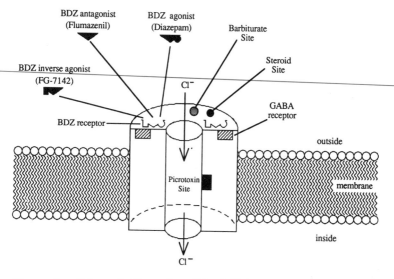

FIGURE 6-3. Schematic illustration of the GABA$_A$ receptor complex and the sites of action of different agents on the receptor.

of the chemical structures of the benzodiazepine agonists and inverse agonists have produced a small number of antagonists, the most important of which is RO15-1788 or flumazenil. This compound has very few, if any, pharmacological effects when administered alone; however, it readily reverses all the effects of benzodiazepine agonists and inverse agonists and has proved to be a very useful experimental tool.

It is now common textbook knowledge that the major inhibitory neurotransmitter receptor, the GABA$_A$ receptor, is a multi-subunit receptor–channel complex that can be allosterically modulated by the two important classes of drugs mentioned above—the benzodiazepines and the barbiturates. The primary structure of the GABA$_A$ receptor, described in 1987, revealed that this receptor is a member of a large superfamily of ligand-gated ion channels that includes the nicotinic–cholinergic, inotropic glutamate, and glycine receptors. The current understanding of the molecular structure of the

GABA$_A$ receptor–ion channel complex is that it is believed to be a heteropentameric glycoprotein of approximately 275 kD composed of a combination of multiple polypeptide subunits (cf. Fig. 6-4). Five distinct classes of polypeptide subunits (α, β, γ, δ, and ρ) have been cloned, and multiple isoforms of each have been shown to exist so that the total number of identified subunits now stands at 15. The existence of a family of at least 15 genes coding for diverse subunits ($α_{1-6}$, $β_{1-4}$, $γ_{1-3}$, δ, $ρ_{1-2}$) provides the basis for the extraordinary structural diversity of GABA$_A$ receptors. The subunit composition of the GABA$_A$ receptors appears to vary from one brain region to another and even between neurons in a given region, but the exact composition of most native GABA$_A$ receptors is unknown. The distribution

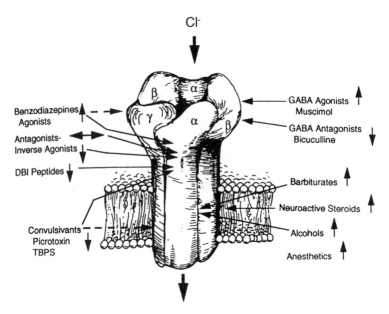

FIGURE 6-4. Schematic illustration of the GABA$_A$ receptor structure containing two α and β subunits and a single γ subunit to form an intrinsic Cl-ion channel. Putative ligands and drugs known to interact at one of the major sites associated with the GABA$_A$ receptors and to either positively or negatively modulate GABA-gated Cl-ion conductance are also illustrated. (From S. M. Paul, 1995.)

of mRNA in the CNS determined by in situ hybridization is very different for each subunit subtype. It is noteworthy that a recently cloned α-subunit (α_6), which confers unique pharmacology (binding of the partial inverse agonist RO-15-4513, a putative alcohol antagonist) to a recombinantly expressed $GABA_A$ receptor, is only expressed in a single type of neuron, the cerebellar granule neuron. Thus it is now becoming clear that the heterogeneity of the $GABA_A$ receptor subunit isoforms appears to confer diversity of pharmacological and perhaps physiological response characteristics upon the $GABA_A$ receptor. For example, it has been shown that coexpression of an additional γ-subunit is necessary for the potentiation of GABA responses by benzodiazepines. In addition, coexpression of individual γ-subunit variants (γ_1, γ_2, or γ_3) with α- and β-subunits results in varying degrees of modulation by benzodiazepine receptor ligands (agonist, antagonist, inverse agonist). Photoaffinity-labeling studies have further suggested that the benzodiazepine binding site resides on the α-subunit while the GABA binding site itself resides on the β-subunit. Finally, it appears that the α-subunit heterogeneity determines the diversity of physiological and pharmacological response characteristics of native $GABA_A$ receptors even though expression of the γ-subunits is essential for conferring the modulatory actions of benzodiazepines on $GABA_A$ receptors. Thus, when coexpressed with β_1, the α_1-subunit yields a receptor with a relatively high affinity for GABA. By contrast, coexpression of the α_2 or α_3-subunit with the β_1-subunit results in $GABA_A$ receptors with far lower affinities for GABA. Thus, the subunit composition of a given receptor may actually determine the local "response" of the $GABA_A$ receptor to synaptically released GABA. These subtle differences in subunit organization may result in subpopulations of $GABA_A$ receptors that have different regional and cellular locations, each with differential sensitivity to GABA and allosteric modulators.

This extraordinary heterogeneity of $GABA_A$ receptors clearly provides a hitherto unexplored diversity in the function of receptor subtypes affecting their sensitivity to GABA, modulation by allosteric effectors, adaptation to stimulus conditions, distribution within a neuron and between neurons, ontogenetic development, and alterations in pathological states. Several studies have suggested that phosphorylation of the GABA$_A$ receptor channels may also be of importance for both short-term and long-term regulation of GABA$_A$ receptor function and expression. At present, however, the physiological significance and specific consequences of phosphorylation of GABA$_A$ receptor channels is uncertain.

An increased understanding of the benzodiazepine GABA receptor chloride channel complex has already resulted in the development of selective anxiolytic and anticonvulsant agents that lack significant sedative and muscle relaxant action, properties that often limit the usefulness of traditional agents such as benzodiazepines and barbiturates. A better understanding of the molecular characteristics and regulation of the multiple allosteric sites of the supramolecular complex and the endogenous substances that may physiologically subserve these sites should not only contribute to our understanding of the possible etiology of anxiety and seizure disorders but also aid in the development of more effective and specific therapeutic agents. Once the functional properties of the GABA$_A$ subunits and their subtypes are more clearly defined, it should be possible to use this knowledge in the rational screening and/or design of new, clinically useful subtype-specific agents.

GABA$_B$ Receptor

Besides the GABA$_A$ receptor, there are at least two other binding sites for GABA in the brain: the GABA$_B$ receptor and the GABA transporter. The GABA$_B$ receptor, as mentioned above, is present at lower levels in the CNS than the GABA$_A$ receptor and is not linked to a chloride channel.

Since its discovery in 1980, much progress has been made in GABA$_B$ receptor pharmacology. Selective agonists and antagonists have been developed, and a functional role for this receptor as a mediator of slow inhibitory postsynaptic potentials in many brain regions has emerged. GABA$_B$ receptor activation also plays a role in attenuating the release of amines, excitatory amino acids, neuropeptides and hormones, as well as that of GABA via an interaction with autoreceptors. Whereas $GABA_A$ receptors are directly associated with a Cl^- channel, the $GABA_B$ receptors seem to be coupled to Ca^{2+} or K^+ channels via second messenger systems. The inhibitory action of $GABA_B$ receptor activation appears to be mediated through either increases in potassium conductance or decreases in calcium conductance.

This receptor can be distinguished pharmacologically from the $GABA_A$ receptor by its selective affinity for the agonist baclofen and its lack of affinity for muscimol and bicuculline (see Table 6-1). The $GABA_B$ receptor is believed to be linked through GTP-sensitive proteins to a calcium channel. Activation of the $GABA_B$ presynaptic receptors by baclofen decreases calcium conductance and transmitter release. Postsynaptic $GABA_B$ receptors are indirectly coupled to K^+ channels via G proteins, and they mediate late inhibitory postsynaptic potentials (IPSPs). Unlike the $GABA_A$ receptor, the $GABA_B$ receptor is not modulated by the benzodiazepines or barbiturates. Pharmacological studies have demonstrated that blockade of $GABA_B$ receptors does not result in the profound behavioral sequelae observed following administration of $GABA_A$ antagonist (e.g., seizures). These observations suggest that, unlike $GABA_A$ receptors, which are believed to be in a continuous tonically activated state, $GABA_B$ receptors may only be activated under certain physiological conditions.

With regard to the functions of the $GABA_B$ receptor in the brain, it seems premature to assign a physiological or pathological role. However, the discovery of selective $GABA_B$ antagonists that cross the blood–brain barrier has aided in evaluating the functions of this receptor (see Fig. 6-5). With the recent development of a potent, orally effective $GABA_B$ antagonist, CGP54626, it became possible to evaluate better the physiological role of this receptor. In vivo and in vitro studies clearly demonstrated that blockade of $GABA_B$ receptors with this agent increased neurotransmitter (GABA and glutamate) release, reduced late IPSPs of CA1 hippocampal pyramidal neurons, and led to an increase in neuronal excitability. Behavioral studies in several species have suggested that $GABA_B$ receptor

TABLE 6-1. Subdivision of GABA Receptors

| Receptor Class | Pharmacology | | | Channels | Second Messengers |
	Agonists	Antagonists	Modulators		
GABA$_A$	GABA	*Competitive* Bicuculline GABAzine *Noncompetitive* Picrotoxin TBPS	Benzodiazepines	Cl$^-$	None
	Muscimol Isoguvacine		Barbiturates Steroids DBI peptides		
GABA$_B$	GABA R(+) Baclofen	Phaclofen CGP-36742* 3-APPA	None	$\uparrow$K$^+$ $\downarrow$Ca^{2+}	Adenylate cyclase Phosphatidyl inositol turnover

TBPS, t-butylbicyclophosphosthianate; 3-APPA, 3-aminopropylphosphinic acid.

*See Figure 6-5.

blockade can improve cognition in rats (social learning), mice (passive avoidance), and rhesus monkeys (conditional spatial color test.

Cloning of the gene for the GABA$_B$ receptor should further facilitate unraveling its functions and provide insight concerning the pos-

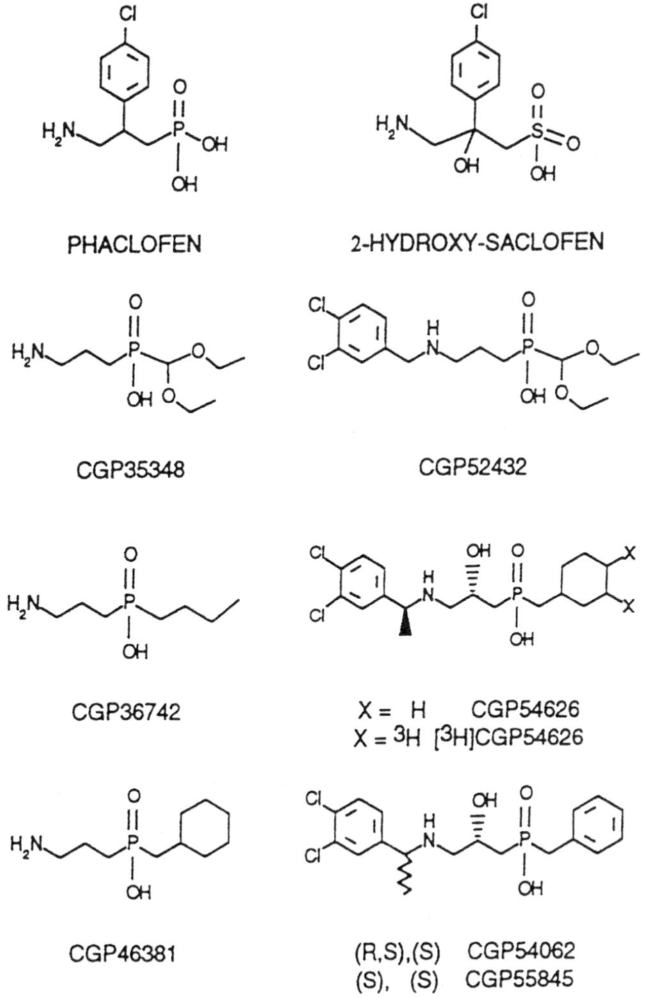

PHACLOFEN

2-HYDROXY-SACLOFEN

CGP35348

CGP52432

CGP36742

X = H   CGP54626
X = 3H [3H]CGP54626

CGP46381

(R,S),(S) CGP54062
(S), (S) CGP55845

FIGURE 6-5. Chemical structure of GABA$_B$ receptor antagonists.

sible changes that may be associated with disease states. To date, attempts to clone the GABA$_B$ receptor by expression cloning in oocytes or by microsequencing a portion of the purified protein have proven unsuccessful.

GABA Transporter

The other GABA binding site in brain is the GABA uptake transporter, which is entirely dependent upon the presence of sodium and is eliminated by reuptake blockers such as nipicotic acid. The purification of rat brain GABA transporter protein provided a reagent for the molecular cloning of a GABA transporter cDNA. A complimentary DNA clone encoding a transporter for GABA has recently been isolated from rat brain and its functional properties examined in *Xenopus* oocytes. This cloned and expressed GABA transporter was found to have a high affinity for GABA, to exhibit sodium and chloride dependence, and to have a pharmacology similar to that of neuronal GABA transporter. The expressed protein also shares antigenic determinants with the native rat brain GABA transporter.

Coexistence with Classic and Nonclassic Transmitters

During the last several years, an increasing number of neurons containing coexisting messenger molecules have been described. GABA neurons are no exception. GABA has been shown to coexist with such classic transmitters as 5-HT, DA, ACh, glycine, and histamine, as well as with a number of peptides including CCK, NPY, VIP, Substance P, and others. The actual function of such coexistent molecules, especially the peptides, is often very difficult to evaluate in the CNS, so that work on the functional implications of multiple signaling through a single synapse by coexistent transmitters has proceeded slowly. Clearly, the realization that GABA can coexist with a large number of other messenger molecules provides a new dimension to the complexity of this neuronal system and the way it can influence follower cells.

PHARMACOLOGY OF GABAERGIC NEURONS

Drugs can influence GABAergic function by interaction at many different sites, both pre- and postsynaptic (Fig. 6-6). Drugs can influence presynaptic events and modify the amount of GABA that ultimately reaches and interacts with postsynaptic GABA receptors. In most cases, presynaptic drug effects do not involve an interaction with "GABA receptors." The most extensively studied presynaptic drug actions involved inhibitory effects exerted on enzymes involved in GABA synthesis (GAD) and degradation (GABA-T) and the neuronal reuptake of GABA. The major exception is the interaction of drugs with GABAergic autoreceptors to modulate both the physiological activity of GABA neurons and the release and synthesis of GABA in a manner analogous to the role played by dopamine autoreceptors in the regulation of dopaminergic function.

A great deal of emphasis has been directed recently to the study of drug interactions with GABA receptors. Drugs interacting at the level of GABA receptors can be assigned into two general categories, GABA antagonists and GABA agonists.

Figure 6-3 depicts possible sites of drug interaction in a hypothetical GABAergic synapse. The structures of compounds that act at GABAergic synapses are depicted in Figs. 6-6 and 6-7. Picrotoxinin is the active component of picrotoxin.

GABA Antagonists

The action of GABA at the receptor–ionophore complex may be antagonized by GABA antagonists either directly by competition with GABA for its receptor or indirectly by modification of the receptor or by inhibition of the GABA-activated ionophore. The two classic GABA antagonists (Fig. 6-6) bicuculline and picrotoxin appear to act by different means. Bicuculline acts as a direct competitive antagonist of GABA at the receptor level, while picrotoxin acts as a noncompetitive antagonist, presumably due to its ability to block GABA-activated ionophores. Although early studies raised some doubts concerning the usefulness of bicuculline as a selective GABA

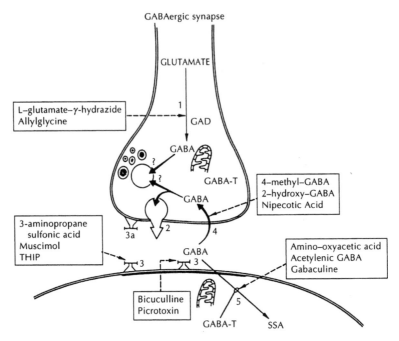

FIGURE 6-6. Schematic illustration of a GABAergic neuron indicating possible sites of drug action.

Site 1: *Enzymatic synthesis:* Glutamic acid decarboxylase (GAD-1) is inhibited by a number of various hydrazines. These agents appear to act primarily as pyridoxal antagonists and are therefore very nonspecific inhibitors. L-glutamate——hydrazide and allylglycine are more selective inhibitors of GAD-1, but these agents are also not entirely specific in their effects.

Site 2: *Release:* GABA release appears to be calcium dependent. At present no selective inhibitors of GABA release have been found.

Site 3: *Interaction with postsynaptic receptor:* Bicuculline and picrotoxin block the action of GABA at postsynaptic receptors. 3-Aminopropane sulfonic acid and the hallucinogenic isoxazole derivative, muscimol, appear to be effective GABA agonists at postsynaptic receptors and autoreceptors.

Site 3a: *Presynaptic autoreceptors:* Possible involvement in the control of GABA release.

Site 4: *Reuptake:* In the brain GABA appears to be actively taken up into presynaptic endings by a sodium-dependent mechanism. A number of compounds will inhibit this uptake mechanism, such as 4-methyl-GABA and 2-hydroxy-GABA, but these agents are not completely specific in their inhibitory effects.

Site 5: *Metabolism:* GABA is metabolized primarily by transamination by GABA-transaminase (GABA-T), which appears to be localized primarily in mitochondria. Amino-oxyacetic acid, gabaculline, and acetylenic

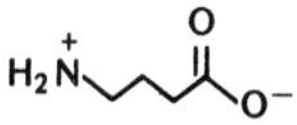

GABA

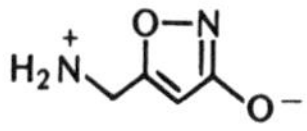

Muscimol

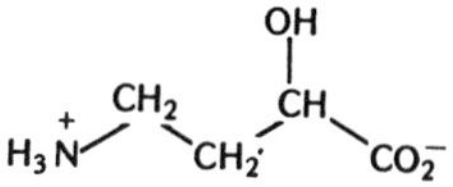

2-hydroxy-GABA

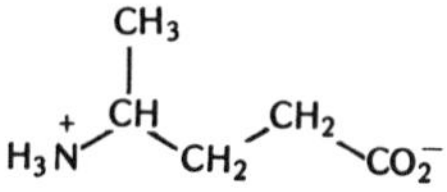

4-methyl-GABA

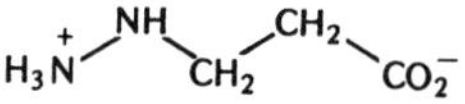

3-hydrazinopropionic acid

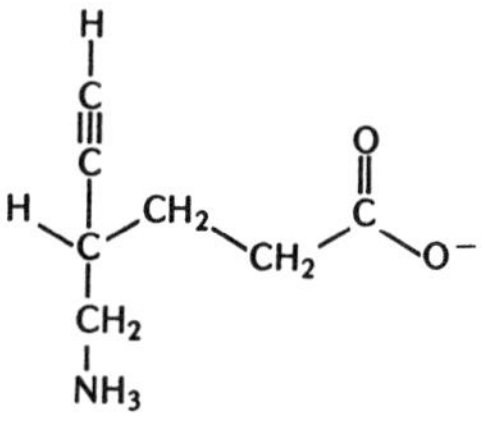

Acetylenic GABA

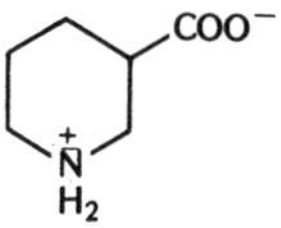

(±)-Nipecotic acid

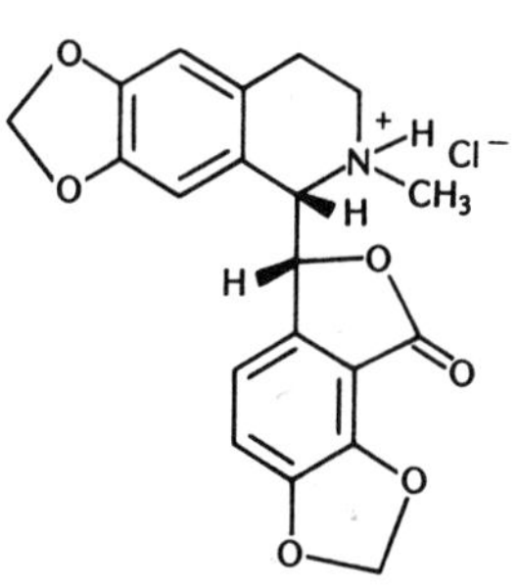

Bicuculline
Hydrochloride

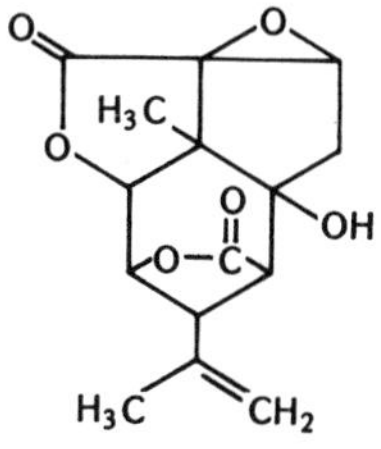

Picrotoxinin

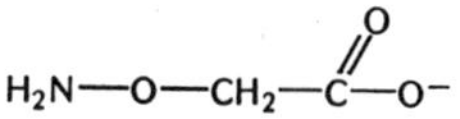

Amino-oxyacetic acid

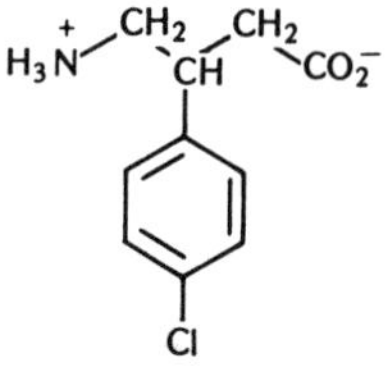

Lioresal
(Baclofen)

FIGURE 6-7. Structures of compounds that act at GABAergic synapses.

TABLE 6-2. GABA$_A$ Receptor Pharmacology

Agent	Activity
Muscimol, GABA	Selective agonists
Bicuculline	Competitive antagonist
Picrotoxin, PTZ, TBPS	Noncompetitive antagonists
Penicillin	Channel blocker
Diazepam, clonazepam, flurazepam	Benzodiazepine receptor agonists
DMCM, βCCM, FG-7142	Benzodiazepine receptor inverse agonists
Flumazenil	Benzodiazepine receptor antagonist
Pentobarbital, phenobarbital	Barbiturate receptor agonists
Alphaxalone, allopregnanalone	"Anesthetic steroid receptor" agonist

antagonist, these expressions of skepticism have been largely resolved and appear primarily related to the instability of bicuculline at 37°C and physiological pH. Under normal physiological conditions bicuculline is hydrolyzed to bicucine, a relatively inactive GABA antagonist with a short half-life of several minutes. The quaternary salts now used for most electrophysiological experiments (bicuculline methiodide and bicuculline methochloride) are much more water soluble and stable over a broad pH range of 2–8. It should be noted, however, that these quaternary salts are not suitable for systemic administration because of their poor penetration into the CNS.

GABA Agonists

Electrophysiological studies have demonstrated that there are a wide variety of compounds that are capable of directly activating bicuculline-sensitive GABA receptors. These agonists can be readily subdivided into two groups based on their ability to penetrate the blood–brain barrier, dictating whether they will be active or inactive following systemic administration. Agents such as 3-aminopropane-sulfonic acid, β-guanidinoproprionic acid, 4-aminotetrolic acid, *trans*-4-aminocrotonic acid, and *trans*-3-aminocyclopentane-1-carboxylic acid are all effective direct-acting GABA agonists. However, entry of these agents into the brain following systemic administration is minimal. In addition, compounds such as *trans*-4-aminotetrolic acid and 4-aminocrotonic acid also inhibit GABA-T and GABA uptake; therefore, their action is not totally attributable to their direct agonist properties.

In contrast to this class of direct-acting GABA agonists, the second group listed below readily passes the blood–brain barrier and is active following systemic administration. Muscimol (3-hydroxy-5-aminomethylisoxazole) is the agent in this group that has been most extensively studied. Some other agents in this group include (5)-(−)-5-(1-aminoethyl)-3-isoxazole, THIP (a bicyclic muscimol analogue), SL-76002 (a[chloro-4-phenyl]fluro-5-hydroxy-2 benzilide-neamino-4H butyramide), and kojic amine (2-aminomethyl-3- hydroxy-4H-pyran-4-one).

In addition to classification based on their ability to penetrate the blood–brain barrier, GABAergic substances may be further divided into those compounds which stimulate GABA receptors directly or those that indirectly cause an activation of GABA receptors by several different mechanisms. For example, agents such as muscimol, isoguvacine, and THIP are true GABA mimetic agents that interact directly with GABA receptors. Indirectly acting GABA mimetics act to facilitate GABAergic transmission by increasing the amount of endogenous GABA which reaches the receptor or by altering in some manner the coupling of the GABA receptor-mediated change in chloride permeability. Thus many drugs often classified as indirect GABA agonists act presynaptically to modify GABA release and metabolism rather than by interacting directly with GABA receptors. For this reason, drugs like gabaculine (a GABA-T inhibitor), nipecotic acid (a GABA uptake inhibitor), and baclofen (an agent that, in addition to many other actions, causes release of GABA from intracellular stores) are often classified incorrectly as GABA agonists. The benzodiazepines mentioned earlier also appear to potentiate the action of tonically released GABA at the receptor by displacement of an endogenous inhibitor of GABA receptor binding, allowing more endogenous GABA to reach and bind receptors. Thus, benzodiazepines are sometimes also classified as GABA agonists. A GABA-like action can also be elicited by agents that bypass GABA receptors and influence GABA ionophores. It has been suggested that pentobarbital acts at the level of the GABA ionophore, but it is unclear whether its CNS depressant effects are explainable by this action.

The structures of some of the more potent and widely used direct-acting GABA agonists are illustrated in Figs. 6-7 and 6-8. Included are muscimol, isoguvacine, THIP, and (+)-*trans*-3-aminocyclopentane carboxylic acid. Useful therapeutic effects have not yet been obtained by use of agents of this sort that have direct GABA mimetic effects (such as muscimol), that inhibit the active reuptake of GABA (such as nipecotic acid), or that alter the rate of synthesis or degradation of GABA (such as amino-oxyacetic acid and gabaculine). However, useful therapeutic effects are achieved with the anxiolytic benzodiazepines (such as Valium (diazepam) and Librium (chlordiazepoxide), which may exert their actions by facilitating GABAergic transmission.

If the anatomical distribution and functional properties of $GABA_A$ receptor subtypes and subunits become clearly defined, this knowledge may enable the development of therapeutically useful subtype-specific agonists that can be directed to modify GABAergic function in selective brain areas.

Endogenous Modulators

The large number of recognition sites associated with $GABA_A$ receptors has led to the speculation that a host of endogenous regulatory factors exists. A number of candidates have been identified but,

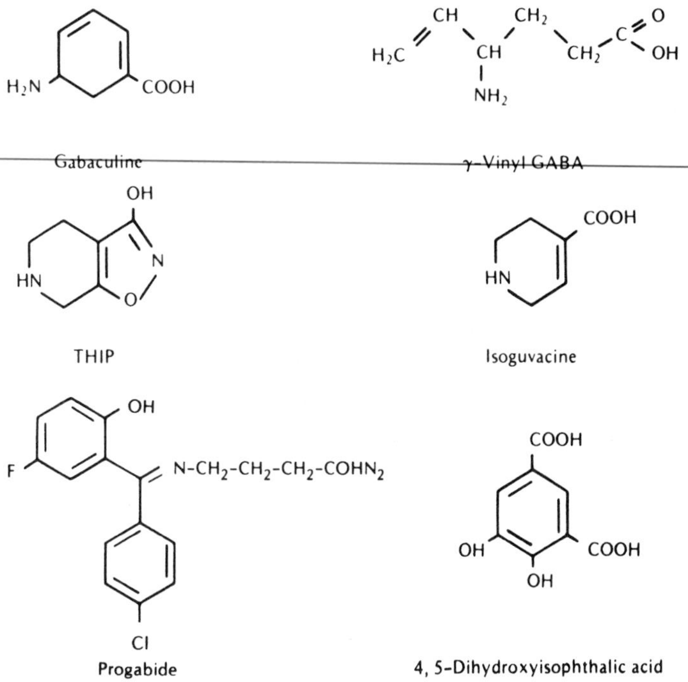
H₂N
COOH
Gabaculine
CH
CH₂
H₂C
CH
CH₂
C
O
OH
NH₂
γ-Vinyl GABA
OH
N
HN
O
THIP
COOH
HN
Isoguvacine
OH
F
N-CH₂-CH₂-CH₂-COHN₂
Cl
Progabide
COOH
OH
COOH
OH
4, 5-Dihydroxyisophthalic acid

FIGURE 6-8. Structures of compounds that act at GABAergic synapses.

with the exception of the neurosteroids and the diazepam-binding inhibitor (DBI), there is little compelling evidence that any play an important role in modulating $GABA_A$ receptor function in vivo. DBI is an endogenous peptide that has been purified to homogeneity from rat and human brain and its structure determined by recombinant DNA technology. Several lines of evidence suggest that DBI functions as a precursor of a family of allosteric modulatory peptides of the $GABA_A$ receptor, causing a negative modulation of the GABA-operated Cl^- ion fluxes (see Chapter 12). Consistent with this action, DBI appears to have anxiogenic properties similar to those associated with inverse benzodiazepine agonists in experimental animals. A similar endogenous peptide has been found in human brain and cerebrospinal fluid. It is noteworthy that the content of human DBI immunoreactivity is elevated in the cerebrospinal fluid of severely depressed patients.

The other postulated endogenous ligands for the GABA$_A$ receptor include two naturally occurring reduced steroid metabolites of deoxycorticosterone and progesterone (allotetrahydro-DOC and allopregnanalone, respectively). These neurosteroids are formed in brain and bind with high affinity to GABA$_A$ receptors, eliciting a barbiturate-like action that potentiates GABA-elicited Cl$^-$ conductance. These are among the most potent known endogenous ligands of GABA$_A$ receptors found in the CNS. Their plasma and brain (cortex and hypothalamus) levels are increased dramatically in rats following exposure to stress. Plasma levels of allopregnanolone are also elevated during the third trimester of pregnancy and decrease dramatically following parturition. The observations that brain and plasma levels of allopregnanolone and allotetrahydro-DOC increase rapidly in the brain after stress (4–20-fold within less than 5 minutes) suggest that these neurosteroids may have a physiological role in stress and anxiety. In addition, conditions that may lead to large increases in neuroactive steroid levels, such as puberty, pregnancy, or menstrual cycle, could also alter neurochemical and behavioral adaptations to stress. To date, however, none of these putative natural ligands (neurosteroids or DBI) have been unequivocally demonstrated to subserve a physiological function or modulate a pathological state.

Neurotransmitter Role in the Mammalian Central Nervous System

The evidence from the mammalian CNS to support a role of GABA as an inhibitory transmitter is not as complete as that generated from the crustacean. With substances such as the amino acids, which may play a dual role (both in metabolism and as neurotransmitters), it is only to be expected that it will be more difficult to obtain conclusive evidence as to their role as neurotransmitters.

Neurophysiological and biochemical studies have indicated that the majority of GABA found in brain may serve a neurotransmitter function. In recent years enough evidence has accumulated to allow the conclusion that GABA functions as an inhibitory transmitter in the mammalian CNS. GABA is considered an inhibitory transmitter even though this substance can induce both hyperpolarizing and depolarizing responses. Both actions are thought to be a result of GABA receptor-mediated change in chloride conductance. The hyperpolarization of neurons produced by an increase in membrane permeability to chloride ions appears identical to the changes induced by glycine-like amino acids. The depolarization induced by GABA of primary afferent terminals is also though to be associated with an increased permeability to chloride ions, but this action is not produced by glycine-like amino acids. The depolarizing response to GABA predominates in spinal cord cells while the GABA-induced hyperpolarization is typical of cerebral cortical cells. These effects establish possible relationships between GABA and postsynaptic (hyperpolarizing) inhibition and between GABA and presynaptic inhibition.

The elementary criteria for identification of a compound as a putative transmitter have already been mentioned. In short, the agent in question must be produced, stored, released, exert its appropriate action, and then be removed from its site of action. There is sufficient evidence for the production, storage, and pharmacological activity of GABA consistent with its suggested role as an inhibitory transmitter, but it has not yet been possible to demonstrate an association of GABA with specific inhibitory pathways in the cortex. Until recently there was also no evidence that GABA could be released from mammalian brain under physiological conditions. Several reports now claim to have demonstrated the spontaneous release of GABA from the surface of the brain. However, only one group of investigators has claimed that the amount of GABA released from the cortex is dependent on the activity of the brain. This group has found that in cats showing an aroused EEG, either following cervical cord section or in awake animals receiving local anesthesia, small amounts of GABA will leak out of the cerebral cortex and can be recovered by a superfusion technique (cortical cup), provided the pia–arachnoid membrane has been punctured. Jasper and coworkers demonstrated that this release of GABA into cortical cups occurs about three times more rapidly from the brains of cats showing an EEG sleep-like pattern with marked spindle activity following midbrain section than in cats showing an EEG awake pattern. When a continuous waking state was maintained by periodic stimulation of the brain stem reticular formation, no measurable amount of GABA could be found in the perfusates. Iversen and coworkers have also demonstrated that GABA is released from the surface of the posterior lateral gyri of cats during periods of cortical inhibition produced by either stimulation of the ipsilateral geniculate or by direct stimulation of the cortex.

The vertebrate neurons most clearly identified by physiological, chemical, and immunochemical criteria as GABAergic are the cerebellar Purkinje cells. When GABA is applied iontophoretically to Deiters' neurons, it induces IPSP-like changes and therefore mimics the action of the inhibitory transmitter released from axon terminals of the Purkinje cells. GABA is also known to be concentrated in Purkinje cells, which exert a monosynaptic inhibitory effect on neurons of Deiters' nucleus. Obata and Takeda observed that stimulation of the cerebellum, which presumably activates the Purkinje cell axons, which have their terminals in cerebellar subcortical nuclei adjacent to the fourth ventricle, induces about a threefold increase in the amount of GABA released into the ventricular perfusate. However, the extremely high voltage employed, the relatively crude perfusion system, and the lack of critical identification of GABA make the significance of these otherwise interesting observations on release somewhat difficult to interpret, although they are consistent with the hypothesis that GABA is an inhibitory transmitter in this particular pathway.

In addition, other experiments have indicated that labeled GABA can be released from brain slices by electrical stimulation or by the addition of high levels of potassium to the incubation medium. However, one should bear in mind that all the above-mentioned release experiments are somewhat gross in nature and do not provide evidence of the neuronal system involved in the release of GABA. Indeed, these experiments do not even indicate that this release is in any way associated with an inhibitory synaptic event. In fact, at the present time no suitable test system in the CNS has been developed that will be as appropriate as the lobster nerve–muscle preparation. The student is advised against holding his or her intellectual breath until it has.

Most of the investigations analyzing the mechanism of GABA's depressive action on the CNS have been conducted on spinal cord motoneurons. Until quite recently, GABA had not shown much evidence of closely mimicking the inhibitory transmitter. In fact, some years ago Curtis and his associates discounted the hypothesis that GABA might be the main inhibitory transmitter in the mammalian CNS. They noted that the action of GABA iontophoresed onto spinal neurons differed significantly from that of the inhibitory transmitter; whereas natural postsynaptic inhibition was associated with membrane hyperpolarization, experiments with coaxial pipettes failed to detect any hyperpolarization of motoneurons during the administration of GABA. Since at that time there was no evidence that GABA had a hyperpolarizing action in other parts of the CNS, Curtis concluded that GABA could not be the main central inhibitory transmitter. Recently, other investigators have demonstrated that in the cortical neurons of cat GABA imitates the action of the cortical inhibitory transmitter, at least qualitatively; it usually raises the membrane potential and increases conductance of cortical neurons, just like the normal inhibitory synaptic mechanism. When the latter inhibitory effect is artificially reversed by the administration of chloride into the neuron, the action of GABA is also reversed in a similar way. Similar results have been obtained on neurons in Deiters' nucleus. Therefore, at least as far as the cortex and Deiters' nucleus are concerned, there appears to be much support for the hypothesis that GABA is an inhibitory transmitter.

There appears to be no pharmacologically sensitive mechanism for the rapid destruction of GABA similar to the cholinesterase mechanism for destruction of acetylcholine. Thus, amino-oxyacetic acid or hydroxylamine given intravenously or iontophoresed does not appear to prolong significantly the duration of action of iontophoretically applied GABA. The exception is Deiters' nucleus, where the hyperpolarization produced by GABA on the neurons in Deiters' nucleus and the IPSP are prolonged in both cases by hydroxylamine, an agent of dubious pharmacological specificity. No functional reuptake mechanism for GABA within a mammalian system has been conclusively demonstrated either. However, GABA is actively and efficiently taken up by brain slices and synaptosomes, and this implies that some uptake phenomena are functional in the intact brain for recycling GABA and terminating its action. This mechanism could also explain why it has been difficult to collect GABA released from neural tissue. Unfortunately, no drug has been discovered that will effectively block the uptake of GABA into nervous tissue without exerting a number of other unwanted pharmacological actions. Therefore, it is difficult to evaluate the functional significance of this uptake mechanism. The development of a selective inhibitor of GABA uptake might prove to be quite useful in the study of GABA systems in the CNS. Some progress in the development of more specific agents has been made recently, and the cloning of the "GABA" transporter should aid in this pursuit.

Several other potentially useful approaches can be identified: (1) It would be helpful to have a specific inhibitor of GAD, which would readily gain access to the CNS so that GABA formation could be selectively blocked. At present, no really specific inhibitors for GAD have been developed that will inhibit this enzyme in vivo without also inhibiting GABA-T to some extent, as well as a number of other B_6-dependent enzymes. (2) Another real advance would be the development of a histochemical method for the visualization of GABA or GABA-related enzymes at both the light and electron microscope levels. The recently developed immunohistochemical method for GAD and for GABA is an important step in this direction. This methodology is already providing useful cytological formation in mapping out GABA-containing and GAD-containing systems in the CNS.

A number of different techniques have been used to glean knowledge concerning the neuronal pathways in the brain using GABA as a transmitter. These include the binding of specific GABA receptor agonists in different brain regions and the autoradiographic localization of labeled GABA taken up through the high-affinity uptake process, the calcium-dependent, depolarization-induced release of GABA from specific brain regions, the measurement of GABA concentration and GAD activity in different brain regions, and the changes in these various parameters after surgical or chemical lesions of afferent pathways. However, the most powerful technique has been the immunohistochemical localization of GAD-like and GABA-like activities in the brain by means of specific antibodies directed against GAD and GABA. In the most recent approach, antibodies have been raised in rabbits against GABA conjugated to bovine serum albumin. This antisera has been used to map GABA-like neurons in the brain. In general, the distributions of GABA-like and GAD-like activity parallel each other, and detailed maps of GABA- and GAD-containing neurons in brain are being produced. The GABA antiserum appears to provide better resolution of GABA-containing cell bodies, while the GAD antiserum provides better resolution of GAD-containing terminal fields. GAD localization studies have already given additional support to a role for GABA as an inhibitory transmitter in certain cerebellar, cortical, and striatonigral neurons by localizing GABA synthesis to specific nerve terminals.

Studies employing the above techniques have demonstrated that in the CNS of mammals, GABA is found primarily in inhibitory interneurons with short axons. For example, the majority of GAD-like and GABA-like activities found in the cerebral cortex, hippocampus, and spinal cord is present in inhibitory interneurons. At the present time, only two long axonal projections from one brain region to another that utilize GABA as a transmitter are known. These are the two systems alluded to above: the Purkinje cells of the cerebellum and their projections to vestibular and cerebellar nuclei and a system of neurons in the striatum with axons projecting to the substantia nigra. As noted in Table 6-1, this latter region of brain (substantia nigra) contains the highest known concentration of GABA. It seems likely that these two immunocytochemical approaches will continue to be employed successfully and will ultimately permit a detailed anatomical description of the GABA pathways throughout the CNS. Once the anatomy of the GABA-containing neuronal systems in the mammalian CNS is more clearly defined, it should be possible to learn a great deal more about the functional aspects of these systems.

In summary, numerous biochemical observations have been made about brain GABA systems over the years which are consistent with a neurotransmitter role for this substance in mammalian brain. Similar to other neurotransmitters or neurotransmitter candidates, GABA and its biosynthetic enzyme (GAD) have a discrete nonuniform distribution in the brain. The brain contains a high-affinity, sodium-dependent transport system, and storage of GABA can be demonstrated in selected synaptosomal populations. The release of endogenous or radioactively labeled exogenously accumulated GABA can be evoked by the appropriate experimental conditions. Most recently, the presence of GABA-containing neurons has been verified, and the anatomical distribution of GABAergic neurons mapped out by use of autoradiographic and immunocytochemical techniques. However, the most compelling evidence that GABA plays a neurotransmitter role in mammalian brain has been provided by intracellular recording studies demonstrating that GABA causes a hyperpolarization of neurons similar to that evoked by the naturally occurring transmitter substance.

GLYCINE

As an Inhibitory Transmitter

Structurally, glycine is the simplest amino acid and is found in all mammalian body fluids and tissue proteins in substantial amounts.

Although glycine is not an essential amino acid, it is an essential intermediate in the metabolism of protein, peptides, one-carbon fragments, nucleic acids, porphyrins, and bile salts. It is also believed to play a role as a neurotransmitter in the CNS. Over the past two decades, numerous neurochemical studies have attempted to separate and distinguish between the general "metabolic" and "transmitter" functions of glycine within the CNS. These studies have confirmed glycine's role as a neurotransmitter in the spinal cord and have suggested that this amino acid may play a similar role in more rostral portions of the CNS and also in the retina. Thus, glycine appears to have a more circumscribed function in the CNS than GABA since the inhibitory role for this substance is restricted to the spinal cord, lower brain stem, and perhaps the retina. Glycine also appears to be an exclusively vertebrate transmitter, making it unique among the transmitter substances. However, within the last several years very little progress has been achieved in developing pharmacological tools that act selectively on glycine systems or in generating more information concerning glycine metabolism in neuronal tissue.

Our knowledge of the metabolism of glycine in nervous tissue is still somewhat rudimentary despite the fact that the process has been studied extensively in other tissues. For example, we still do not know whether biosynthesis is important for the maintenance of glycine levels in the spinal cord or whether the neurons depend upon the uptake and accumulation of preformed glycine. As indicated in Figure 6-9, glycine can be formed from serine by a reversible folate-dependent reaction catalyzed by the enzyme, serine *trans*-hydroxy-methylase. Serine itself can also be formed in nerve tissue from glucose via the intermediates 3-phosphoglycerate and 3-phosphoserine. It is also conceivable that glycine might be formed from glyoxylate via a transaminase reaction with glutamate. Although not established definitively, it seems likely that serine serves as the major precursor of glycine in the CNS and that serine hydroxy-methyl transferase and D-glycerate dehydrogenase are the best candidates for the rate-limiting enzymes involved in the biosyn-

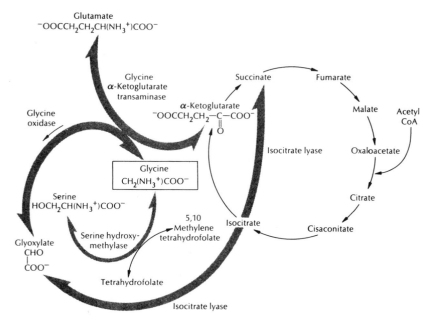

FIGURE 6-9. Possible metabolic routes for the formation and degradation of glycine by nervous tissue. (Modified after Roberts and Hammerschlag, in: *Basic Neurochemistry*, Little, Brown, Boston, 1972, p. 151)

thesis of glycine. Not only is our knowledge of the metabolism of glycine in nervous tissue minimal, but at the present time only scanty information is available on the factors regulating the concentration of glycine in the CNS. Glycine in the spinal cord is labeled only slowly from radioactive glucose via one of the glycolytic intermediates indicated in Figure 6-9. It is also possible to label glycine by administration of labeled glyoxylate, which is readily transmitted to glycine by nervous tissue.

Substantial evidence now exists to indicate that glycine may play a role as an inhibitory transmitter in the mammalian spinal cord. From the evidence available we will try to assess how well this compound fulfills the criteria necessary to categorize this substance as a CNS inhibitory transmitter. First, we can say that this amino acid is

TABLE 6-3. Amino Acid Concentration of Cat Spinal Cord and Roots (μmoles/g)

Amino Acids	Spinal Gray	Dorsal Root	Ventral Root
Alanine	0.62	0.36	0.30
Arginine	0.14	0.07	0.06
Aspartate	2.14	0.79	1.13
Cystathionine	3.02	0.02	0.01
GABA	0.84	Trace	Trace
Glutamate	4.48	3.33	2.09
Glutamine	5.48	1.46	1.10
Glycine	4.47	0.28	0.32
Leucine	0.16	0.06	0.05
Lysine	0.16	0.07	0.05
Serine	0.41	0.32	0.23
Threonine	0.21	0.18	0.15

SOURCE: M. H. Aprison and R. Werman (1968), *Neurosciences Research 1*, 157, Academic Press, New York.

found in relatively high concentrations in the spinal cord compared with other amino acids. Table 6-3 illustrates the occurrence of glycine and other free amino acids in the spinal cord, and from these data it is apparent that glycine is more concentrated in the spinal gray matter than in the spinal white matter. The concentration found in the spinal gray matter is much higher than the level in whole brain or spinal roots. This high level of glycine in the ventral horn, together with the comparatively low content in ventral root fibers, initially suggested that glycine may be associated with the inhibitory interneurons, and subsequently this was shown to be the case. From iontophoretic studies the compounds in Table 6-3 can be divided into four main groups:

1. Excitatory   glutamate, aspartate
2. Inhibitory   GABA, glycine, alanine, cystathionine, and serine

3. Inactive   glutamine, leucine, threonine, and lysine
4. Untested   arginine and methyl histidine

Of the amino acids in group 2, cystathionine and serine appear to have a weaker and relatively sluggish action on spinal neurons compared with that of GABA and glycine. Therefore, in view of the distribution and the relative inhibitory potency of the compounds in this group, it was suggested by Aprison and Werman that GABA and glycine were the most likely candidates for spinal inhibitory transmitters. Glycine administered by iontophoretic techniques was consistently found to diminish the firing and excitability of both spinal motoneurons and interneurons. This is interesting in itself since glycine is quite ineffective as an inhibitor of cortical neurons. In addition, the hyperpolarization and the changes in membrane permeability produced by glycine seem to be quite similar to those produced by the spinal inhibitory transmitter. Alteration in K^+ or Cl^- ion concentrations affect the inhibitory postsynaptic potentials and the glycine-induced potentials in the same fashion. Strychnine, a compound that has been shown to reduce spinal postsynapic inhibition, also blocks the effects of glycine on spinal motoneurons. It was originally reported that strychnine reversibly blocks the action of the natural inhibitory synaptic transmitter(s), as well as glycine and β-alanine, but does not have any effect on the hyperpolarizing action of GABA in cats anesthetized with pentobarbital. However, similar experiments on decerebrated, unanesthetized cats have indicated that the quantitative dichotomy between glycine and β-alanine and GABA originally reported does not usually hold. Instead, only a quantitative difference in the interaction of strychnine and the amino acids above is reported. It is of interest that, except for experiments with strychnine, there has been little pharmacological manipulation of neurons that release glycine. In all these experiments, strychnine antagonized glycine more potently than GABA. However, GABA also inhibits these spinal neurons and is present endogenously, and in the course of distinguishing between the relative merits of GABA and glycine as inhibitory transmitter candidates in the spinal cord, it was of interest to see if a loss of function of inhibitory interneurons would be accompanied by a change in one or both of these compounds. Anoxia of the lumbosacral cord produced by clamping the thoracic aorta seems to destroy the interneurons preferentially, while leaving 80% or more of the motoneurons intact. Thus, the presynaptic cells responsible for inhibition (and other functions) are lost, and therefore the concentration of transmitter associated with those cells would be expected to be decreased. When the effect of anoxia on GABA and glycine content of the cat spinal cord was analyzed, glycine was the only potential inhibitory amino acid markedly decreased. Thus the distribution of glycine in the spinal cord of the cat appears to be related to the inhibitory interneurons. Some preliminary results with perfused toad spinal cord have indicated that glycine release appears to occur during stimulation of the dorsal roots. However, the release of glycine from interneurons in the cord has not yet been demonstrated. In summary, glycine thus satisfies many of the criteria sufficiently to warrant its consideration as a possible inhibitory transmitter in the cat spinal cord.

1. It occurs in the cat spinal cord associated with interneurons.

2. When administered iontophoretically, it hyperpolarizes motoneurons to the same equilibrium potential as postsynaptic inhibition.

3. The permeability changes of the postsynaptic membrane induced by glycine appear to be similar to those associated with postsynaptic inhibition.

4. Strychnine, a drug that blocks the action of glycine, also blocks postsynaptic inhibition.

5. Stimulation of dorsal roots causes release of glycine from perfused spinal cord.

The actions of synaptic glycine are terminated by reuptake into the nerve terminals by Na^+/Cl^--dependent, high-affinity glycine transporters. In the spinal cord and brain stem the specific uptake of glycine has been demonstrated in regions exhibiting high densities of inhibitory glycine receptors. Two glycine transporter proteins have been cloned and shown to be expressed in brain as well as in peripheral tissues. These glycine transporters are both members of the large family of Na^+/Cl^--dependent neurotransmitter transporters (see Chapter 10), and both share approximately 50% sequence identity with the GABA transporter. The glycine transporters have been named GLYT-1 and GLY-2 in the order in which they were reported. These transporters have very similar kinetics and pharmacological properties but differ in the distribution of their transcripts measured by in situ hybridization. The distribution of GLYT-1 mRNA closely parallels the distribution of the glycine receptor, suggesting that the GLYT-1 is primarily a glial transporter, whereas the GLYT-2 is primarily associated with neurons.

The strychnine-sensitive glycine receptor has also been cloned and expressed and appears to be structurally quite homologous to the subunits of other ligand-gated ion channels, including the $GABA_A$ receptor. The native glycine receptor appears to be a pentameric structure, and photoaffinity labeling reveals that both glycine and strychnine binding sites are located on the α-subunit. Several glycine receptor α-subunit variants have recently been identified ($α_{1-4}$) and shown to differ in their pharmacological properties and level of expression. The expression of $α_1$- and $α_2$-subunits is developmentally regulated, with a switch from the neonatal $α_2$-subunit (strychnine insensitive) to the adult $α_1$ form (strychnine sensitive) at about 2 weeks postnatally in the mouse. It is interesting that the timing of this switch corresponds with the development of spasticity in the mutant spastic mouse, prompting the speculation that insufficient expression of the adult strychnine-sensitive isoform may underlie some forms of spasticity.

To date, very little appears to be known about the factors controlling the release of glycine from the spinal cord. Again, as with GABA, the efficient uptake process may explain why it is difficult to detect glycine release from the CNS. The main problems (as with

GABA, glutamate, etc.) is that there is no distinct neuronal pathway that may be isolated and stimulated; thus, all the induced activity is very generalized, making the significance of any demonstrable release (metabolite or excess transmitter) very difficult to interpret.

In summary, probably the most critical missing piece of evidence to establish an even more complete identification of the inhibitory role of glycine in the spinal cord is the demonstration that glycine is the main substance contained in the terminals of the interneurons that synapse on the motoneurons and that it is glycine that is released from these terminals when direct inhibition is produced. Some partial support for the localization of glycine has been provided by autoradiographic localization of glycine uptake sites as visualized by electron microscopy. However, demonstration of discretely evoked release of glycine from spinal cord interneurons has been more difficult to obtain.

Recently, a new role has been proposed for glycine that is distinct from its established role as an inhibitory transmitter in lower brain stem areas and in spinal cord mediated by a strychnine-sensitive chloride conductance. Several groups have shown that nanomolar concentrations of glycine increase the frequency of NMDA-receptor channel opening in a strychnine-insensitive manner and have suggested a mechanism involving allosteric regulation of the NMDA-receptor complex through a distinct glycine-binding site (Fig. 6-12). This action can be mimicked with glycine agonists and blocked by glycine antagonists. This concept is supported by the existence of strychnine-insensitive glycine binding sites that have an anatomical distribution identical to that of the NMDA receptor. When compared with the effects of benzodiazepines on the GABA receptor, the enhancement of NMDA responses observed with glycine is much greater. This suggests that the main effect of glycine is to prevent desensitization of the NMDA receptor during prolonged exposure to agonists. Glycine appears to accomplish this by an acceleration of the recovery of the receptor from its desensitized state rather than by blocking the onset of desensitization. It has been speculated that glycine facilitation of synaptic responses medi-

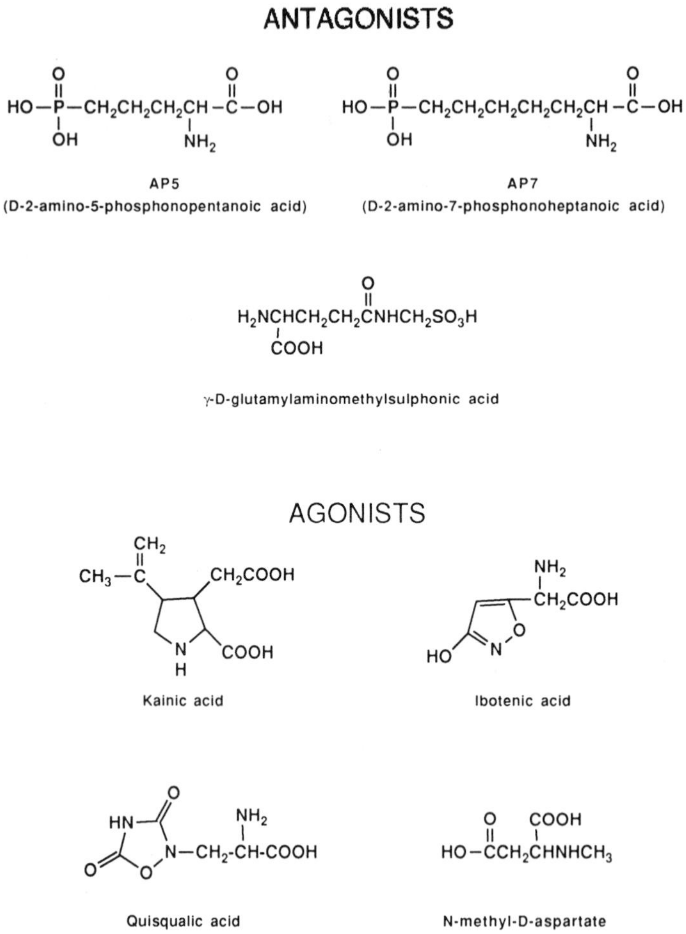

ANTAGONISTS

$$HO-\overset{\overset{O}{\|}}{\underset{\underset{OH}{|}}{P}}-CH_2CH_2CH_2\overset{\overset{}{}}{\underset{\underset{NH_2}{|}}{CH}}-\overset{\overset{O}{\|}}{C}-OH$$

AP5
(D-2-amino-5-phosphonopentanoic acid)

$$HO-\overset{\overset{O}{\|}}{\underset{\underset{OH}{|}}{P}}-CH_2CH_2CH_2CH_2\overset{\overset{}{}}{\underset{\underset{NH_2}{|}}{CH}}-\overset{\overset{O}{\|}}{C}-OH$$

AP7
(D-2-amino-7-phosphonoheptanoic acid)

$$H_2N\overset{}{\underset{\underset{COOH}{|}}{CH}}CH_2CH_2\overset{\overset{O}{\|}}{C}NHCH_2SO_3H$$

γ-D-glutamylaminomethylsulphonic acid

AGONISTS

Kainic acid

Ibotenic acid

Quisqualic acid

N-methyl-D-aspartate

FIGURE 6-10. Structures of several conformationally restricted analogs of glutamate and several antagonists.

ated by NMDA receptors may be a common regulatory mechanism for excitatory synapses, which raises the question of the processes that regulate extracellular glycine concentration in brain regions where NMDA receptors play a critical role in excitatory transmission. An important question is whether endogenous glycine antagonists (such as kynurenate) play a role in regulating neuronal function where NMDA receptors are involved.

GLUTAMIC ACID

It has been recognized for many years that certain amino acids such as glutamate and aspartate occur in uniquely high concentrations in the brain and that they can exert very powerful stimulatory effects on neuronal activity. Thus, if any amino acid is involved in regulation of nerve cell activity, as an excitatory transmitter or otherwise, it seems unnecessary to look beyond these two candidates.

The excitatory potency of glutamate was first demonstrated in crustacean muscle and later by direct topical application to mammalian brain. However, except for the invertebrate model, where substantial evidence has accumulated to support a role for glutamate as an excitatory neuromuscular transmitter, its status as a neurotransmitter in mammalian brain was uncertain for many years. This is probably in part explainable by the fact that glutamate (and also aspartate) is a compound that is involved in intermediary metabolism in neural tissue. For example, glutamate plays an important function in the detoxification of ammonia in the brain, is an important building block in the synthesis of proteins and peptides including glutathione, and also plays a role as a precursor for the inhibitory neurotransmitter GABA. Thus it has been extremely difficult to dissociate the role this amino acid plays in neuronal metabolism, and as a precursor for GABA, from its possible role as a transmitter substance. The transport of circulating glutamate to the brain normally plays only a very minor role in regulating the levels of brain glutamate. In fact, the influx of glutamate from the blood across the blood–brain barrier is much lower than the efflux of glutamate from the brain.

In brain L-glutamate is synthesized in the nerve terminals from two sources: from glucose via the Krebs cycle and transamination of α-oxoglutatrate (Fig. 6-1) and from glutamine that is synthesized in glial cells, transported into nerve terminals, and locally converted by glutaminase into glutamate (see Fig 6-11). In the glutamate-containing nerve terminals, glutamate is stored in synaptic vesicles, and on depolarization of the nerve terminal it is released by a calcium-dependent exocytotic process. The action of synaptic glutamate is terminated by a high-affinity uptake process via the plasma membrane glutamate transporter on the presynaptic nerve terminal and/or on glial cells. The glutamate taken up into glial cells is converted by glutamine synthetase into glutamine, which is then transported via a low-affinity process into the neighboring nerve terminals, where it serves as a precursor for glutamate. In astrocytes glutamine can also be oxidized, (via the Krebs cycle) into α-ketoglutarate, which can be actively transported into the neuron to

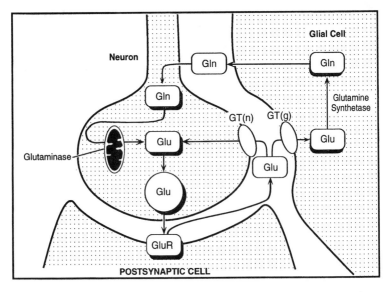

FIGURE 6-11. Pathways for glutamate utilization and metabolism. Glutamate (Glu) released into the synaptic cleft is recaptured by neuronal-type ($GT_{[n]}$) and glial-type ($GT_{[g]}$) Na^+-coupled glutamate transporters. Glial glutamate is converted to glutamine (Gln) by the enzyme glutamine synthetase (GS). Gln is present at high concentrations in the CSF (about 0.5 mM) and can enter the neuron to help replenish glutamate after hydrolysis by mitochondrial glutaminase (GA). (Modified from D. Nicholls, 1994.)

replace the α-ketoglutarate lost during the synthesis of neuronal glutamate.

Despite its ubiquitous occurrence in high levels throughout the CNS, the unequal regional distribution of glutamate in the spinal cord (higher in dorsal than corresponding ventral regions) is one of the main observations in support of the proposal that this common amino acid functions as an excitatory transmitter released from primary afferent nerve endings.

Even though neuronal systems believed to utilize glutamate or aspartate as transmitter substances are being described in the CNS, because of their role in intermediary metabolism it seems quite unlikely that it will be possible to map these systems accurately by simply tracking the presence of glutamate or aspartate or their synthesizing enzymes, as has been done in the past for the monoamines. The recent development of antibodies against excitatory amino acids, especially antisera that can distinguish between glutamate and aspartate with a high degree of selectivity, has facilitated the anatomical mapping of excitatory amino acid pathways. The development of reliable methods for combining anterograde labeling of primary afferent terminals with immunocytochemistry has helped to identify afferent nerve terminals enriched in glutamate or aspartate and to dissociate the role played by these amino acids in neurotransmission from their general role in metabolism. Nerve terminal enrichment in a specific excitatory amino acid provides the most direct anatomical evidence that a pathway uses a particular amino acid as a neurotransmitter. The elegant studies of Rustioni and coworkers employing immunocytochemistry at the electron microscopic level have identified primary afferent fibers terminating in spinal laminae of the lumbar spinal cord with nerve terminals enriched in glutamate and/or aspartate, providing direct anatomical evidence that these primary afferents appear to use these excitatory amino acids as neurotransmitters. Use of these techniques in conjunction with other less direct approaches, including mapping the localization of excitatory amino acid receptors, has provided strong support for a neurotransmitter role for excitatory amino acids in the mammalian CNS.

A unique approach to investigating the possible role of glutamate as a transmitter in mammalian CNS has been the employment of a virus that selectively destroys granule cells in the cerebellum. After viral infection had destroyed more than 95% of the granule cells, a marked decrease in cerebellar glutamate was observed, but no significant decrease in other amino acids, including aspartate, was detected. The high-affinity, sodium-dependent uptake of glutamate was also significantly reduced following destruction of the granule cells. The selective decline of glutamic acid levels and uptake does not prove that glutamic acid is a neurotransmitter in the granule cells, although it is consistent with this hypothesis, especially since the granule cell transmitter is known to be excitatory.

A recent technical advance that has aided in the elucidation of the excitatory effects of glutamate and related amino acids is the discovery of several conformationally restricted analogs of glutamic acid that exhibit marked potency and specificity in depolarizing central neurons. These agents include quisqualic acid (isolated from seeds), ibotenic acid (isolated from mushrooms), and kainic acid (isolated from seaweed). (See Fig. 6-10.) Of these agents, kainic acid has been the most extensively studied. Kainic acid is about 50-fold more potent in depolarizing neurons in the mammalian CNS than glutamate and thus is one of the most potent of the amino acid neuroexcitants. Extensive studies with this substance have revealed that direct injection of kainic acid in discrete brain regions results in selective destruction of the neurons that have their cell bodies at or near the site of injection. Axons and nerve terminals appear to be more resistant to the destructive effects of kainic acid than the cell soma. Although it was originally believed that the selective neurotoxic effects of kainic acid were due simply to excessive excitation of all neuronal soma exposed to this agent, it is now clear that the mechanism of its neurotoxic action is more complex. The neurotoxic action appears to involve an interaction between specific receptors for kainic acid and certain types of presynaptic input. For example, studies in the striatum and hippocampus indicate that kainic acid toxicity depends on an intact glutaminergic innervation of the injected brain region. Kainic acid is now being used extensively as a research tool in neurobiology to produce selective lesions in the brain of experimental animals and to explore the physiological pharmacology of excitatory transmission. Although the status of glutamate and aspartate as neurotransmitters has suffered through many cycles of acceptability and nonacceptability in the past several decades, the rush to clone the glutamate receptors and the extensive research on long-term potentiation demonstrating a function for glutamate have stopped the doubters from speaking. In spite of considerable evidence suggesting that glutamate may be an excitatory transmitter in the CNS, little was known about the biosynthesis and release of the pool of releasable "transmitter glutamate." Utilizing the molecular layer of the dentate gyrus of the hippocampal formation to provide a definitive system in which the major input appears to be glutaminergic, Cotman and coworkers have addressed these questions. Glutamate was shown to be released by depolarization from slices of the dentate gyrus in a Ca^{2+}-dependent manner, and lesions of the major input to the dentate gyrus originating from the entorhinal cortex diminished this release and the high-affinity uptake of glutamate. Glutamate biosynthesis in the releasable pools was rapidly regulated by the activity of glutaminase and by the uptake of glutamine. These properties are all consistent with the properties expected of a neurotransmitter, and the observations strengthened the premise that glutamate may be an important neurotransmitter in the molecular layer of the dentate gyrus. Furthermore, these studies demonstrate that the regulation of glutamate synthesis and release share many properties in common with other transmitters. For example, similar to acetylcholine synthesis, the synthesis of glutamate is regulated in part via the accumulation of its major precursor, glutamine, and newly synthesized glutamate, like acetylcholine, is preferentially released. In addition, the synthesis of glutamate is regulated by end-product inhibition. This is similar to the mechanism by which the rate-limiting enzyme in catecholamine synthesis, tyrosine hydroxylase, is regulated in catecholaminergic neurons by dopamine and norepinephrine. It is interesting that these similarities are demonstrable despite the involvement of glutamate in general brain metabolism.

The vesicle hypothesis describing the quantal release of acetyl-choline at the neuromuscular junction was introduced in the mid-1950s. Since then the concept of vesicular storage and release of acetylcholine has become firmly established and has been extended to a number of other synapses and neurotransmitters. However, there was no direct experimental evidence for the participation of synaptic vesicles in the storage and release of excitatory amino acid neurotransmitters until recently. The concept has received strong support from the studies of Jahn and Veda and coworkers on isolated synaptic vesicles from mammalian brain. These studies have shown that vesicles are capable of glutamate uptake and storage and that they have a specific carrier for L-glutamate.

Plasma Membrane Glutamate Transporter

Most of the molecular biological research on excitatory amino acid (EAA) transmission has focused on receptors rather than the process of transmitter inactivation. Inactivation is an especially important process in EAA neurotransmission. The rapid removal of glutamate from the synapse by high-affinity uptake not only serves to terminate the excitatory signal and recycle the glutamate but also plays an important role in the maintenance of extracellular levels of glutamate below those that could induce excitotoxic damage. Three glutamate transporter genes have been identified that share 50% sequence homology and exhibit minimal homology with other eukaryotic proteins, including the super family of neurotransmitter transporters that mediate the uptake of GABA, glycine, choline, and the biogenic amines. Two of these glutamate transporters appear to be of glial origin while the third is found in neurons. However, each individual transporter exhibits a specific distribution in brain. All three transporters demonstrate a strong Na^+ dependence and are enantioselective (i.e., D-aspartate, L-aspartate, and L-glutamate are substrates, whereas D-glutamate is not). The transporters are also inhibited by well-characterized uptake blockers including β-threo-hydroxy-aspartate, dihydrokainate, and L-trans-2,4-pyrrolidine decarboxylate. Neurons and glial cells appear to possess a similar plasma membrane glutamate uptake carrier that serves to terminate the postsynaptic action of neurotransmitter glutamate and to maintain the extracellular glutamate concentrations below levels that may damage neurons. It is not known whether synaptically released glutamate is removed from the synaptic cleft by uptake into the presynaptic neuron, by uptake into associated glial cells, or mainly by diffusion out of the synapse down a concentration gradient maintained by the glial cell uptake. The normal process for termination of the synaptic action of glutamate may involve all three processes to varying degrees. Radiotracer studies with labeled glutamate applied to brain indicate that most of the label is taken up into glial cells. However, this may be an artifact of the method employed since synaptically released as compared to exogenous applied glutamate may be better positioned to the uptake carrier in the presynaptic nerve membrane than to the glial transporter.

The studies on the dentate gyrus as well as many others have demonstrated that lesion studies coupled with biochemical measures of alterations in high-affinity, sodium-dependent uptake of glutamate in brain synaptosomes provide a useful technique for mapping out glutaminergic neuronal systems in the brain, although this technique cannot be used to differentiate between glutamate- and aspartate-containing neurons. Retrograde transport of D-aspartate after microinjection in terminal regions is another method that may be of value in tracing glutamate pathways. The success and specificity of this method is also dependent on the specificity of the uptake process and therefore is not capable of discriminating between glutamate- and aspartate-containing neurons.

It is somewhat disturbing that glutamate, aspartate, and synthetic derivatives of these dicarboxylic acids result in almost universal activation of unit discharge; these endogenous excitants appear to be somewhat ubiquitous in the nervous system without the expected asymmetric distribution. It has also been reported that glutamate does not bring the cell membrane potential to the same level as the natural excitatory transmitter. In addition, both the D and the naturally occurring L isomers of EAA are active, although in the case of glutamate the D isomer is often reported to be somewhat less active. These findings have led some investigators to suggest that the response to amino acids represents a nonspecific receptivity of the neuron to these agents and is therefore not necessarily indicative of a transmitter function.

However, based on the current evidence, glutamate appears to have satisfied four of the main criteria for classification as an excitatory neurotransmitter in the mammalian CNS: (1) It is localized presynaptically in specific neurons where it is stored in and released from synaptic vesicles; (2) it is released by a calcium-dependent mechanism by physiologically relevant stimuli in amounts sufficient to elicit postsynaptic responses; (3) a mechanism (reuptake and specific transporters) exists that will rapidly terminate its transmitter action; and (4) it demonstrates pharmacological identity with the naturally occurring transmitter.

The most clearcut evidence that EAA can act physiologically as excitatory neurotransmitters at a given synapse comes from experiments in which intracellular recordings of pre- and postsynaptic events can be made. In studies of this nature, the criterion of identical and parallel change induced by antagonists on synaptic events elicited by stimulation and those elicited by the action of the exogenously administered putative transmitter substance is of critical importance in identifying the excitatory amino transmitter involved. In many situations, however, such studies are not feasible, and more indirect studies like those summarized above have been utilized.

It should be clearly stated that quite often at excitatory synapses the actual molecule acting at the postsynaptic receptor has not been definitively identified, even though pharmacological analysis indicates that the synaptic response is mediated by a particular excitatory amino acid receptor. Thus, a critical link in establishing a neurotransmitter role for an EAA within a specific pathway is the demonstration and characterization of the EAA receptor that mediates synaptic transmission at that synapse.

Excitatory Amino Acid Receptors

Our understanding of EAA transmitters and their function and regulation has been greatly enhanced by studies directed at the identification, characterization, localization, and isolation of receptors for these amino acids. In fact, in the last decade, progress on the definition of receptor subtypes and the availability of more selective agonists and antagonists have produced a quantum leap in knowledge about EAA at synaptic sites throughout the vertebrate CNS.

Until the mid-1980s, we were content with two major classes of EAA receptors—the NMDA (N-methyl-D-aspartate) receptors and the non-NMDA receptors, the latter composed at that time of kainate and quisqualate receptors. With the development of more selective agonists and antagonists, however, the classes of EAA receptors have expanded to at least five different types (NMDA, kainate, AMPA, AP4, and ACPD) in the CNS, each displaying distinct physiological characteristics. Three of these receptors have been defined by the depolarizing excitatory actions of select agonists (NMDA; kainate; and quisqualate, or α-amino-3-hydroxy-5-methyl-ioxyzole-4-propionic acid—also called AMPA) and the blockade of the effects of these agonists by selective antagonists. A fourth—the AP4 receptor (1-2-amino-4-phosphonobutyrate)—appears to represent an inhibitory autoreceptor. The fifth receptor, activated by trans-1-aminocyclopentane-1-3-dicarboxylic acid (ACPD) modifies inositol phosphate (IP) metabolism and has been called a metabotrophic receptor. A summary of representative agonists and antagonists for each of these EAA receptor classes is given in Table 6-4.

Synaptic transmission in synapses using EAA does not appear to follow the simple model of fast-acting synaptic transmission mediated by a single receptor class. In fact, individual synapses that use EAA may not be restricted to distinct receptors but rather may have a combination of receptors and thereby exhibit different input/output properties and second messenger responses.

In recent years, a specific subtype of EAA receptor, the NMDA receptor, has become a major focus of attention because of evidence suggesting that it may be involved in a wide range of both neurophysiological and pathological processes as important and diverse as memory acquisition (see Chapter 13), developmental plasticity, epilepsy, and the neurotoxic effects of brain ischemia.

TABLE 6-4. Classification of Excitatory Amino Acid Receptors in the Mammalian CNS

Currently Accepted Name	NMDA		AMPA	Kainate	Metabotropic
	Glutamate Site	Glycine Site			
Subtype selective agonists	NMDA	Glycine D-serine R(+)HA-966(partial) L-687,414 (partial)	AMPA S(−)-5-Flu Quisqualic acid	Kainic acid Domoic acid	L-AP4 ACPD LCCG-1
Subtype selective antagonists	D(−)-AP-5 D(−)-AP-7 CGS-19755 CGP-37849 CGP-40116 CPP, (±)-D-D-CPPene	7-Chlorokynurenic 5,7-Dichlorokynurenic MNQX L-689,560	NBQX GYKI 52466		MCPG

Channel blockers	MK-801 Phencyclidine (PCP)			
Receptor selective agonists	NMDA	AMPA	Kainic acid	L-AP4
Receptor selective antagonists	D(−) AP-5	CNQX DNQX	CNQX DNQX	MCPG
Effector pathways	(Na$^+$/K$^+$/Ca^{2+})	(Na$^+$/K$^+$/Ca^{2+})	(Na$^+$/K$^+$/Ca^{2+})	IP$_3$,DAG

ACPD, 1-aminocyclopentane-1,3-dicarboxylic acid; AMPA, α-amino-3-hydroxy-5-methylisoxazole-4-propionic acid; AP-5, 2-amino-5-phosphonopentanoic acid; AP-7, amino-7 phosphonoheptanoic acid; CNQX, 6-cyano-7-nitroquinoxaline-2,3-dione; CPP, 3-(2-carboxypiperazin-4-yl)-propyl-1-phosphonic acid; D-CCPene, D-3-(2-carboxypiperazin-4-yl)-propyl-1-phospho-nene; DNQX, 6,7-dinitroquinoxaline-2,3-dione; HA-966, 3-amino-1-hydroxypyrrolidone-2; L-AP4, L-2-amino-4-phos-phonobutanoic acid; L-687,414, (R(+)-cis-β-methyl-3-amino-1-hydroxypyrrolid-2-one; LCCG-I, (2S,3S,4S)-α-(car-boxycyclopropyl)glycine; MCPG, α-methyl-4-carboxyphenylglycine; MNQX, 5,7-dinitroquinoxaline-2,3-dione; NBQX, 2,3-dihydro-6-nitro-7-sulphamoyl-benzo(f)quinoxaline; NMDA,N-methyl-D-aspartic acid; S(−)-5-Flu, S(−)-5-fluorowillardine.

NMDA Receptor–Ionophor Complex

The NMDA receptor is a ligand-gated ion channel composed of two different protein subunits named NMDAR1 and NMDAR2. NMDAR1 can exist in seven splice variants, and there are four different genes encoding variants of NMDAR2 (NMDAR2A, B, C, D). At present, it is not clear how many NMDAR1 and NMDAR2 subunits are present in each functional NMDA receptor. This receptor complex has been extensively characterized physiologically and pharmacologically and is widely distributed in mammalian brain and

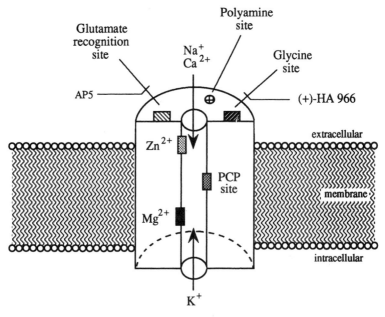

FIGURE 6-12. Schematic illustration of the NMDA receptor and the sites of action of different agents on the receptor. The NMDA receptor gates a cation channel that is permeable to Ca^{2+} and Na^+ and is gated by Mg^{2+} in a voltage-dependent fashion; K^+ is the counterion. The NMDA receptor channel is blocked by PCP and MK801 and the complex is regulated at two modulatory sites by glycine and polyamines; AP5 and CPP are competitive antagonists at the NMDA site.

spinal cord, with particularly high receptor densities found in hippocampus and cerebral cortex. NMDA receptors appear to have a pivotal role in long-term depression (LTD), long-term potentiation (LTP), and developmental plasticity. However, overactivation or prolonged stimulation of NMDA receptors can damage and eventually kill target neurons via a process referred to as excitotoxicity. Like the $GABA_A$ receptor, the NMDA receptor is a complex molecular entity endowed with a number of distinct recognition sites for endogenous and exogenous ligands, each with discrete binding domains. At present, there appear to be at least six pharmacologically distinct sites through which compounds can alter the activity of this receptor (see Fig. 6-12): (1) A transmitter binding site that binds L-glutamate and related agonists, promoting the opening of a high conductance channel that permits entry of Na and Ca into target cells. L-glutamate is virtually ineffective unless the site that binds glycine, the strychnine-insensitive glycine modulatory site (2), is also occupied. This glycine site is distinct from the glycine binding site on the strychnine-sensitive glycine inhibitory receptor (see discussion of glycine receptors, above), which is present in high density in brain stem and spinal cord. (3) A site within the channel that binds phencyclidine (PCP site) and related noncompetitive antagonists (MK-801, ketamine). These agents act most effectively when the receptor is activated (i.e., "open channel block"). (4) A voltage-dependent magnesium (Mg^{2+}) binding site. (5) An inhibitory divalent cation site near the mouth of the channel that binds zinc (Zn^{2+}) to produce a voltage-independent block. (6) A polyamine regulatory site whose activation by spermine and spermidine facilitates NMDA receptor-mediated transmission.

In addition, two distinct binding sites are also apparently associated with the transmitter recognition site, one that preferentially binds agonists and one that preferentially binds antagonists. It is of interest that quinolinic acid, a metabolite of tryptophan and thus a natural brain constituent, may be a specific antagonist for a particular subtype of NMDA receptor since it shows regional variation in potency. A number of glycine agonists and antagonists have already been identified for the glycine regulatory site, even though the glycine binding component of the NMDA receptor was discovered only recently.

The glycine modulatory site has attracted a great deal of interest as a potential site for the action of new antiepileptic drugs or agents that might be useful in preventing ischemic brain damage. D-serine is a potent agonist at this site, and (+)HA-966 is a selective antagonist. Glycine in submicromolar concentrations increases the frequency of the NMDA receptor channel opening in a strychnine-insensitive manner, and, even though brain and CSF concentrations of glycine are in the millimolar range, glycine is effective in vivo, indicating that the glycine site is not saturated. Thus, conditions that alter the extracellular concentration of glycine or compete with its binding site can dramatically alter NMDA receptor-mediated responses.

Polyamines such as spermine and spermidine function as allosteric modulators of NMDA receptors and potentiate NMDA currents in the presence of saturating concentrations of glutamate and glycine, but in contrast to glycine their presence is not a requirement for NMDA receptor activation. Under pathological conditions such as brain ischemia or trauma in which the production of polyamines is dramatically increased, it has been suggested that polyamines may mediate or potentiate the excitotoxic mechanisms responsible for the neuronal damage produced. This idea is supported by the findings that phenylethanolamines, e.g., ifenprodil and eliprodil (SL820715), which are potent antagonists of the polyamine modulatory site of the NMDA receptor complex in a number of biochemical models, exhibit effective neuroprotective action in ischemia and trauma.

Attention has recently been directed to kynurenate, a tryptophan metabolite that has been shown to block the high-affinity glycine binding site. Kynurenate can be thought of as an endogenous neuroprotective agent that is released from glial cells following the transamination of L-kynurenine. Manipulation of endogenous kynurenate formation may provide further insight into the role it plays in modulating glycine's action on the NMDA receptor.

Quinolinic Acid

Quinolinic acid is an endogenous excitatory neurotoxin, synthesized from L-tryptophan via the kynurenine pathway, that has the potential of mediating NMDA-induced neurotoxicity and dysfunction. Its endogenous occurrence in normal brain is relatively low (ca. 50–100 pmol/g). Toxicity induced by exogenous quinolinic acid can be prevented or reversed by noncompetitive or competitive NMDA antagonists, suggesting that the neurotoxicity produced by locally administered quinolinic acid is mediated through the NMDA subtype of EAA receptors. Although quinolinic acid was first identified in human brain in 1983, knowledge of the possible role played by this toxin in neuropathology was uncertain while speculation abounded. The observation by Heyes and coworkers in the late 1980s that quinolinate levels were dramatically elevated in AIDS was a major advance toward understanding the possible pathobiology of quinolinate. Further studies have demonstrated that the motor signs of AIDS dementia correlate strikingly with the levels of quinolinate in the cerebrospinal fluid. Treatment with the antiretroviral agent AZT decreased the viremia and associated dementia and lowered the cerebrospinal fluid levels of quinolinate. Because cerebrospinal fluid levels of quinolinate are often higher than those found in blood, it has been argued that the origin of this neurotoxin may be via intracerebral synthesis, although this has not yet been documented. The highest reported levels of cerebrospinal fluid quinolinate are often obtained in AIDS patients with opportunistic infections. It is now appreciated that elevated levels of cerebrospinal fluid quinolinate are found in human patients and in nonhuman primates with inflammatory neurologic diseases, but it is unclear whether the mechanisms underlying the increased levels of quinolinate induced by bacterial and viral infection are similar. It also remains to be determined if the bacteria are actually involved in the synthesis of some of the increased quinolinate. No specific organism has been identified as responsible for elevating quinolinic acid since this neurotoxin is elevated in a variety of bacterial and viral infections. The discovery of an effective inhibitor of quinolinic acid synthesis will help elucidate the importance of this neurotoxin in inflammatory and infectious disease.

Non-NMDA Receptors

Both AMPA and kainaic acid (KA) receptors mediate fast excitatory synaptic transmission and are associated primarily with voltage-independent channels that gate a depolarizing current primarily carried by the influx of Na^+ ions. While these receptors are easily distinguished from NMDA receptors, they are more difficult to distinguish from each other. Molecular biological studies have confirmed the existence of AMPA and KA classes of non-NMDA receptors but have revealed a considerable degree of heterogeneity within these two families. Some pharmacological discrimination can be achieved, with AMPA and quisqualate being the preferred agonist for AMPA receptors and Domoate and Kainate the preferred agonist for KA receptors. The most selective and potent non-NMDA antagonists available are a series of dihydroxyquinoxaline derivatives including NBQX, CNQX, and DNQX (see Table 6-4). However, these agents competitively block both types of non-NMDA receptors, although NBQX appears to exhibit the best selectivity for AMPA receptors. Very few KA receptor-selective compounds have been identified. Recently, a new class of 2,3-benzodiazepine derivatives (most notably GYKI 52466) has been shown to block AMPA-induced responses noncompetitively and to attenuate ischemic neuronal damage effectively in animal models, highlighting the fact that non-NMDA receptors also play an important role in CNS pathology.

Metabotropic Glutamate Receptors

The metabotropic glutamate receptors (mGluRs) constitute a family of EAA receptors that are linked to G proteins and second messenger systems and are distinct from the inotropic EAA receptors that form ion channels and are comprised of the NMDA, AMPA, and kainate subtypes discussed above. This more recently characterized group of receptors (mGluRs) is coupled to a variety of signal transduction pathways via guanine nucleotide binding proteins (G proteins), producing alterations in intercellular second messengers and generating slow synaptic responses. This is in clear contrast to the inotropic glutamate receptors that are directly coupled to cation-specific ion channels and that mediate fast excitatory synaptic responses. The widespread distribution of metabotropic receptors in the CNS, coupled with the prevalence of glutamate as a neurotransmitter, indicates that this system is a major modulator of second messengers in the mammalian CNS. Recent molecular cloning studies have revealed the existence of at least seven different subtypes of mGluRs, termed $mGluR_1$ to $mGluR_7$, which share the common structure of a large extracellular domain preceded by the seven-member spanning domains. Members of the mGluR family can be divided into three subgroups according to their sequence similarities, signal transduction properties, and pharmacological profiles to agonists (i.e., relative potencies when expressed in cell lines of glutamate, quisqualate, ACPD, and AP4). The first subgroup comprised of $mGluR_1$ and $mGluR_5$, is coupled to the stimulation of PI hydrolysis/Ca^{2+} signal transduction. The second group, $mGluR_2$ and $mGluR_3$, is negatively coupled through adenylate cyclase to cAMP formation. The third group, $mGluR_4$, $mGluR_6$, and $mGluR_7$, is also negatively linked to adenylate cyclase activity but shows a different agonist preference from that of $mGluR_2$ and $mGluR_3$. As a group, the mGluRs are widely expressed throughout the brain, but the individual subtypes appear to show some differential distribution. The pharmacology of the individual subtypes expressed in Chinese hamster ovary or in baby hamster kidney cells show some interesting differences (cf. Table 6-4). LAP_4 is a potent agonist of $mGluR_4$, $MGluR_6$, and $mGluR_7$, but has little effect on the other receptor subtypes. L-CCG-I, on the other hand, activates $mGluR_2$ at concentrations that have little or no effect on $mGluR_1$ and $mGluR_4$. No agonist yet identified appears to be specific for any single metabotropic receptor subtype.

Considerable experimental evidence indicates that mGluRs are involved in the regulation of synaptic transmission in the CNS.

However, the lack of specific antagonists has limited the precise characterization of the role of individual mGluRs in glutamatergic transmission and has severely hampered progress in identifying the physiological and pathological roles of mGluRs. The recent discovery that phenylglycine derivatives are selective antagonists of mGluRs has permitted a more rigorous testing of the physiological role of this receptor subclass in brain function and dysfunction. Data are already emerging that suggest a role in both synaptic transmission and synaptic plasticity. New compounds currently being developed are more potent and show greater subtype specificity. With the availability of these new subtype-selective antagonists, the next few years should witness major advances in our knowledge of the roles played by metabotropic glutamate receptors in physiological and pathological processes.

The function and distribution of the four classes of excitatory amino acid receptors are summarized in Table 6-5. The NMDA receptor is an essential component in the generation of long-term potentiation (LTP). LTP results in an increase in synaptic efficacy that has been proposed as an underlying mechanism involved in memory and learning (see Chapter 13).

In addition to the roles excitotoxic mechanisms may play in various chronic neurodegenerative disorders like Huntington's disease and viral diseases like AIDS, two chronic neurological syndromes have been linked to dietary consumption of amino acid toxins of plant origin. Neurolathyrism, a spastic disorder occurring in East Africa and Southern Asia, is associated with dietary consumption of the chick pea *Lathyrus sativus*. β-N-oxalylamino-L-alanine (BOAA) has been identified as the responsible toxin in this plant. This amino acid acts as an agonist at AMPA receptors. Guam disease, also referred to as ALS/parkinsonism/dementia, is thought to be related to the consumption of flour prepared from the seeds of the cycad *Cycas circinalis*, which contains the amino acid, β-N-methyl-amino-L-alanine (BMAA). Although BMAA is a neutral amino acid that is not directly excitatory or toxic in vitro, in the presence of bicarbonate it becomes excitotoxic and acts as an agonist at AMPA and NMDA receptors.

TABLE 6-5. Distribution and Function of Excitatory Amino Acid
Receptors in the Mammalian CNS

Receptor Type	Distribution/Function
NMDA	Widely distributed in mammalian CNS (enriched in hippocampus, cerebral ~~cortex). Demonstrated most easily by~~ pharmacological antagonism under MG^{2+}-free or depolarizing conditions or in binding experiments. Usually recognized as a slow component in repetitive activity generated primarily by non-NMDA receptors. Important in synaptic plasticity.
AMPA	Widespread in CNS; parallel distribution to NMDA receptors. Involved in the generation of fast component of EPSPs in many central excitatory pathways.
Kainate	Concentrated in a few specific areas of CNS, complementary to NMDA/AMPA distribution (e.g. stratum lucidum region of hippocampus). Difficult to distinguish from AMPA receptors pharmacologically due to nonspecificity of kainate in electrophysiological experiments. However, present specificity (in absence of AMPA and NMDA receptors) on dorsal root C fibers.
ACPD (metabotropic)	Linked to IP_3 formation. Activated by glutamate, quisqualate, ibotenate and trans-1-aminocyclopentane-1,3-dicarboxylate (ACPD) but not by AMPA, NMDA, or kainate. Not antagonized by NMDA or non-NMDA antagonists but sensitive to pertussis toxin. May be involved in developmental plasticity.

SOURCE: Modified from Watkins et al. (1990).

Neurotoxicity has also been observed following ingestion of domoic acid. Domoic acid is an analog of KA that is about three times as potent. This substance is synthesized by seaweed and can be consumed in toxic amounts by eating mussels that have fed on the seaweed. An outbreak of domoic poisoning occurred in 1987 in western Canada. Consumption of this neurotoxin by humans can damage the hippocampus and produce dementia.

With the availability of more specific pharmacological agents, it should be possible to evaluate in more detail the involvement of excitatory amino acid pathways in normal brain function and in neuropathological conditions. The participation of NMDA receptors in long-term potentiation provides a strong link between these systems and the mechanisms of learning and memory. NMDA and other EAA receptors also appear to play a role in cell damage caused by hypoglycemia, hypoxia, seizures, and other disturbances associated with excess excitatory amino acids.

SELECTED REFERENCES

GABA

Barbaccia, M. L., P. Guarneri, A. Berkowich, C. Wambebe, A. Guidotti, and E. Costa (1989). Studies on the endogenous modulator of GABA$_A$ receptors in human brain and CSF. In *Allosteric Modulation of Amino Acid Receptors: Therapeutic Implications* (E. A. Barnard and E. Costa, eds.), pp. 125–138, Raven Press, New York.

Bowery, N. G. (1993). GABA$_B$ receptor pharmacology. *Annu. Rev. Pharmacol Toxicol. 33*, 109–147.

Erlander, M. G. and A. J. Tobin (1991). The structural and functional heterogeneity of glutamic acid decarboxylase: A review. *Neurochem. Res. 16*, 215–226.

Haefely, W. (1990). The GABA–benzodiazepine interaction fifteen years later. *Neurochem. Res. 15*, 169–174.

Jursky, F., S. Tamura, A. Tamura, S. Mandiyan, H. Nelson, and N. Nelson (1994). Structure, function and brain localization of neurotransmitter transporters. *J. Exp. Biol. 196*, 283–295.

Lüddens, H. and W. Wisden (1991). Function and pharmacology of multiple GABA_A receptor subunits. *Trends Pharmacol. Sci. 12*, 49–51.

Macdonald, R. L. and Olsen, R. W. (1994). GABA_A receptor channels. *Annu. Rev. Neurosci. 17*, 569–602.

Paul, S. M. (1995). GABA and Glycine. In *Psychopharmacology: The Fourth Generation of Progress* (Bloom and Kupfer, eds.), pp. 87–94, Raven Press, New York.

Purdy, R. H., A. L. Morrow, P. H. Moore, Jr. and S. M. Paul (1991). Stress-induced elevations of γ-aminobutyric acid type A receptor-active steroids in the rat brain. *Proc. Natl. Acad. Sci. U.S.A. 88*, 4553–4557.

Roberts, E. (1986). GABA: the road to neurotransmitter status. In *Benzodiazepine/GABA receptors and chloride channels: Structural and functional properties* (Olsen and Venter, eds.), pp. 1–39, Alan R. Liss, New York.

Scott, R. H., K. G. Sutton, and A. C. Dolphin (1993). Interactions of polyamines with neuronal ion channels. *TINS 16(4)*, 153–160.

Zorumski, C. F. and K. E. Isenberg (1991). Insights into the structure and function of GABA–benzodiazepine receptors: Ion channels and psychiatry. *Am. J. Psychiatry 148*, 162–173.

Glycine

Becker, C.-M. (1990). Disorders of the inhibitory glycine receptor: The spastic mouse. *FASEB J. 4*, 2767–2774.

Betz, H. (1992). Structure and function of inhibitory glycine receptors. *Q. Rev. Biophys. 25*, 381–394.

Ottersen, O. P. and J. Storm-Mathisen (eds.) (1990). *Glycine Neurotransmission*. John Wiley & Sons, New York.

Vandenberg, R. J., C. A. Handford, and P. R. Schofield (1992). Distinct agonist- and antagonist-binding sites on the glycine receptor. *Neuron*, 491–496.

Excitatory Amino Acids

Birse, E. F., et al. (1993). Phenylglycine derivatives as new pharmacological tools for investigating the role of metabotropic glutamate receptors in the central nervous system. *Neuroscience 52*, 481–488.

Cotman, C. W., J. S. Kahle, S. E. Miller, J. Ulas, and R. J. Bridges (1995). Excitatory amino acid neurotransmission. In *Psychopharmacology: The Fourth Generation of Progress* (Bloom and Kupfer, eds.), pp. 75–85, Raven Press, New York.

Hayashi, Y., N. Sekiyama, S. Nakanishi, D. E. Jane, D. C. Sunter, E. F. Birse, P. M. Udvarhelyi, and J. C. Watkins (1994). Analysis of agonist and antagonist activities of phenylglycine derivatives for different cloned metabotropic glutamate receptor subtypes. *Neurosci. 14(5),* 3370–3377.

Heyes, M. P., K. Saito, J. S. Crowley, L. E. Davis, M. A. Demitrack, M. Der, L. A. Dilling, J. Elia, M.J.P. Kruesi, A. Lackner, S. A. Larsen, K. Lee, H. L. Leonard, S. P. Markey, A. Martin, S. Milstein, M. M. Mouradian, M. R. Pranzatelli, B. J. Quearry, A. Salazar, M. Smith, S. E. Strauss, T. Sunderland, S. W. Swedo, and W. W. Tourtellotte (1992). Quinolinic acid and kynurenine pathway metabolism in inflammatory and non-inflammatory neurological disease. *Brain 115,* 1249–1273.

Ishii, T., K. Moriyoshi, H. Sugihara, K. Sakurada, et al. (1993). Molecular characterization of the family of the N-methyl-D-aspartate receptor subunits. *J. Biol. Chem. 268,* 2836–2843.

Kemp, J. A., and P. D. Leeson (1993). The glycine site of the NMDA receptor—five years on. *Trends Pharmacol. Sci. 14,* 20–25.

Lodge, D. and D. Schoepp (1993). Excitatory amino acids. *Trends Pharmacol. Sci. Receptor Nomenclature Supplement,* 4th ed., insert.

Mayer, M. L., M. Benveniste, K. K. Patneau, L. J. Vyklicky (1992). Pharmacologic properties of NMDA receptors. *Ann. N.Y. Acad. Sci. 648,* 194–204.

Meldrum, B. and J. Garthwaite (1990). EAA pharmacology: Excitatory amino acid neurotoxicity and neurodegenerative disease. *Trends Pharmacol. Sci. 11,* 379–386.

Nakanishi, S. (1992). Molecular diversity of glutamate receptors and implications for brain functions. *Science 258,* 597–603.

Nicholls, D. G. (1994) *Proteins, Transmitters and Synapses.* Blackwell Science, Cambridge, MA.

Reinhard, J. F. Jr., J. B. Erickson, and E. M. Flanagan (1994). Quinolinic acid in neurological disease: Opportunities for novel drug discovery. *Adv. Pharmacol. 30,* 85–127.

Rogawski, M. A. (1993). Therapeutic potential of excitatory amino acid antagonists: Channel blockers and 2,3-benzodiazepines. *Trends Pharmacol. Sci. 14,* 325–331.

Scatton, B. (1993). The NMDA receptor complex. *Fundam. Clin. Pharmacol. 7,* 389–400.

Schoep, D. D. and P. J. Conn (1993). Metabotropic glutamate receptors in brain functional and pathology. *Trends Pharmacol. Sci. 14,* 13–20.

Seeburg, P. H. (1993). The molecular biology of mammalian glutamate receptor channels. *Trends Pharmacol Sci. 14,* 297–303.

Tracey, D. J., S. DeBiasi, K. Phend, and A. Rustioni (1991). Aspartate-like immunoreactivity in primary afferent neurons. *Neuroscience* 40(3), 673–686.

Valtschanoff, J. G., K. D. Phend, P. S. Bernardi, R. J. Weinberg, and A. Rustioni (1994). Amino acid immunocytochemistry of primary afferent terminals in the rat dorsal horn. *J. Comp. Neurol. 346, 237–252.*

Watkins, J. and G. Collingridge (1994). Phenylglycine derivatives as antagonists of metabotropic glutamate receptors. *TIPS 15*, 333–342.

Watkins, J. C., P. Krogstgaard-Larsen, and T. Honoré (1990). Structure–activity relationships in the development of excitatory amino acid receptor agonists and competitive antagonists. *Trends Pharmacol. Sci. 11*, 25–33.

7 | Acetylcholine

The neurophysiological activity of acetylcholine (ACh) has been known since the turn of the century and its neurotransmitter role since the mid-1920s. With this long history, it is not surprising when students assume that everything is known about the subject. Unfortunately, the lack of sophisticated methods for determining the presence of ACh in cholinergic tracts and terminals, which only recently has been overcome, has left this field far behind the biogenic amines. The structural formula of ACh is presented below:

$$(CH_3)_3N^+ - CH_2CH_2 - O - \overset{\overset{\displaystyle O}{\|}}{C} - CH_3$$

ASSAY PROCEDURES

ACh may be assayed by its effect on biological test systems or by physiochemical methods. Popular bioassay preparations include the frog rectus abdominis, the dorsal muscle of the leech, the guinea pig ileum, the blood pressure of the rat (or cat), and the heart of *Venus mercenaria*. In general, bioassays tend to be laborious, subject to interference by naturally occurring substances, and on occasion to behave in a mysterious fashion (e.g., the frog rectus abdominis is not as sensitive to ACh in the summer months as in the winter). Nevertheless, bioassays currently represent one of the most sensitive (0.01 pmol in the toad lung) and, under properly controlled conditions, the most specific procedures for determining ACh. It is probably fair to say that the neurochemically oriented investigator's natural fear and distrust of a bioassay have hampered progress in elucidating the biochemical and biophysical aspects of ACh. This statement is supported by a consideration of the plethora of information on norepinephrine. This neurotransmitter can also be bioassayed, but it was only after the development of sensitive fluorometric and radiometric procedures for determining components of the adrenergic nervous system that the information explosion occurred.

Until about 1965 physiochemical methods for determining ACh were so insensitive as to be virtually useless in measuring endogenous levels of ACh. Since then, however, papers have been published on enzymatic, fluorometric, gas chromatographic, chemiluminescent, and radioimmunoassay techniques that approach the sensitivity and specificity of the bioassays. Currently, the most popular assays are the GC–mass spectrometric procedure of Jenden and coworkers, the radiometric procedure of McCaman and Stetzler, the chemiluminescent assay of Israel and Lesbats, and the HPLC-electrochemical procedure of Potter and colleagues. Ricny and coworkers have modified and combined several existing methods to develop a highly sensitive (0.2 pmol) determination of the transmitter, and the sensitivity has been further increased in the procedure of Flentge et al.

SYNTHESIS

Acetylcholine is synthesized in a reaction catalyzed by choline acetyltransferase:

$$\text{Acetyl CoA} + \text{choline} \rightleftharpoons \text{ACh} + \text{CoA}$$

Before entering into a discussion of choline acetyltransferase, we should take note of Fig. 7-1, which depicts the possible sources of acetyl CoA and choline. In theory, the acetyl CoA for ACh synthesis may arise from glucose, through glycolysis and the pyruvate dehydrogenase system; from citrate, either by a reversal of the condensing enzyme (citrate synthetase) or by the citrate cleavage enzyme (citrate lyase); or from acetate through acetatethiokinase. In brain slices, homogenates, acetone powder extracts, and preparations of nerve-ending particles, glucose, or citrate are the best sources for ACh synthesis, with acetate rarely showing any activity. In lobster axons, the electric organ of the *Torpedo*, corneal epithelium, and frog neuromuscular junction, acetate appears to be the preferred sub-

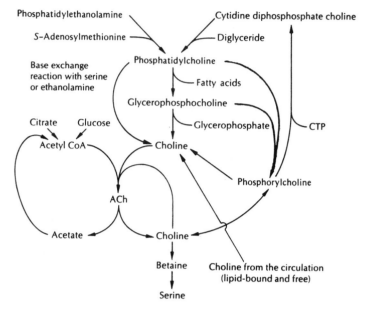

FIGURE 7-1. Acetylcholine metabolism.

strate. However, these systems are all in vitro and do not necessarily reflect the situation in vivo. Regardless of its source, acetyl CoA is primarily synthesized in mitochondria. Since, as detailed below, choline acetyltransferase appears to be in the synaptosomal cytoplasm, another still unsolved problem is how acetyl CoA is transported out of the mitochondria to participate in ACh synthesis. A probable carrier for acetyl CoA is citrate, which can diffuse into the cytosol and produce acetyl CoA via citrate lyase; a possible carrier is acetyl carnitine, and another possibility is a Ca^{2+}-induced leakage of acetyl CoA from mitochondria.

Although choline can be synthesized de novo in brain by successive methylations of ethanolamine, the extent is too minor to be of significance. Rather, choline is transported to the brain both free and in phospholipid form (possibly as phosphatidylcholine) by the blood. Following the hydrolysis of ACh, about 35%–50% of the liberated choline is transported back into the presynaptic terminal by a sodium-dependent, high-affinity active transport system, to be re-utilized in ACh synthesis (see below). As outlined in Fig. 7-1, the remaining choline may be catabolized or become incorporated into phospholipids, which can again serve as a source of choline. Regarding the possibility that membrane phosphatidylcholine is a reservoir for choline. Wurtman has speculated that an increased catabolism of this phospholipid to supply choline for ACh synthesis may be a primary cause of neurodegenerative diseases such as Alzheimer's disease. It should be noted that lesioning in the CNS or denervation in the superior cervical ganglion produces a decrease in choline content. A curious observation is that when brain cortical slices are incubated for 2 hours in a Krebs-Ringer medium, choline accumulates to a level about 10 times its original concentration. Similarly, a rapid postmortem increase in choline has been observed. The precise source of this choline is unknown; a probable candidate is phosphatidylcholine.

CHOLINE TRANSPORT

Recent studies of choline transport have produced a number of significant findings:

1. Choline crosses cell membranes by two processes referred to as high-affinity and low-affinity transport. High-affinity transport with a K_m for choline of 1–5 μM is saturable, carrier mediated, dependent on sodium, and stimulated by chloride. It is also dependent on the membrane potential of the cell or organelle so that any agent (e.g., K^+) that depolarizes the cell will concurrently inhibit high-affinity transport. Low-affinity choline transport with a K_m of 40–80 μM appears to operate by a passive diffusion process, linearly dependent on the concentration of choline, and virtually nonsaturable.

2. In contrast to the other neurotransmitters, ACh is taken up in terminals only via low-affinity transport; it is only choline that exhibits high-affinity kinetics.

3. Current evidence suggests that the high-affinity transport of choline is specific for cholinergic terminals and is not present in aminergic nerve terminals. Furthermore, it is kinetically (but not physically) coupled to ACh synthesis. It has been calculated that about 50%–85% of the choline that is transported by the high-affinity process is utilized for ACh synthesis. Low-affinity transport, on the other hand, is found in cell bodies and in tissues such as the corneal epithelium, and it is thought to function in the synthesis of choline-containing phospholipids. On the other hand, tissues that do not synthesize ACh (e.g., fibroplasts, erythrocytes, photoreceptor cells) exhibit high-affinity choline transport which is coupled to phospholipid synthesis.

4. Hemicholinium-3 is an extremely potent inhibitor of high-affinity transport (K_m of 0.05–1 μM) but a relatively weak (K_m of 10–120 μM) inhibitor of low-affinity transport.

5. There are three obvious mechanisms for regulating the level of ACh in cells: feedback inhibition by ACh on choline acetyltransferase, mass action, and the availability of acetyl CoA and/or choline. Of these three possibilities, the major regulatory factor seems to be high-affinity choline transport. This view derives from early observations that choline is rate limiting in the synthesis of ACh, coupled with findings in a number of laboratories. Using the septal-hippocampal pathway, a known cholinergic tract, Kuhar and his associates have shown that changes in impulse flow induced via electrical stimulation or pentylenetetrazol administration (both of which increase impulse flow) or via lesioning or the administration of pentobarbital (both of which decrease neuronal traffic) will alter high-affinity transport of choline into hippocampal synaptosomes. In their studies, procedures that activated impulse flow increased the V_{max} of choline transport, while agents that stopped neuronal activity decreased V_{max}. In neither situation was the K_m changed, a result to be ex-

pected since the concentration of choline outside the neuron (5–10 μM) normally exceeds the K_m for transport (1–5 μM). Recent evidence, however, is accumulating to suggest that this relationship between impulse traffic and choline transport does not occur in all brain areas (e.g., in the striatum, where cholinergic interneurons abound). In addition, the endogenous concentration of ACh is also implicated in regulating the level of the transmitter in the brain. Thus, in several studies an increase in choline uptake following depolarization of a preparation has been attributed to the release of endogenous ACh on depolarization. Other studies, however, suggest that this increased choline uptake is not related to ACh release but rather to an increase in Na^+, K^+-ATPase activity.

6. In contrast to the other neurotransmitters, the choline transporter has not yet been cloned.

CHOLINE ACETYLTRANSFERASE

With respect to the cellular localization of choline acetyltransferase (CAT), the highest activity is found in the interpeduncular nucleus, the caudate nucleus, retina, corneal opithelium, and central spinal roots (3000–4000 μg ACh synthesized/g/hr). In contrast, dorsal spinal roots contain only trace amounts of the enzyme, as does the cerebellum.

Intracellularly, after differential centrifugation in a sucrose medium, CAT is found in mammalian brain predominantly in the crude mitochondrial fraction. This fraction contains mitochondria, nerve ending particles (synaptosomes) with enclosed synaptic vesicles, and membrane fragments. When this fraction is subjected to sucrose density gradient centrifugation, the bulk of the CAT is found to be associated with nerve ending particles. When these synaptosomes are ruptured by hypo-osmotic shock, synaptosomal cytoplasm can be separated from synaptic vesicles. In a solution of low ionic strength, the CAT is adsorbed to membranes and to vesicles, but in the presence of salts at physiological concentration the enzyme is solubilized and remains in the cytoplasm. In vivo the enzyme is most likely present in the cytoplasm of the nerve ending particle. However, a particulate form of the enzyme has been found at nerve terminals and, more recently, at synaptic vesicle membranes. This CAT, which when solubilized has the same kinetic characteristics as the soluble form, occurs in a much lower concentration but with a higher specific activity. Since a kinetic coupling has been observed between uptake of choline and acetylation, it is conceivable that this membrane-bound CAT is the physiologically relevant form of the enzyme. Support for this contention derives from three experimental findings: (1) Homocholine cannot be acetylated with a purified soluble CAT but can form acetylhomocholine when a lysed synaptosomal preparation is used. (2) Choline mustard aziridinium ion, an inhibitor of choline transport as well as of CAT, is a much more potent inhibitor of membrane-bound CAT than the soluble form of the enzyme. (3) A monoclonal antibody raised against *Torpedo* terminal membranes inhibits CAT and ACh release.

A cell-free system of CAT was first described by Nachmansohn and Machado in 1943. Since that time, the enzyme from squid head ganglia, human placenta, *Drosophila melanogaster*, *Torpedo californica*, and brain has been purified and some of its characteristics have been defined.

When highly purified from rat brain, CAT has a molecular weight of 67–75 kD: It has an apparent Michaelis constant (K_m) for choline of 7.5×10^{-4} M and for acetyl CoA of 1.0×10^{-5} M. Recent estimates suggest an equilibrium constant of 13. The enzyme is activated by chloride and is inhibited by sulfhydryl reagents. A variety of studies on the substrate specificity of the enzyme indicate that various acyl derivatives both of CoA and of ethanolamine can be utilized by the enzyme. The major gap in our knowledge of CAT is that as yet we do not know of any useful (i.e., potent and specific) direct inhibitor of the enzyme. Styrylpyridine derivatives inhibit the enzyme but suffer from the fact that they are light sensitive, somewhat insoluble, and possess varying degrees of anticholinesterase activity. Hemicholinium inhibits the synthesis of ACh indirectly by preventing the transport of choline across cell membranes. Genomic aspects of CAT have been recently reviewed by Wu and Hersh.

ACETYLCHOLINESTERASE

Everybody agrees that ACh is hydrolyzed by cholinesterases, but nobody is sure just how many cholinesterases exist in the body. All cholinesterases will hydrolyze not only ACh but other esters. Conversely, hydrolytic enzymes such as arylesterases, trypsin, and chymotrypsin will not hydrolyze choline esters. The problem in deciding the number of cholinesterases that exist is that different species and organs sometimes exhibit maximal activity with different substrates. For our purposes we will divide the enzymes into two rigidly defined classes: acetylcholinesterase (also called "true" or specific cholinesterase) and butyrylcholinesterase (also called "pseudo" or nonspecific cholinesterase; the term "propionylcholinesterase" is sometimes used since in some tissues propionylcholine is hydrolyzed more rapidly than butyrylcholine). Although their molecular forms are similar, the two enzymes are distinct entities, encoded by specific genes. Current evidence suggests that in lower forms butyrylcholinesterase predominates, gradually giving way to acetylcholinesterase with evolution. When distinguishing between the two types of cholinesterases, at least two criteria should be used because of the aforementioned species or organ variation.

The first criterion is the optimum substrate. Acetylcholinesterase hydrolyzes ACh faster than butyrylcholine, propionylcholine, or tributyrin; the reverse is true with butyrylcholinesterase. In addition, acetyl-—methyl choline (methachol) is only split by acetylcholinesterase. That this criterion for distinguishing between the two esterases is not inviolate and must be used along with other indices is illustrated by the fact that chicken brain acetylcholinesterase will hydrolyze acetyl-—methyl choline but will also hydrolyze propionylcholine faster than ACh. Also the beehead enzyme will not hydrolyze either ACh or butyrylcholine but will split acetyl-β-methyl choline.

A second criterion for differentiating the cholinesterases is the substrate concentration versus activity relationship. Acetylcholinesterase is inhibited by high concentrations of ACh so that a bell-shaped substrate concentration curve results. This is observed also when butyrylcholine or propionylcholine is used. In contrast, butyrylcholinesterase is not inhibited by high substrate concentrations, so that the usual Michaelis-Menten type of substrate concentration curve is obtained. The reason for this difference is that in acetylcholinesterase there is at least a two-point attachment of substrate to enzyme, whereas with butyrylcholinesterase the substrate is attached at only one site.

The type of cholinesterase found in a tissue is often a reflection of the tissue. This fact is used as a discriminating index between cholinesterases. In general, neural tissue contains acetylcholinesterase, while glial cells and nonneural tissue usually contain butyrylcholinesterase. However, this is a generalization, and some neural tissue (e.g., autonomic ganglia) contains both esterases, as do some extraneural organs (e.g., liver, lung). In the blood, erythrocytes contain only acetylcholinesterase, while plasma contains butyrylcholinesterase. But plasma has primary substrates varying from species to species. Because of its ubiquity, cholinesterase activity cannot be used as the sole indicator of a cholinergic system in the absence of additional supporting evidence. To generalize on this point, until neuron-specific transmitter-degradating enzymes are discovered, it is a neurochemical commandment that, in order to delineate a neuronal tract, one must always assay an enzyme involved in the synthesis of a neurotransmitter and not one concerned with catabolism.

A final criterion that may be applied to differentiate between the esterases is their susceptibility to inhibitors. Thus the organophosphorous anticholinesterases such as diisopropyl phosphorofluoridate (DFP) and iso-OMPA are more potent inhibitors of butyrylcholinesterase whereas WIN8077 (Ambinonium) is about 2000 times better an inhibitor of acetylcholinesterase. The compound BW284C51 (1,5-bis-(4-allyldimethylammoniumphenyl) penta-3-

one dibromide) is presumed to be a specific reversible inhibitor of acetylcholinesterase.

In discussing the various techniques that are used to classify the cholinesterases, we touched on some aspects of the molecular properties of the enzymes. Because very little work has been done on butyrylcholinesterase and because no physiological role for this enzyme (or enzymes) has been demonstrated, we will focus our attention on acetylcholinesterase. In sucrose homogenates of mammalian brain subjected to differential centrifugation, acetylcholinesterase is found in both the mitochondrial and the microsomal fractions. The latter, consisting of endoplasmic reticulum and plasma cell membranes, exhibits a higher specific activity. This localization of the enzyme is supported by electron microscopic and histochemical studies that fix the activity at membranes of all kinds in both the CNS and the peripheral nervous system.

Both cholinesterases occur in several molecular forms that are classified as either globular or asymmetric. The globular forms G_1, G_2, and G_4 (Fig. 7-2) exist as monomers, dimers, and tetramers. Elongated forms that contain as many as 12 subunits and are attached to a collagen tail are classified as asymmetric. Regardless of the form, both cholinesterases occur in both a water-soluble and a membrane-bound state.

With its turnover time of 150 μs, equivalent to hydrolyzing 5000 molecules of ACh per molecule of enzyme per second, acetylcholinesterase ranks as one of the most efficient enzymes extant.

With respect to the topography of the enzyme, the twin-hatted diagram of the anionic and esteratic sites has been reproduced countless times and need not be presented again here. However, some discussion is in order since this was the first enzyme to be dissected at a molecular level. For this initiation into molecular biology we owe a debt of gratitude to Nachmansohn and his colleagues, particularly Wilson.

The active center of acetylcholinesterase has two main subsites. The first is an anionic site that attracts the positive charge in ACh; the second, about 5 Å distant, is an esteratic site that binds the car-

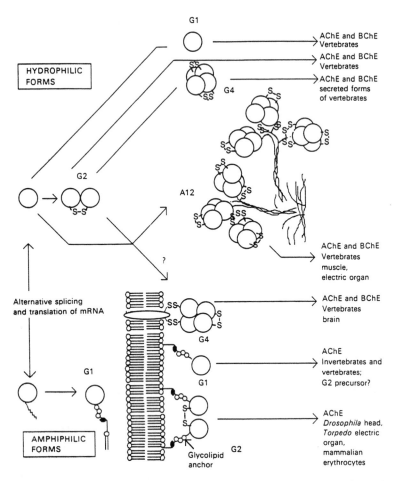

FIGURE 7-2. Schematic model of the molecular polymorphism of AChE and BChE. Open circles designate catalytic subunits. Disulphide bonds are indicated by S-S. Hydrophilic forms are G1, G2 and G4 forms. The asymmetric A12 forms have three hydrophilic G4 heads linked to a collagen tail via disulphide bonds. The G4 amphiphilic forms of brain are anchored into a phospholipid membrane through a 20-kDa anchor. The G2 amphiphilic forms of erythrocytes have a glycolipid anchor. In *Torpedo* AChE hydrophilic forms and amphiphilic G2 forms are produced by alternative splicing, so that the proteins are identical at 535 amino acids but are nonidentical at their C-termini [6].

SOURCE: From Chatonnet and Lockridge (1989).

bonyl carbon atom of ACh. Current information suggests that the anionic site contains at least one carboxyl group, possibly from glutamate, and the esteratic site involves a histidine residue adjacent to serine. The overall reaction is written as follows:

$$E - OH + ACh \underset{k_{-1}}{\overset{k_1}{\rightleftharpoons}} (E - OH - ACh) \overset{k_2}{\rightharpoonup}$$

$$E - OAC + choline \overset{k_3}{\underset{H_2O}{\rightharpoonup}} E - OH + ACO^- + H^+$$

Information on the architecture of the active center has been derived not only from kinetic studies using model compounds but from a group of inhibitors known as the anticholinesterases. (The pharmacology of these agents will be discussed in the last section of the chapter.) The anticholinesterases are classified as reversible and irreversible inhibitors of the enzyme. Like ACh, both types of inhibitor acylate the enzyme at the esteratic site. However, in contrast to ACh or to a reversible inhibitor such as physostigmine, the irreversible inhibitors, which are organophosphorous compounds, irreversibly phosphorylate the esteratic site. This phosphorylation has been shown to occur on the hydroxyl group of serine when DFP was incubated with purified acetylcholinesterase. Although the organophosphorous agents (referred to as nerve gases, although they are actually oils) are classed as irreversible anticholinesterases, there is a slow detachment of the compounds from the enzyme. Wilson observed that hydroxylamine speeded up this dissociation and regenerated active enzyme. He then set about designing a nucleophilic agent with a spatial structure that would fit the active center of acetylcholinesterase, and he did in fact produce a compound that is very active in displacing the inhibitor. This is 2-pyridine aldoxime methiodide (PAM), which has been used with moderate success in treating poisoning from organophosphorous compounds used as insecticides. PAM, with its quaternary ammonium group, does not penetrate the blood–brain barrier well enough to overcome central actions of the anticholinesterase. For this reason, atropine is usually used as an antidote along with PAM. It will be recalled that atropine blocks the effect of ACh at neuroeffector sites and has nothing to do with acetylcholinesterase. Currently, the oxime of choice in dephosphorylating organophosphorous-inactivated cholinesterase is HI-6 (1 [2-{hydroxyimino} methyl pyridinium]-2-[4-carboxyamidopyridinium] dimethyl ether dichloride).

Although the hydrolysis of Substance P by purified serum cholinesterase has been described, evidence now suggests that this activity may be due to contamination by dipeptidyl peptidase. Recent reviews of the cholinesterases by Taylor and Radic and by Massoulié et al. are recommended.

Uptake, Synthesis, and Release of ACh

Superior Cervical Ganglion, Brain, and Skeletal Muscle

To date the only major and thorough studies of ACh turnover in nervous tissue were done originally by MacIntosh, Birks, and colleagues and more recently by Collier using the superior cervical ganglion of the cat. By using one ganglion to assay the resting level of ACh and by perfusing the contralateral organ, these investigators have determined the amount of transmitter synthesized and released under a variety of experimental conditions including electrical stimulation, the addition of an anticholinesterase to the perfusion fluid, and perfusion media of varying ionic composition. Their results may be summarized as follows:

1. During stimulation, ACh turns over at a rate of 8%–10% of its resting content every minute (i.e., about 24–30 ng/min). At rest, the turnover rate is about 0.5 ng/min. Since there is no change in the ACh content of the ganglion during stimulation at physiological frequencies, it is evident that electrical stimulation not only releases the transmitter but also stimulates its synthesis.
2. Choline is the rate-limiting factor in the synthesis of ACh.

3. In the perfused ganglion, Na^+ is necessary for optimum synthesis and storage, and Ca^{2+} is necessary for the release of the neurotransmitter.

4. Newly synthesized ACh appears to be more readily released on nerve stimulation than depot or stored ACh.

5. About half of the choline produced by cholinesterase activity is reutilized to make new ACh.

6. At least three separate stores of ACh in the ganglion are inferred from these studies: "surplus" ACh, considered to be intracellular, which accumulates only in an eserine-treated ganglion and which is not released by nerve stimulation but is released by K^+ depolarization; "depot" ACh, which is released by nerve impulses and accounts for about 85% of the original store; and "stationary" ACh, which constitutes the remaining 15% that is nonreleasable.

7. Choline analogs such as triethylcholine, homocholine, and pyrrolcholine are released by nerve stimulation only after they are acetylated in the ganglia.

8. Increasing the choline supply in the plasma during perfusion of the ganglion only transiently increases the amount of ACh that is releasable with electrical stimulation, despite an accumulation of the transmitter in the ganglion.

9. The compound AH5183 (Vesamicol), shown by the Parsons' lab to inhibit ACh transport into synaptic vesicles, ultimately blocks release of ACh from the stimulated ganglia. This finding supports the contention that there is vesicular release of the transmitter in the periphery.

As stated above, this work on the superior cervical ganglion represents the most complete information on the turnover of ACh in the nervous system. With respect to regulation of ACh turnover in cholinergic terminals of skeletal muscle, Vaca and Pilar, using chick iris, have elaborated on the work of Potter with the rat phrenic nerve–diaphragm preparation. With the iris and the diaphragm preparation, the same relationship of high-affinity choline uptake,

ACh synthesis, and regulation by endogenous ACh has been demonstrated as previously described with CNS and autonomic nervous system preparations.

As noted earlier, ACh is not taken up into cholinergic terminals by a high-affinity transport system. However, as first described by Parsons' laboratory, ACh is transported into synaptic vesicles via a proton-pumping ATPase activity. A glycosylated ATPase pumps protons out of vesicles and drives ACh via a separate transporter into vesicles in exchange for the protons. This uptake is blocked by a compound referred to as vesamicol. Interestingly, it has been shown recently that this vesicular transporter and choline acetyltransferase arise from the same gene locus or an alternative splicing.

Microdialysis is a relatively new technique that is now being used to determine ACh regulation in freely moving rats. The procedure is to implant stereotaxically a cannula covered with a dialysis membrane that is perfused with a physiological solution. The dialysate is led into an HPLC system for determining ACh. This technique offers considerable promise for measuring basal and evoked ACh release under various experimental conditions.

Brain Slices, Nerve-Ending Particles (Synaptosomes), and Synaptic Vesicles

In 1939 Mann, Tennenbaum, and Quastel demonstrated the synthesis and release of ACh in cerebral cortical slices. In the succeeding 56 years these observations have been repeatedly confirmed but only moderately extended. The major finding of interest in all these studies is that in the usual incubation medium the level of ACh in the slices reaches a limit and cannot be raised. In a high K^+ medium the total ACh is increased substantially because much of its leaks into the medium from the slices. The experiments again suggest that the intracellular concentration of the neurotransmitter may play a role in regulating its rate of synthesis, in addition to the high-affinity uptake system for choline. This concept of a feedback mechanism is supported by the findings that the administration of drugs such as morphine, oxotremorine, or anticholinesterases at the most only succeed in a doubling of the original level of ACh in the brain. Regardless of the dose of the drug, no higher level can be obtained. Much of the current neurochemical work on the release of ACh from the brain involves the use of nerve ending particles (synaptosomes). This preparation, independently developed by DeRobertis and by Whittaker, is derived from sucrose density gradient centrifugation of a crude mitochondrial fraction of brain. Although synaptosomes represent presynaptic terminals with enclosed vesicles and mitochondria, some postsynaptic fragments are often attached to them. A disadvantage of the preparation is its heterogeneity; the usual synaptosome fraction is a mixture of cholinergic, noradrenergic, serotonergic, and other terminals. In addition, as judged by electron microscopy and enzyme markers, the purity is around 60%; the contaminants may include glial cells, ribosomes, and membrane fragments that may be axonal, mitochondrial, or perikaryal. On the other hand, the advantage of these preparations is that they can be isolated easily and that synaptic vesicles can be collected by hypoosmotically shocking the synaptosomes. With respect to the disposition of ACh in synaptosomes, roughly half of the transmitter is found in vesicles and the other half in synaptosomal cytoplasm. The cytosolic localization could be artifactual, resulting from the preparation methods.

Following the discovery of these presynaptically localized vesicles that contained ACh, the conclusion was almost unavoidable that these organelles are the source of the quantal release of transmitter as described in the neurophysiological experiments of Katz and his collaborators. Thus, the obvious interpretation has been that as the nerve is depolarized, vesicles in apposition to the terminal fuse with the presynaptic membrane, and ACh is released into the synaptic cleft to interact with receptors on the postsynaptic cell to change ion permeability. Synapsin I, a phosphoprotein that is localized in vesicles, may mediate the translocation of vesicles to the plasma membrane. Other vesicular proteins that have been implicated in the exocitotic process are synaptotagmins, synaptophysins, and synaptobrevins (see Chapter 3). Synaptobrevin is of particular interest in that both tetanus toxin and botulinum toxin type B, which are zinc endopeptidases, inhibit ACh release by cleaving synaptobrevin. At any rate, the subsequent sequence of events is not clearly pictured, but in some fashion presynaptic membrane is pinocytotically recaptured, and vesicles are resynthesized and simultaneously or subsequently repleted with ACh; the neuron is then ready for the next quantal release of transmitter.

Another possibility to explain ACh release comes from the laboratories of Israel and Dunant. The Israel group has isolated a lipoprotein from presynaptic terminals of the *Torpedo* electric organ that they refer to as a mediatophore. A 15 kDa subunit of this protein exhibits a vacuolar ATPase activity. When the mediatophore is incorporated into ACh-loaded proteoliposomes, the addition of calcium and the calcium ionophore A23187 triggers the release of ACh. The Dunant group transfected mediatophore cDNA into neuroblastoma cells and restored quantal ACh release. Dunant and Israel suggest that the vesicles release ACh to the cytosol and that the mediatophore, acting as a gate around a calcium channel, will release the transmitter. The action of calcium is terminated by uptake into synaptic vesicles, where it is ultimately excreted.

CHOLINERGIC PATHWAYS

The identification of cholinergic synapses in the peripheral nervous system has been relatively easy, and we have known for a long time now that ACh is the transmitter at autonomic ganglia, at parasympathetic postganglionic synapses, and at the neuromuscular junction. In the CNS, however, until relatively recently, technical difficulties have limited our knowledge of cholinergic tracts to the motoneuron collaterals to Renshaw cells in the spinal cord. With respect to the aforementioned technical difficulties, the traditional approach has been to lesion a suspected tract and then assay for ACh, choline acetyltransferase, or high-affinity choline uptake at the presumed terminal area. Problems with lesioning include the difficulty in making discrete, well-defined lesions and in interrupting fibers of passage. This latter problem is illustrated by the discovery that a habenula-interpeduncular nucleus projection that, based on lesioning of the habenula, was always quoted as a cholinergic pathway is not: It turned out that what was lesioned were cholinergic fibers that passed through the habenula. Thus, although the interpeduncular nucleus has the highest choline uptake and choline acetyltransferase activity of any area in the brain, the origin of this innervation remains largely unknown. A quantum leap in technology for tracing tracts in the CNS has occurred in the past several years. Through the use of histochemical techniques (originally developed by Koelle and coworkers) that stain for regenerated acetylcholinesterase after DFP treatment (Butcher, Fibiger), autoradiography with muscarinic receptor antagonists (Rotter, Kuhar), and immunohistochemical procedures with antibodies to choline acetyltransferase (McGeer, Salvaterra, Cuello, Wainer), a clear picture of cholinergic tracts in the CNS is now emerging. The well-documented tracts are depicted in Fig. 7-3. There is additional information that in the striatum and the nucleus accumbens septi, only cholinergic interneurons are found. Also, intrinsic cholinergic neurons recently have been reported to exist in the cerebral cortex, colocalized with VIP (vasoactive intestinal polypeptide) and often in close proximity to blood vessels.

With respect to other neurotransmitter functions of ACh, there are a few bits of information that suggest that ACh may participate in circuits involved with pain reception. Thus, the findings that nettles *(Urtica urens)* contain ACh and histamine, that high concentrations of ACh injected into the brachial artery of humans have been shown to result in intense pain, and that ACh applied to a blister produced a brief but severe pain, all indicate a relationship between ACh and pain. That ACh may act as a sensory transmitter in thermal receptors, taste fiber endings, and chemoreceptors has also been suggested, based on the excitatory activity of the compound on these sensory nerve endings.

CELLULAR EFFECTS

A variety of actions of ACh that may be viewed as cellular effects rather than neurotransmitter activity have been described. These in-

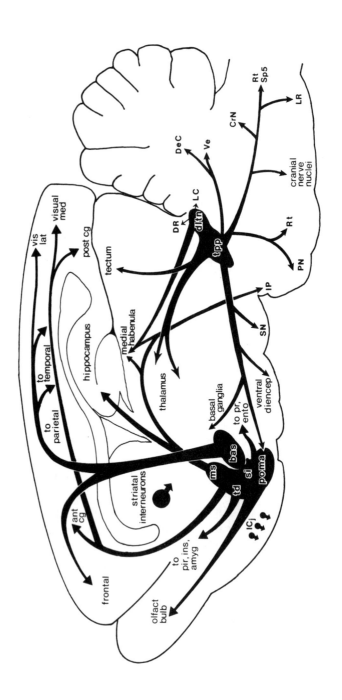

FIGURE 7-3. Schematic representation of the major cholinergic systems in the mammalian brain. As illustrated, central cholinergic neurons exhibit two basic organizational schemata: (a) local circuit cells (i.e., those that morphologically are arrayed wholly within the neural structure in which they are found) exemplified by the interneurons of the caudate-putamen nucleus, nucleus accumbens, olfactory tubercle, and *Islands of Calleja complex (ICj)* and (b) projection neurons (i.e., those that connect two or more different regions). Of the cholinergic projection neurons that interconnect central structures, two major subconstellations have been identified: (a) the basal forebrain cholinergic complex composed of ChAT-positive neurons in the *medial septal nucleus (ms), diagonal band nuclei (td), substantia innominata (si), magnocellular preoptic field (poma),* and *nucleus basalis (bas)* and projecting to the entire nonstriatal telencephalon; and (b) the pontomesencephalotegmental cholinergic complex composed of ChAT-immunoreactive cells in the *pedunculopontine (tpp)* and *laterodorsal (dltn) tegmental nuclei* and projecting ascendingly to the thalamus and other diencephalic loci and descendingly to the pontine and medullary *reticular formations (Rt), deep cerebellar (DeC)* and *vestibular (Ve) nuclei,* and cranial nerve nuclei. Not shown are the somatic and parasympathetic cholinergic neurons of cranial nerves III–VII and IX–XII and the cholinergic α- and γ-motor and autonomic neurons of the spinal cord. Additional abbreviations: amyg, amygdada; ant cg, anterior cingulate cortex; CrN, dorsal cranial nerve nuclei; diencep, diencephalon; DR, dorsal raphe nucleus; ento, entorhinal cortex; frontal, frontal cortex; IP, interpeduncular nucleus; ins, insular cortex; LC, locus ceruleus; LR, lateral reticular nucleus; olfact, olfactory; pir, piriform cortex; PN, pontine nuclei; pr, perirhinal cortex; parietal, parietal cortex; post cg, posterior cingulate cortex; SN, substantia nigra; Sp5, spinal nucleus of cranial nerve five. (From Butcher and Woolf [1986] and Woolf and Butcher [1989].)

clude ciliary movement in the gill plates of *Mytilis edulis,* ciliary motility of mammalian respiratory and esophageal tracts, a hyperpolarizing effect on atrial muscle, limb regeneration in salamanders, protein production in the silk gland of spiders, the induction of sporulation in the fungus *Trichoderma,* protoplasmic streaming in slime molds, and photic control of circadian rhythms and seasonal reproductive cycles.

There is also a variety of tissues and organisms, such as human placenta, *Lactobacillus plantarum, Trypanosoma rhodesience,* and the fungus *Claviceps purpurea* in which ACh is found but where nothing is known of its action. One of the most interesting situations is the corneal epithelium, which contains the highest concentrations of ACh of any tissue in the body. Some evidence suggests that ACh may be involved in sodium transport in this tissue.

All the activities of ACh and its localization in nonnervous tissue that we have noted above suggest that this agent may be a hormone as well as a neurotransmitter. It is worthwhile for the student to keep in mind that all known neurotransmitters may possess this dual function. These two activities have already been shown to occur with the biogenic amines. Even when a neurotransmitter is found in nervous tissue, its action may satisfy the criteria for defining a hormone or a modulator rather than the currently strict criteria for a "classic" neurotransmitter.

CHOLINERGIC RECEPTORS

As noted in Chapter 4, cholinergic receptors fall into two classes, muscarinic (Table 7-1) and nicotinic (Table 7-2). At last count, five muscarinic receptors (M_1–M_5) had been cloned. All of them exhibit a slow response time (100–250 ms), are coupled to G proteins, and either act directly on ion channels or are linked to a variety of second messenger systems. M_1, M_3, and M_5 are coupled to PI hydrolysis; M_2 and M_4 are coupled to cAMP. When activated, the final effect can be to open or close K channels, Ca channels, or Cl channels, depending on the cell type. With this array of channel activity, therefore, stimulation of muscarinic receptors will lead to ei-

TABLE 7-1. Muscarinic Receptors

Currently Accepted Name	M_1	M_2	M_3	M_4	M_5
Subtype selective agonists	McN-A343 (ganglion) Pilocarpine (relative to M_3 and M_5) L-689,660	Bethanechol (relative to M_4)	L-689,660	McN-A343 (relative to M_2)	
Subtype selective antagonists	Pirenzepine Telezepine 4-DAMP	Methoctramine AF-DX-116 Gallamine (noncompetitive) Himbacine	Hexahydro-sila-difenidol p-Fluorohexahydro-sila difenidol 4-DAMP	Tropicamide Himbacine 4-DAMP	4-DAMP
Signal transduction mechanisms	IP_3/DAG	↓ cAMP	IP_3/DAG	↓ cAMP	

4-DAMP, 4-Diphenylacetoxy-N-methylpiperidine.

SOURCE: Modified from Kebabian, J. W. and Neumeyer, J. L., eds (1994), *RBI Handbook of Receptor Classification.*

ther depolarization or hyperpolarization. As noted in Table 7-1, second messenger systems have been described following activation of the muscarinic receptors. Knowing the messengers is fine, but it doesn't tell us anything about the message (i.e., what is the ultimate effect?).

The pharmacological antagonists that have been used to define three of the muscarinic subtypes are pirenzepine, which has a high affinity for M_1; AFDX-116, with a high affinity for M_2; and 4-DAMP, which exhibits the highest affinity for M_3. It should be noted that most antagonists do not show more than a fivefold selectivity for one subtype over all other subtypes. The two classic muscarinic antagonists atropine and QNB do not distinguish the subtypes but block all equally well. What is needed are selective muscarinic agonists that will define the various subtypes. More importantly, it would be a major advance if we knew whether each subtype subserved a specific function since, if this were the case, it could lead to the development of specific drugs devoid of side effects.

Until relatively recently the identification of nicotinic cholinergic receptors in the CNS was an enigma. Using labeled α-bungarotoxin, nicotine, mecamylamine, or dihydro-βerythroidine, each investigation yielded mystifying results in which the antagonist could not be easily displaced by ACh or by unlabeled ganglionic or neuromuscular antagonists but on occasion was displaced by muscarinic agonists and antagonists. A major problem has been the low density of nicotinic compared to muscarinic receptors in the brain.

Conversely, much is known about the properties of the nicotinic cholinergic receptor of *Torpedo* and *Electrophorus* electric organs. This reflects the abundance of the receptor in this tissue and the availability of two snake toxins, α-bungarotoxin and *Naja naja siamensis*, that specifically bind to the receptor and have facilitated its isolation and purification. In the past 4 years, however, research with monoclonal antibodies and cDNAs has yielded considerable information about the mammalian nicotinic receptors (Table 7-2). At least seven different functional receptors have been identified and can be tentatively differentiated by CNS, ganglionic, and muscle types as well as pre- and postsynaptic localizations in the CNS.

Cholinergic nicotinic receptors from muscle or from electric organ contain four different subunits (α, β, γ, and δ) with a stoichiometry of two α-subunits and each of the other three. In contrast, neuronal nicotinic receptors contain (as of 1995) only two kinds of subunits (α and β) with the α occurring in at least seven different forms and the β in three. When one considers the variety of combinations of these α- and β-subunits that are theoretically possible, there appears to be a surfeit of nicotinic receptors in the brain. Whether future electrophysiological studies can assign a specific function to each of them is an open question. Reviews of this problem by Patrick et al. and by Papke are recommended.

As noted above, the nicotinic ACh receptor from electric organs is a pentameric integral membrane protein composed of four glycosylated polypeptide chains designated α, β, γ, and δ, with a stoichiometry of two α-subunits to one each of the other three. All subunits traverse the membrane and, when viewed face on, resemble a five-petal rosette with a central pit. ACh binds to the α-subunits and produces a conformational change in the channel that selectively allows cations rather than anions with a diameter of about 0.65 nm to pass through. The molecular weight of the receptor complex is about 255,000. Desensitization of the receptor increases when it is phosphorylated by cAMP protein kinase, protein kinase C, or tyrosine kinase.

An unexpected windfall resulting from the isolation of the nicotinic cholinergic receptor has been the demonstration that when this lipoprotein as isolated from *Electrophorus* is injected into rabbits, all the signs of myasthenia gravis appear. This finding, coupled with autoradiographic studies of muscle biopsies of myasthenics using labeled α-bungarotoxin as a tag where a marked deficiency of receptors is observed, has led to a better understanding of the disease at a molecular level. It is now clear that myasthenia gravis is an autoimmune disease in which a circulating antibody appears to be involved in an increased rate of degradation and damage, as well as antagonism of the ACh receptor. Since antibodies produced from *Electrophorus* or *Torpedo* ACh receptor result in myasthenic signs in rabbits, guinea pigs, and monkeys, this work also carries the implication

TABLE 7-2. Nicotinic Receptors (Members of the Ligand-Gated Ion Channel Superfamily)

	Muscle type	Neuronal type; α-Bungarotoxin-insensitive	Neuronal type; α-Bungarotoxin-insensitive
Currently accepted name	Muscle type	Neuronal type; α-Bungarotoxin-insensitive	Neuronal type; α-Bungarotoxin-insensitive
Alternate names	C-10 receptor	C-6 receptor, Ganglionic receptor	Neuronal α-bungarotoxin Binding site
Selective nicotinic agonists	Nicotine[a] Dimethylphenylpiperazinium Methylcarbamylcholine[c] (+)-Anatoxin A Cytisine[a] Methylisoarecolone	Nicotine Dimethylphenylpiperazinium Methylcarbamylcholine (+)-Anatoxin A Cytisine Methylisoarecolone	Nicotine Dimethylphenylpiperazinium Methylcarbamylcholine (+)-Anatoxin A Cytisine Methylisoarecolone
Selective nicotinic antagonists	d-Tubocurarine Lophotoxin Dihydro-β-erythroidine	d-Tubocurarine Lophotoxin Dihydro-β-erythroidine	d-Tubocurarine Lophotoxin Dihydro-β-erythroidine

Subtype selective antagonists	α-Bungarotoxin Conotoxin MI	Neosurugatoxin Neuronal bungarotoxin[b]	α-Bungarotoxin Methyllycaconitine[c]
Channel permeability and conductance	$Na^+ > K^+ \gg Ca^{2+}$ 40 or 60 pS	$Na^+ > K^+ \gg Ca^{2+}$ Varies, 5 to 80 pS	$Na^+ > K^+ \gg Ca^{2+}$ Varies
Selective channel blockers		Hexamethonium Mecamylamine Chlorisondamine	

[a]Although nicotine, cytisine, and methylcarbamylcholine activate a variety of nicotinic receptor subtypes, only certain α-bungarotoxin insensitive subtypes bind these ligands with high enough affinity to detect binding.

[b]Also known as κ-bungarotoxin, this toxin blocks only certain subtypes of a α-bungarotoxin-insensitive receptors. In addition, neuronal bungarotoxin recognizes at least some of the neuronal α-bungarotoxin-sensitive subtypes.

[c]Methyllycaconitine blocks other subtypes of receptors but is reported to be a 1000 times more potent at α-bungarotoxin sensitive receptors on neurons.

SOURCE: Modified from Kebabian, J. W. and Neumeyer, J. L., eds (1994), *RBI Handbook of Receptor Classification.*

that the ACh receptor is a phylogenetically conserved protein that can exhibit immunological cross-reactivity.

ACH IN DISEASE STATES

Aside from myasthenia gravis and other autoimmune diseases, such as the Lambert-Eaton myasthenic syndrome (a presynaptic problem involving diminished release of ACh), the role of ACh in nervous system dysfunction is unclear. Certainly a strong case can be made for familial dysautonomia, an autosomal recessive condition affecting Ashkenazi Jews that is in fact diagnosed by a supersensitivity of the iris to methacholine. Huntington's disease, involving a degeneration of Golgi type 2 cholinergic interneurons in the striatum, is reported to be partially ameliorated by physostigmine. The administration of physostigmine to patients with tardive dyskinesia has produced mixed results.

In Alzheimer's disease, characterized behaviorally by a severe impairment in cognitive function and neuropathologically by the appearance of neuritic plaques and neurofibrillary tangles, a cholinergic dysfunction has been implicated. As discussed more fully in Chapter 13, this is based on the fact that a variety of cholinergic abnormalities have been found. However, efforts to treat patients with choline, lecithin, or anticholinesterases such as tetrahydroaminoacridine (Tacrine) have met with little success. In animal models with imposed cognitive deficits, a restoration of memory has been achieved with synthetic cholinergic agonists such as YM796, DuP996, L-689,660, and WEB 1881 FU. Since there are no animal models of Alzheimer's disease, it is still unknown if any of these agents will produce a clinical benefit in humans. At any rate, the major focus currently is on the β-amyloid peptide that is found in the plaques of Alzheimer's disease brain. Considerable evidence suggests that aberrant proteolytic processing of this β-amyloid protein is associated with degenerating nerve terminals. Thus, current research is directed to developing specific protease inhibitors as well as exploring techniques for delivering neurotrophic factors such as NGF to the brain.

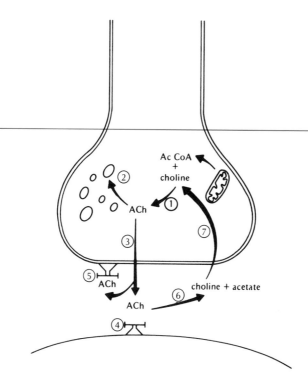

FIGURE 7-4. Sites of drug action at cholinergic synapses.

Site 1: ACh synthesis can be blocked by styryl pyridine derivatives such as NVP.

Site 2: ACh transport into vesicles is blocked by vesamicol (AH5183).

Site 3: Release is promoted by β-bungarotoxin, black widow spider venom, and La³⁺. Release is blocked by botulinum toxin, cytochalasin B, collagenase pretreatment, and Mg²⁺.

Site 4: Postsynaptic receptors are activated by cholinometic drugs and anticholinesterases. Nicotinic receptors, at least in the peripheral nervous system, are blocked by rabies virus, curare, hexamethonium, or dihydro—erythroidine; n-methylcarbamylcholine and dimethylphenyl piperazinium are nicotinic agonists. Muscarinic receptors are blocked by atropine, pirenzepine, and quinuclidinyl benzilate.

Site 5: Presynaptic muscarinic receptors may be blocked by AFDX-116 (an M_2 antagonist), atropine or quinuclidinyl benzilate. Muscarinic agonists (e.g., oxotremorine) will inhibit the evoked release of ACh by acting on these receptors.

Site 6: Acetylcholinesterase is inhibited reversibly by physostigmine (eserine) or irreversibly by DFP, or soman.

Site 7: Choline uptake competitive blockers include hemicholinium-3, troxypyrrolium tosylate, or AF64A (noncompetitive).

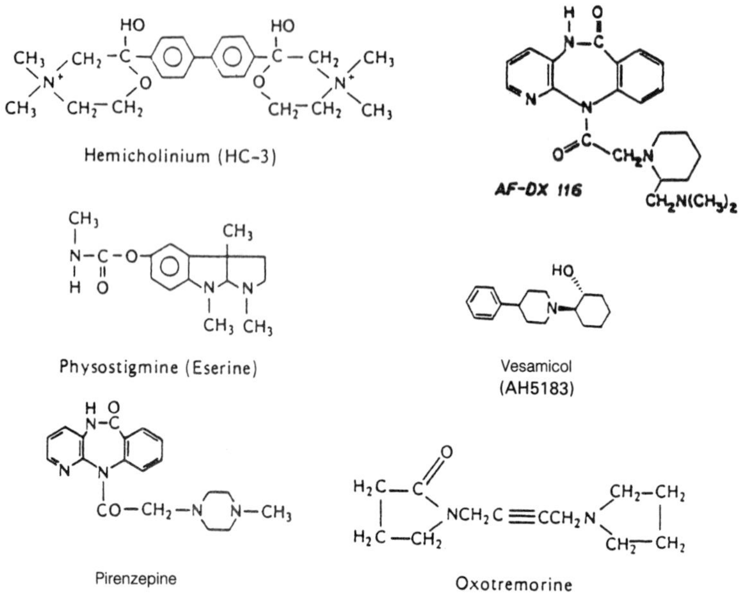

FIGURE 7-5. Structures of some drugs that affect the cholinergic nervous system.

Sites of drugs that affect the synthesis, release, and neurotransmitter activity of ACh are shown in Fig. 7-4. Structures of drugs that affect the cholinergic nervous system are depicted in Fig. 7-5.

SELECTED REFERENCES

Benishin, C. G. and P. T. Carroll (1983). Multiple forms of choline-O-acetyltransferase in mouse and rat brain: Solubilization and characterization. *J. Neurochem. 41*, 1016.

Bonner, T. I. (1989). The molecular basis of muscarinic receptor diversity. *Trends Neurosci. 12*, 148.

Butcher, L. L. and N. J. Woolf (1986). Central cholinergic systems: Synopsis of anatomy and overview of physiology and pathology. In *The Biological Substrates of Alzheimer's Disease* (A. B. Scheibel and A. F. Wechsler, eds.), pp. 73–86, Academic Press, New York.

Chatonnet, A. and O. Lockridge (1989). Comparison of butyryl-cholinesterase and acetylcholinesterase. *Biochem. J. 260, 625.*

Collier, B., A. Tandon, M.A.M. Prado, and M. Bachoo (1993). Storage and release of acetylcholine in a sympathetic ganglion. *Prog. Brain Res.* *98,* 183.

Cooper, J. R. (1994). Unsolved problems in the cholinergic nervous system. *J. Neurochem. 63,* 395.

Claudio, T. (1989) Molecular genetics of acetylcholine receptor-channels. In *Frontiers of Molecular Biology* (D. Glover and D. Hanes, eds.), IRL Press, London.

Cordell, B. (1994). β-Amyloid formation as a potential therapeutic target for Alzheimer's disease. *Annu. Rev. Pharmacol. Toxicol. 34,* 69.

Deneris, E. S., J. Boulter, J. Connoly, E. Wada, D. Goldman, L. W. Swanson, J. Patrick, and S. Heinemann (1989). Genes encoding neuronal nicotinic acetylcholine receptors. *Clin. Chem. 35,* 731.

Dunant, Y., and M. Israel (1993). Ultrastructure and biophysics of acetylcholine release: Central role of the mediatrophore. *J. Physiol. (Paris)* *87,* 179.

Eder-Colli, L., S. Amato, and Y. Froment (1986). Ampiphilic and hydrophilic forms of choline-O-acetyltransferase in cholinergic nerve endings of the Torpedo. *Neuroscience 19,* 275.

Eder-Colli, L., P. A. Briand, M. Pellegrini, and Y. Dunant (1989). A monoclonal antibody raised against plasma membrane of cholinergic nerve terminals of the *Torpedo* inhibits choline acetyltransferase activity and acetylcholine release. *J. Neurochem. 53,* 1419.

Ericson, J. P., H. Varoqui, M.K.H. Schafer, W. Modi, M. F. Diebler, E. Weihe, J. Rand, L. Eiden, T. I. Bonner, and T. B. Usdin (1994). Functional identification of a vesicular acetylcholine transporter and its expression from a "cholinergic" gene locus. *J. Biol. Chem. 269,* 21929.

Flentge, F., K. Venema, T. Koch, and J. Korf (1992). An enzyme-reactor for electrochemical monitoring of choline and acetylcholine: Applications in high-performance liquid chromatography, brain tissue, microdialysis, and cerebrospinal fluid. *Anal. Biochem. 204,* 305.

Goyal, R. K. (1989). Muscarinic receptor subtypes. *N. Engl. J. Med. 321,* 1022.

Israel, M. and B. Lesbats (1981). Chemiluminescent determination of acetylcholine and continuous detection of its release from *Torpedo* electric organ synapses and synaptosomes. *Neurochem. Int. 3,* 81.

Jenden, D. J., M. Roch, and R. A. Booth (1973). Simultaneous measurement of endogenous and deuterium-labeled tracer variants of choline and acetylcholine in subpicomole quantities by gas chromatography/mass spectrometry. *Anal. Biochem. 55,* 438.

Kawashima, L., H. Ishikawa, and M. Mochizuki (1980). Radioimmunoassay for acetylcholine in the rat brain. *J. Pharmacol. Methods* 3, 115.

Klein, J., R. Gonzalez, A. Koppen, and K. Löffelholz (1993). Free choline and choline metabolites in rat brain and body fluids: Sensitive determination and implications for choline supply to the brain. *Neurochem. Int.* 22, 293.

Kuhar, M. J. (1981). Autoradiographic localization of drug and neurotransmitter receptors in brain. *Trends Neurosci.* 4, 10.

Lindstrom, J., R. Schoepfer, W. G. Conroy, and P. Whiting (1990). Structural and functional heterogeneity of nicotinic receptors. *Ciba Found. Symp.* 152.

Massoulié J., L. Pezzementi, S. Bon, E. Krejci, and F. M. Vallette (1993). Molecular and cellular biology of cholinesterases. *Prog. Neurobiol.* 41, 39.

McCaman, R. E. and J. Stetzler (1977). Radiochemical assay for ACh: Modifications for sub-picomole measurements. *J. Neurochem.* 28, 669.

Papke, R. L. (1993). The kinetic properties of neuronal nicotinic receptors: Genetic basis of functional diversity. *Prog. Neurobiol.* 41, 509.

Parsons, S. M., C. Prior, and I. G. Marshall (1993). Acetylcholine transport, storage, and release. *Int. Rev. Neurobiol.* 35, 279.

Patrick, J., P. Séquéla, S. Vernino, M. Amador, C. Luetje, and J. A. Dani (1993). Functional diversity of neuronal nicotinic acetylcholine receptors. In *Prog. Brain Res.* 98, 113.

Potter, L. T. (1970). Synthesis, storage and release of [^{14}C]acetylcholine in isolated rat diaphragm muscles. *J. Physiol.* 206, 145.

Potter, P. E., M. Hadjiconstantinou, J. L. Meek, and N. H. Neff (1984). Measurement of acetylcholine turnover rate in brain: An adjunct to a simple HPLC method for choline and acetylcholine. *J. Neurochem.* 43, 288.

Ricny, J., J. Coupek, and S. Tucek (1989). Determination of acetylcholine and choline by flow-injection with immobilized enzymes and fluorometric or luminometric detection. *Anal. Biochem.* 176, 221.

Roghani, A., J. Feldman, S. A. Kohan, A. Shirzadi, C. B. Gundersen, N. Brecha, and R. H. Edwards (1994). Molecular cloning of a putative vesicular transporter for acetylcholine. *Proc. Natl. Acad. Sci. U.S.A.* 91, 10620.

Rotter, A., N. J. Birdsall, A.S.V. Burgen, P. M. Field, E. C. Hulme, and G. Raisman (1979). Muscarinic receptors in the central nervous system of the rat. *Brain Res. Rev.* 1, 141.

Salvaterra, P. M. and J. E. Vaughn (1989). Regulation of choline acetyltransferase. *Int. Rev. Neurobiol.* 31, 81.

Sastry, B.V.R. and C. Sadarangivivad (1979). Cholinergic systems in nonnervous tissue. *Pharmacol. Rev.* 30, 65.

Schimerlik, M. L. (1989). Structure and regulation of muscarinic receptors. *Annu. Rev. Physiol. 51*, 217.

Schmidt, B. M. And R. J. Rylett (1993). Basal synthesis of acetylcholine in hippocampal synaptosomes is not dependent upon membrane-bound choline acetyltransferase activity. *Neuroscience 54*, 649–656.

Silinsky, E. M. (1985). The biophysical pharmacology of calcium-dependent acetylcholine secretion. *Pharmacol. Rev. 37*, 81.

Simon, R., S. Atweh, and M. J. Kuhar (1976). Sodium-dependent high affinity choline uptake: A regulator step in the synthesis of acetylcholine. *J. Neurochem. 26*, 909.

Taylor, P. and Z. Radic (1994). The cholinesterases: from genes to proteins. *Ann Rev. Pharmacol. Toxicol.* 34, 281.

Tucek, S. (1993) Short-term control of the synthesis of acetylcholine. *Prog. Biophys. Mol. Biol. 60*, 59.

Vaca, K. and G. Pilar (1979). Mechanisms controlling choline transport and acetylcholine synthesis in motor nerve terminals during electrical stimulation. *J. Gen. Physiol. 73*, 605.

Wainer, B. H., A. I. Levey, E. J. Mutson, and M.-M. Mesulam (1984). Cholinergic systems in mammalian brain identified with antibodies against choline acetyltransferase. *Neurochem. Int. 6*, 163.

Woolf, N. J. and L. L. Butcher (1989). Cholinergic systems: Synopsis of anatomy and overview of physiology and pathology. In *The Biological Substrates of Alzheimer's Disease* (A. B. Scheibel and A. F. Wechsler, eds.), pp. 73–86, Academic Press, New York.

Wu, D. and L. B. Hersh (1994). Choline acetyltransferase: Celebrating its fiftieth year. *J. Neurochem. 62*, 1653.

Wurtman, R. J. (1992). Choline metabolism as a basis for the selective vulnerability of cholinergic neurons. *Science 14*, 117.

8 | Norepinephrine and Epinephrine

Norepinephrine and epinephrine belong to a class of compounds known as catecholamines. The term "catecholamine" refers generically to all organic compounds that contain a catechol nucleus (a benzene ring with two adjacent hydroxyl substituents) and an amine group (Fig. 8-1). In practice, the term usually means dihydroxyphenylethyl amine (dopamine, DA) and its metabolic products norepinephrine (NE) and epinephrine (E). Major advances in the understanding of the biochemistry, physiology, and pharmacology of norepinephrine and related compounds have come mainly through the development of sensitive assay techniques and of methods for visualizing catecholamines and their metabolic enzymes or receptors in vivo and in vitro.

Bioassay procedures used to be one of the most sensitive methods for the estimation of epinephrine and norepinephrine in tissue extracts or biological fluids. However, the more recently developed radioenzymatic assays and high-performance liquid chromatographic techniques, coupled with electrochemical detection and mass fragmentographic methods, are more sensitive and specific and have virtually displaced all other assay methods.

In 1946, it was demonstrated by Ulf von Euler in Sweden, and shortly thereafter by Holtz in Germany, that mammalian peripheral sympathetic adrenergic nerves use norepinephrine as a transmitter instead of epinephrine (which is the sympathetic transmitter in the frog). These findings laid the foundation for a new era of research in the field of catecholamines. At that time, it was not technically possible to study the content of transmitter in terminal parts of the adrenergic fiber. Despite this, Euler predicted that norepinephrine was in fact highly concentrated in the nerve terminal region from which it was released to act as a neurotransmitter. This prediction was conclusively documented some 10 years later with the develop-

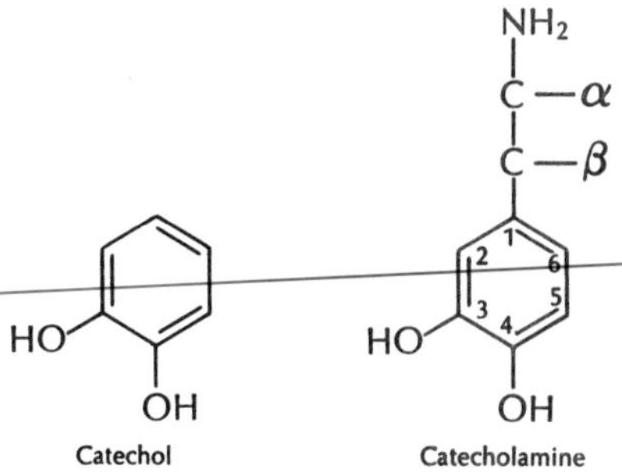

FIGURE 8-1. Catechol and catecholamine structure.

ment of techniques for visualizing catecholamines in freeze-dried tissue sections that helped define the anatomy of the adrenergic neuron.

MORPHOLOGY OF ADRENERGIC NEURON

The morphology of the peripheral noradrenergic neuron became clear following the development by Falck and Hillarp of the fluorescent histochemical method for the visualization of catecholamines. Catecholamines in freeze-dried tissue sections are converted to highly fluorescent derivatives after exposure to dry formaldehyde vapor at 60°–80° C. Under the fluorescent microscope all parts of the adrenergic neuron can be visualized (cell bodies, dendrites, axons, and nerve terminals) but the highest concentration of NE and strongest fluorescence is found in the nerve terminal varicosities. A diffuse innervation pattern is characteristic of the sympathetic nervous system and noradrenergic nerves. A single neuron can give rise to nerve terminal branches with lengths on the order of 10–20 cm, possessing several thousand nerve terminal varicosities (Fig. 8-2). The localization of catecholamines and their precursors within morphologically recognizable microscopic structures has been a great advantage to investigators interested in studying and understanding adrenergic mechanisms. The fluorescent histochemical method has also been used in conjunction with electron mi-

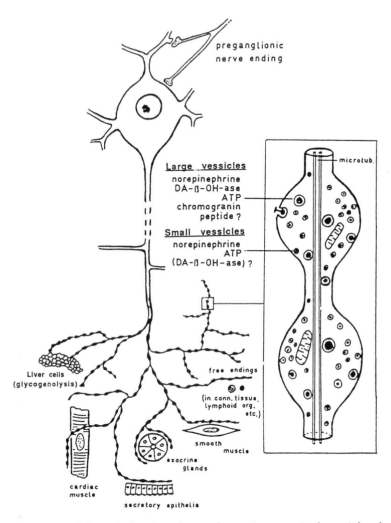

FIGURE 8-2. Schematic drawing of a noradrenergic neuron in the peripheral autonomic nervous system. Examples of innervated structures are illustrated. (Modified after Dahlstrom, 1986)

croscopy quite successfully in the peripheral nervous system where it is possible to correlate morphological changes (fluorescent intensity, content of granular vesicles) and catecholamine content. The lack of a suitable histochemical technique for the visualization of acetylcholine has been a serious handicap for those interested in cholinergic mechanisms.

Shortly after norepinephrine was established as the neurotransmitter substance of adrenergic nerves in the peripheral nervous system, Holtz identified norepinephrine as a normal constituent of mammalian brain. For some years, however, it was thought that the presence of norepinephrine in mammalian brain only reflected vasomotor innervation to the cerebral blood vessels. In 1954, Vogt demonstrated that norepinephrine was not uniformly distributed in the CNS and that this nonuniform distribution did not in any way coincide with the density of blood vessels found in a given brain region. This regional localization of norepinephrine within mammalian brain suggested that norepinephrine might subserve some specialized function, perhaps as a central neurotransmitter. The relative distribution of norepinephrine is quite similar in most mammalian species. The highest concentration is usually found in the hypothalamus and other areas of central sympathetic representation. More norepinephrine is generally found in gray matter than in white matter.

Dopamine is also present in the mammalian CNS and its distribution differs markedly from that of norepinephrine, an early indication that dopamine functions as more than a precursor of norepinephrine in the CNS. In fact, it represents more than 50% of the total catecholamine content of the central nervous system of most mammals. The highest levels of dopamine are found in the neostriatum, nucleus accumbens, and tuberculum olfactorium. The abundance of dopamine in the basal ganglia stimulated intensive research on the functional aspects of this compound and there is now a wealth of evidence that it plays an important role in extrapyramidal motor function (see Chapter 9). Dopamine is also present in the carotid body and superior cervical ganglion where it may play a role other than as a precursor of norepinephrine. The superior cervical ganglion appears to have at least three distinct populations of neurons: cholinergic neurons, noradrenergic neurons; and, small, intensely fluorescent (SIF) cells. The SIF cells are believed to be the small interneurons that contain dopamine. The actual functional significance of these cells is unclear, although it has been suggested that dopamine released from them is responsible for hyperpolarization of the ganglion.

Epinephrine concentration in the mammalian CNS is relatively low, approximately 5%–17% (by bioassay) of the norepinephrine content. Many investigators have suggested that these original estimates are subject to error and in the past have discounted the importance of epinephrine in mammalian brain. However, its presence has now been documented by more sophisticated analytical techniques such as gas chromatography–mass spectrometry and liquid chromatography coupled with electrochemical detection and has been confirmed by immunohistochemical techniques (see below).

The detailed topographical survey of brain catecholamines at different levels of organization within the CNS has provided a framework for organizing and conducting logical experiments concerning the possible functions of these amines. The anatomy, biochemistry, and pharmacology of CNS norepinephrine and epinephrine systems are discussed in detail at the end of this chapter.

LIFE CYCLE OF THE CATECHOLAMINES

Biosynthesis

Catecholamines are formed in brain, chromaffin cells, sympathetic nerves, and sympathetic ganglia from their amino acid precursor tyrosine by a sequence of enzymatic steps first postulated by Blaschko in 1939 and finally confirmed by Nagatsu and coworkers in 1964 when they demonstrated that an enzyme (tyrosine hydorxylase) is involved in the conversion of L-tyrosine to 3,4-dihydroxyphenylalanine (DOPA). This amino acid precursor, tyrosine, is normally present in the circulation in a concentration of about $5-8 \times 10^{-5}$ M. It is taken up from the bloodstream and concentrated within the brain and presumably also in other sympathetically innervated tissue by an active transport mechanism. Once inside the peripheral neuron, tyrosine undergoes a series of chemical transformations resulting in the ultimate formation of norepinephrine—or in the brain—norepinephrine, dopamine, or epinephrine, depending upon the availability of phenylethanolamine-N-methyl transferase and dopamine-β-hydroxylase. This biosynthetic pathway for the formation of catecholamines is illustrated in Figure 8-3. The conversion of tyrosine to norepinephrine and epinephrine was first demonstrated in the adrenal medulla. The availability of radioactive precursors of high specific activity and chromatographic separation techniques has allowed the confirmation of the above-mentioned pathway to norepinephrine in sympathetic nerves, ganglia, heart, arterial and venous tissue, and brain. In mammals, tyrosine can be derived from dietary phenylalanine by a hydroxylase (phenylalanine hydroxylase) found primarily in liver. Both phenylalanine and tyrosine are normal constituents of the mammalian brain, present in a free form in a concentration of about 5×10^{-5} M. However, norepinephrine biosynthesis is usually considered to begin with tyrosine, which represents a branch point for many important biosynthetic processes in animal tissues. It should be emphasized that the percentage of tyrosine utilized for catecholamine biosynthesis as opposed to other biochemical pathways is very minimal ($<2\%$).

Tyrosine Hydroxylase

The first enzyme in the biosynthetic pathway, tyrosine hydroxylase, was the last enzyme in this series of reactions to be identified. It was demonstrated by Udenfriend and his colleagues in 1964 and its properties have been reviewed repeatedly. It is present in the adrenal medulla, in brain, and in all sympathetically innervated tissues studied to date. This enzyme appears to be a unique constituent of catecholamine-containing neurons and chromaffin cells; it completely disappears from renal, salivary gland, vas deferens, and cardiac tissue upon chronic sympathetic denervation. The enzyme is stereospecific, requires molecular O_2, Fe^{2+}, and a tetrahydropteri-

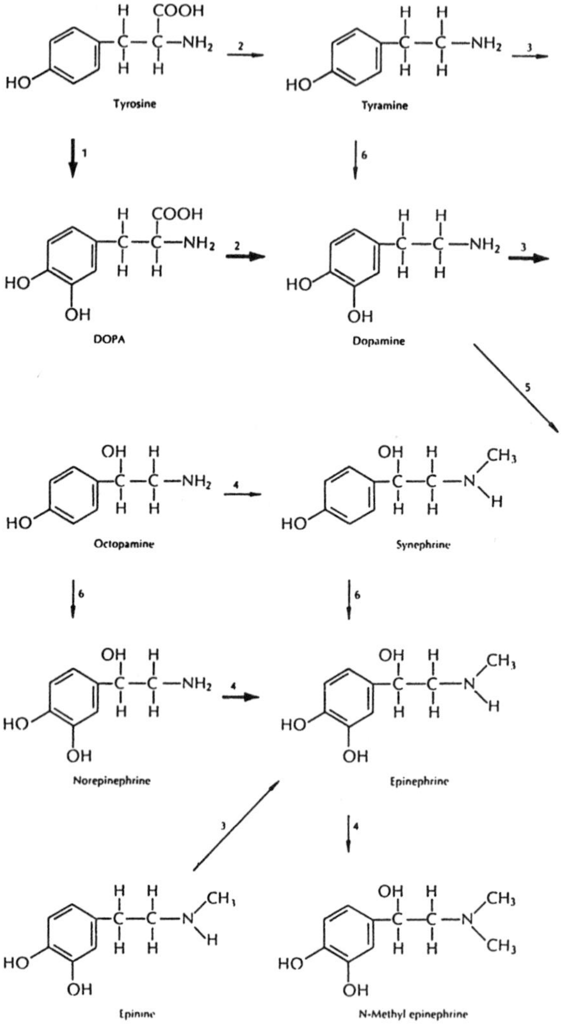

FIGURE 8-3. Primary and alternative pathways in the formation of catecholamine: (1) tyrosine hydroxylase; (2) aromatic amino-acid decarboxylase; (3) dopamine-β-hydroxylase; (4) phenylethanolamine-N-methyl transferase; (5) nonspecific N-methyl transferase in lung and folate-dependent N-methyl transferase in brain; (6) catechol-forming enzyme.

dine cofactor, and shows a fairly high degree of substrate specificity. Thus, this enzyme, in contrast to tyrosinase, oxidizes only the naturally occurring amino acid L-tyrosine and, to a smaller extent, L-phenylalanine. D-tyrosine, tyramine, or L-tryptophan will not serve as substrates for the enzyme. The human gene for tyrosine hydroxylase has been cloned. The single gene encodes multiple mRNAs that are heterogeneous at the 5' end of the coding region. The functional significance of the different messages remains to be determined.

The K_m for the enzymatic conversion of tyrosine to DOPA by purified adrenal tyrosine hydroxylase is about 2×10^{-5} M and in a preparation of brain synaptosomes about 0.4×10^{-5} M. Tyrosine hydroxylation appears to be the rate-limiting step in the biosynthesis of norepinephrine in the peripheral nervous system and is likely to be the rate-limiting step in the formation of norepinephrine and dopamine in the brain as well. In most sympathetically innervated tissues including the brain, the activity of DOPA decarboxylase and that of dopamine-β-oxidase have a magnitude 100–1000 times that of tyrosine hydroxylase. This lower activity of tyrosine hydroxylase may be due either to the presence of less enzyme or to a lower turnover number of the enzyme. Since this enzyme has been demonstrated to be the rate-limiting step in catecholamine biosynthesis, it is logical that pharmacological intervention at this step would cause a reduction of norepinephrine biosynthesis. Many earlier attempts to produce chemical sympathectomy by blockade of the last two steps in the synthesis of norepinephrine have proved largely unsuccessful. On the other hand, studies with inhibitors of tyrosine hydroxylase have proved to be much more successful producing a marked reduction in endogenous norepinephrine and dopamine in brain and norepinephrine in heart, spleen, and other sympathetically innervated tissues. Effective inhibitors of this enzymatic step can be categorized into four main groups: (1) amino acid analogues; (2) catechol derivatives; (3) tropolones; and, (4) selective iron chelators. Some effective amino acid analogues include: α-methyl-ρ-tyrosine and its ester; α-methyl-3-iodotyrosine; 3-iodotyrosine: and, α-methyl-5-hydroxytryptophan. In general, α-methyl-amino acids are more potent than the unmethylated analogues and a marked increase in activity in the case of the tyrosine analogues can also be produced by substituting a halogen at the 3 position of the benzene ring. Most of the agents in this category act as competitive inhibitors of the substrate tyrosine. In this respect, α-methyl-5-hydroxytryptophan appears to be unique since it does not appear to compete with substrate or pteridine cofactor. Its actual mechanism remains unknown at the present time. The potent halogenated tyrosine analogues such as 3-iodotyrosine are about 100 times as active as α-methyl-p-tyrosine in vitro but they are substantially less active in vivo. This is probably due to the very rapid deiodination of these compounds to tyrosine or tyrosine analogues which occurs in vivo. α-methyl-p-tyrosine and its methyl ester have been the inhibitors most widely used to demonstrate the effects of exercise, stress, and various drugs on the turnover of catecholamines and also to lower norepinephrine formation in patients with pheochromocytoma and malignant hypertension.

Dihydropteridine Reductase

Although not directly involved in catecholamine biosynthesis, dihydropteridine reductase is intimately linked to the tyrosine hydroxylase step. This enzyme catalyzes the reduction of the quinonoid dihydropterin that has been oxidized during the hydroxylation of tyrosine to DOPA. Since reduced pteridines are essential for tyrosine hydroxylation, alterations in the activity of dihydropteridine reductase would effectively influence the activity of tyrosine hydroxylase. Thus, this might be a potential site for drug intervention in catecholamine biosynthesis. Dihydropteridines with amine substitution in positions 2 and 4 are effective inhibitors of this enzyme while folic acid antagonists such as aminopterin and methotrexate are relatively ineffective. The distribution of dihydropteridine reductase is quite widespread, the highest activity being found in the liver, brain, and adrenal gland. The distribution of this enzyme activity in the brain does not appear to parallel the catecholamine or serotonin content of brain tissue, suggesting that reduced pterins most likely participate in other reactions besides the hydroxylation of tyrosine and tryptophan.

Dihydroxyphenylalanine Decarboxylase

The second enzyme involved in catecholamine biosynthesis is DOPA-decarboxylase, which was actually the first catecholamine synthetic enzyme to be discovered. Although originally believed to remove carboxyl groups only from L-DOPA, a study of purified enzyme preparations and specific inhibitors has subsequently demonstrated that this DOPA-decarboxylase acts on all naturally occurring aromatic L-amino acids, including histidine, tyrosine, tryptophan, and phenylalanine as well as both DOPA and 5-hydroxytryptophan. Therefore, this enzyme is more appropriately referred to as "L-aromatic amino acid decarboxylase." There is no appreciable binding of this enzyme to particles within the cell since, when tissues are disrupted and the resultant homogenates centrifuged at high speeds, the decarboxylase activity remains associated largely with the supernatant fraction. The exception to this is in the brain where some of the decarboxylase activity is associated with synaptosomes. However, since synaptosomes are in essence pinched-off nerve endings, they would be expected to retain entrapped cytoplasm as well as other intracellular organelles. The DOPA-decarboxylase found in synaptosomal preparations is thought to be present in the entrapped cytoplasm. DOPA-decarboxylase is, relative to other enzymes in the biosynthetic pathway for norepinephrine formation, very active and requires pyridoxal phosphate (vitamin B6) as a cofactor. The apparent K_m value for this enzyme is 4×10^{-4} M. The high activity of this enzyme may explain why it has been difficult to detect endogenous DOPA in sympathetically innervated tissue and brain. It is rather ubiquitous in nature, occurring in the cytoplasm of most tissue including the liver, stomach, brain, and kidney in high levels, suggesting that its function in metabolism is not limited solely to catecholamine biosynthesis. Although decarboxylase activity can be reduced by production of vitamin B6 deficiency in animals, this does not usually result in a significant reduction of tissue catecholamines, although it appears to interfere with the rate of repletion of adrenal catecholamines after insulin depletion. In addition, potent decarboxylase inhibitors also have very little effect on endogenous levels of norepinephrine in tissue. However, these inhibitors have been useful as pharmacological tools (e.g., DOPA accumulation following administration of a decarboxylase inhibitor as an in vivo index of tyrosine hydroxylation).

Dopamine-β-Hydroxylase

Although it has been known for many years that the brain, sympathetically innervated tissue, sympathetic ganglia, and adrenal medulla, could transform dopamine into norepinephrine, it was not until 1960 that the enzyme responsible for this conversion was isolated from the adrenal medulla. This enzyme called dopamine-β-hydroxylase is, like tyrosine hydroxylase, a mixed-function oxidase. It requires molecular oxygen and utilizes ascorbic acid as a cofactor. The K_m of this enzyme for its substrate dopamine is about 5×10^{-3} M. Dicarboxylic acids such as fumaric acid are not absolute requirements, but they stimulate the reaction. Dopamine-β-hydroxylase is a Cu^{2+}-containing protein, with about 2 mole of cupric ion per 1 mole of enzyme. It appears to be associated with the particulate fraction from the heart, brain, sympathetic nerve, and adrenal medulla, and it is believed that this enzyme is localized primarily in the membrane of the amine storage granules. This enzyme usually disappears after chronic sympathetic denervation and therefore is believed to be present largely only in adrenergic neurons or adrenal chromaffin tissue. Dopamine-β-hydroxylase does not show a high degree of substrate specificity and acts in vitro on a variety of substrates besides dopamine, oxidizing almost any phenylethylamine to its corresponding phenylethanolamine (i.e., tyramine → octopamine, → α-methyldopamine → α-methylnorepinephrine). A number of the resultant structurally analogous metabolites can replace norepinephrine at the noradrenergic nerve endings and function as "false neurotransmitters."

Dopamine-β-hydroxylase can be inhibited by a variety of compounds. The most effective are compounds that chelate copper: D-cysteine and L-cysteine; glutathione; mercaptoethanol; and, coenzyme A. Inhibition can be reversed by the addition of N-ethylmaleimide, which reacts with the sulfhydryl groups and interferes with the chelating properties of these substances. Copper chelating agents such as diethyldithiocarbamate and FLA-63 (bis[1-methyl-4-homopiperazinyl-thiocarbonyl]-disulfide) have proved to be effective inhibitors both in vivo and in vitro. Thus, it has been possible to treat animals with disulfuram or FLA-63 and produce a reduction in brain norepinephrine and an elevation of brain dopamine.

Since dopamine-β-hydroxylase obtained from the bovine adrenal medulla can be prepared in a relatively pure form, it has been possible to produce a specific antibody to the enzyme. This antibody inactivates bovine dopamine-β-hydroxylase but does not appear to cross-react with either DOPA-decarboxylase or tyrosine hydroxylase. However, cross-reactivity between dopamine-β-hydroxylase from human, guinea pig, and dog, with the antibody against the bovine enzyme was observed and indicates that the enzymes from these various sources are probably structurally related. By coupling immunochemical techniques with fluorescence and electron microscopy, this antibody has already proved useful in localization of this enzyme in intact tissue. Dopamine-β-hydroxylase has been cloned. Further studies on the molecular structure and expression of the enzyme should yield interesting information.

Phenylethanolamine-N-Methyl Transferase

In the adrenal medulla, norepinephrine is N-methylated by the enzyme phenylethanolamine-N-methyl transferase to form epinephrine. This enzyme is largely restricted to the adrenal medulla, although low levels of activity have been reported in heart and mammalian brain (see Chapter 10). Like the decarboxylase, this enzyme also appears in the supernatant of homogenates. Demonstration of activity requires the presence of the methyl donor S-adenosyl methionine. The adrenal medullary enzyme shows poor substrate specificity and will transfer methyl groups to the nitrogen atom on a variety of β-hydroxylated amines. However, this adrenal enzyme is distinct from the nonspecific N-methyl transferase of rabbit lung, which, in addition to N-methylating phenylethanolamine derivatives, will also react with many normally occurring indoleamines and such diverse structures as phenylisopropylamine, aromatic amines, and phenanthrenes. Interest in the biosynthetic pathway for catecholamines has also led to the cloning of phenylethanolamine-N-methyl-transferase.

Synthesis Regulation

It has been known for a long time that the degree of sympathetic activity does not influence the endogenous levels of tissue norepinephrine; and it has been speculated that there must be some homeostatic mechanism whereby the level of transmitter is maintained at a relatively constant level in the sympathetic nerve endings despite the additional losses assumed to occur during enhanced sympathetic activity.

More than 30 years ago, Euler hypothesized on the basis of experiments carried out in the adrenal medulla that during periods of increased functional activity, the sympathetic neuron must also increase the synthesis of its transmitter substance norepinephrine to meet the increased demands placed upon the neuron. If the sympathetic neuron had the ability to increase transmitter synthesis, this would enable the neuron to maintain a constant steady-state level of transmitter despite substantial changes in transmitter utilization. Some years later, experiments carried out by several laboratories on different sympathetically innervated tissues directly demonstrated that this was, in fact, the case. Electrical stimulation of sympathetic nerves both in vivo and in vitro resulted in an increased formation of norepinephrine in the tissues innervated by these nerves. Further studies demonstrated that the observed acceleration of norepinephrine biosynthesis produced by enhanced sympathetic activity was due to an increase in the activity of the rate-limiting enzyme involved in catecholamine biosynthesis, tyrosine hydroxylase. Emphasis then shifted toward attempting to determine the mechanism by which an alteration in impulse flow within the sympathetic neuron could result in a change in the activity of this potential regulatory enzyme.

It was appreciated at an early stage that the observed enhancement of tyrosine hydroxylase activity that accompanied an increase in sympathetic activity was not the result of new enzyme synthesis but rather was related to an increase in the activity of existing enzyme molecules. The classic in vitro studies by Udenfriend and coworkers, demonstrating that catecholamines could act as feedback inhibitors of tyrosine hydroxylase, provided a reasonable theoretical mechanism by which impulse activity might regulate tyrosine hydroxylase. Since tyrosine hydroxylase is inhibited by catechols and catecholamines, presumably because of their ability to antagonize competitively the binding of the pteridine cofactor to the apoenzyme, it was proposed that free intraneuronal norepinephrine may control its own synthesis by negative feedback inhibition of tyrosine hydroxylase. This concept was supported by many pharmacological studies including the observation that an increase in the endogenous norepinephrine content of peripheral sympathetically innervated tissue or of brain catecholamine content, which is produced by treatment with monoamine oxidase inhibitors (MAOI), results in a marked reduction of catecholamine biosynthesis. In addition, the increase in synthesis observed in vitro during neuronal depolarization is dependent upon the presence of Ca^{2+} and appears inextricably coupled to catecholamine release. This latter observation suggested that the release-induced depletion of transmitter might be a prerequisite for the observed acceleration of catecholamine synthesis.

In the late 1960's the following conceptual model emerged. During periods of increased impulse flow when more transmitter is released and metabolized, a strategic regulatory pool of norepinephrine normally accessible to tyrosine hydroxylase is depleted, end-product inhibition is removed, and tyrosine hydroxylase activity is increased. Likewise, during periods of quiescence when transmitter utilization is decreased, norepinephrine accumulates and tyrosine hydroxylase activity is decreased. Over the past decade, much indirect evidence has amassed to suggest that the release of tyrosine hydroxylase from feedback inhibition is a mechanism that is operative in the sympathetic neuron and in central catecholamine neurons for increasing neurotransmitter synthesis in response to enhanced impulse flow.

In more recent years, it has been apparent that even in peripheral noradrenergic neurons, the regulation of tyrosine hydroxylase is a much more complex process and that in addition to feedback regulation, other processes intimately linked to neuronal activity control norepinephrine synthesis. The concept that the frequency of neuronal depolarization might in some way influence the physical properties of tyrosine hydroxylase is one of the more exciting possibilities that have emerged.

A significant advance in our understanding of the way that short-term changes in impulse flow may regulate transmitter synthesis came from several experiments. It was demonstrated that electrical stimulation of both central and peripheral catecholamine neurons results in an apparent allosteric activation of tyrosine hydroxylase that persists after the stimulation period ends. This poststimulation increase in the activity of tyrosine hydroxylase is maximal after about 10 minutes of continuous stimulation and persists for a measurable period of time after the stimulation is halted. Kinetic studies demonstrated that this activation of tyrosine hydroxylase observed following depolarization is mediated in part by an increased affinity of the enzyme for pteridine cofactor and a decreased affinity for the natural end-product inhibitors norepinephrine and dopamine. The increased affinity for pterin cofactor (whose concentration in tissue is thought to be subsaturating) and the decreased affinity of the enzyme for catecholamine may be important physiologically for the regulation of transmitter synthesis in response to increased impulse flow. These studies provided the first direct experimental evidence that some short-term mechanism in addition to depletion of endogenous catecholamine is operative in altering tyrosine hydroxylase activity during increased neuronal activity.

Numerous in vitro studies have demonstrated that tyrosine hydroxylase can be activated by phosphorylation, providing credence for the speculation that phosphorylation is the major mode by which catecholamine neurons may regulate tyrosine hydroxylase activity in response to altered neuronal activity. The activation of tyrosine hydroxylase by phosphorylation leads to covalent modifications of the enzyme, and phosphorylation/dephosphorylation of tyrosine hydroxylase represents an important mechanism in the regulation of enzyme activity. Tyrosine hydroxylase is phosphorylated by distinct protein kinases at five serine phosphorylation sites, Ser 8, Ser 19, Ser 31, Ser 40, and Ser 153, in the N-terminal region of the enzyme. However, phosphorylation at Ser 8 and Ser 153 does not activate tyrosine hydroxylase. The enzyme is phosphorylated at Ser 19 by Ca^{2+}–calmodulin protein kinase II (Ca/CaMpKII), at Ser 40 by protein kinase A (PKA) and protein kinase C (PKC), (and to a small extent by Ca/CaMpKII), and at Ser 31 by extracellular signal related kinases (ERKs) ERK1 and ERK2, and indirectly by PKC. The phosphorylation of tyrosine hydroxylase by Ca/CaMpKII alone does not increase enzyme activity. Several studies have suggested that an "activator protein" is required for an increase in catalytic activity of the enzyme. Electrical stimulation of the medial forebrain bundle increases phosphorylation at Ser 19, Ser 31, and Ser 40 of the striatal enzyme. A Ca^{2+}-dependent activation of tyrosine hydroxylase by in vitro depolarization of peripheral and CNS tissue has been demonstrated, and this activation is linked to Ca/CaMpKII phosphorylation at the Ser 19 site on the enzyme. In most cases phosphorylation increases the V_{max} of tyrosine hydroxylase (i.e., the maximal catalytic activity of a single enzyme molecule) or the affinity of the enzyme for pterin cofactor, which makes the enzyme more active at subsaturating concentrations of endogenous cofactor. α2-Adrenergic or D2-dopaminergic autoreceptor-mediated reduction of catecholamine synthesis may also be achieved via regulation of the cAMP or Ca^{2+}–calmodulin-dependent protein phosphatases. Although PKC, cAMP-dependent protein kinase, and calcium–calmodulin-dependent kinase all phosphorylate tyrosine hydroxylase, producing a kinetic activation of the enzyme, the question of which of these mechanisms is operative physiologically is still unresolved. The most recent studies favor the involvement of a Ca^{2+}–calmodulin-dependent phosphorylation in the impulse-dependent activation process (Fig. 8-4).

In addition to the mechanisms for immediate adaptation to increased transmitter utilization discussed above, a second mechanism comes into play after prolonged increases in the activity of adrenergic neurons. This latter mechanism is reflected by an increase in the activity of tyrosine hydroxylase measured in vitro under saturating

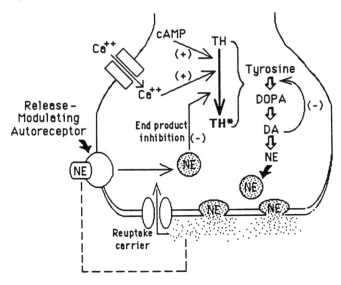

FIGURE 8-4. Model illustrating mechanisms for regulation of transmitter synthesis in noradrenergic neurons.

concentrations of substrate and cofactor. For a while it was unclear whether this increase in activity observed in vitro was due to the disappearance of an inhibitor, the appearance of an activator, or an actual increase in the number of active enzyme sites. In recent years, however, much evidence has accumulated which indicates that there is a measurable increase in the formation of new enzyme molecules. This prolonged neuronal activity is thought to result in an induction of tyrosine hydroxylase. This process of transsynaptic induction does not appear to be specific for tyrosine hydroxylase, since an increase in the formation of dopamine-β-hydroxylase is also observed. No significant increase in the formation of DOPA-decarboxylase has been reported. Based on recent studies employing gene technology, Joh and coworkers proposed that the major enzymes involved in catecholamine biosynthesis—tyrosine hydroxylase, dopamine-β-hydroxylase, and phenylethanolamine-N-methyltransferase—may derive from a common gene and thus may be coregulated (i.e., coinduced by various treatments).

It has become apparent that the regulation of tyrosine hydroxylase activity and tyrosine hydroxylase enzyme protein in norepinephrine neurons differs from that observed in dopamine neurons. For example, the effects of reserpine on tyrosine hydroxylase activity and on tyrosine hydroxylase mRNA in the CNS are restricted to norepinephrine neurons in the locus ceruleus where a significant increase is observed in both activity and mRNA; no effects are observed on tyrosine hydroxylase activity or mRNA in dopamine neurons in the substantia nigra. Exposure of rats to chronic stress also increases levels of tyrosine hydroxylase mRNA and tyrosine hydroxylase protein in the locus ceruleus but does not alter tyrosine hydroxylase in the substantia nigra or ventral tegmental area. Repeated administration of morphine likewise increases tyrosine hydroxylase mRNA levels in the locus ceruleus but not in the substantia nigra. Studies that have evaluated the state of phosphorylation of tyrosine hydroxylase in norepinephrine and dopamine neurons by in situ immunohistochemistry, using specific antibodies that recognize Ser 40–phosphorylated tyrosine hydroxylase, revealed that the tyrosine hydroxylase in norepinephrine and epinephrine neurons under basal conditions is largely phosphorylated, while most of the tyrosine hydroxylase in dopamine neurons is in the nonphosphorylated form. These findings suggest that a large reserve for tyrosine hydroxylase phosphorylation exists under basal conditions in midbrain dopamine neurons in contrast to a much lower reserve in noradrenergic neurons.

Even though tyrosine hydroxylase has been demonstrated to be rate limiting under most circumstances, this reaction may not remain rate limiting under all physiological or pharmacological conditions. Since there is a sequence of reactions, any one of these steps could assume a rate-limiting role depending on the given pathological or pharmacologically induced situations. Thus, reserpine can transform the dopamine-β-hydroxylase step into the rate-limiting step presumably by blocking access of the substrate dopamine to the site of its conversion to norepinephrine.

Although it is well known that alterations in brain tryptophan can influence serotonin biosynthesis, the failure of tyrosine administration to increase levels of norepinephrine in sympathetically innervated tissue or catecholamine levels in brain has led most investigators to assume that tyrosine hydroxylase in catecholamine- containing neurons is normally saturated with its amino-acid substrate tyrosine in vivo and therefore its activity not readily influenced by alterations in the availability of tyrosine. A number of studies by Wurtman, Fernstrom, and their coworkers have demonstrated, however, that under conditions that are believed to cause an increase in impulse flow in peripheral sympathetic or central noradrenergic or dopaminergic neurons, tyrosine hydroxylation and release (evaluated by measuring the accumulation of catecholamine metabolites) can be enhanced by systemic administration of tyrosine (see Chapter 9).

Alternative Biosynthetic Pathways

The question has often been raised as to whether the sequence of enzymatic reactions described in detail above is in fact obligatory. Many attempts have been made to find alternative pathways in vivo and to determine whether they are functionally important. Thus, if labeled tyramine is administered to animals, both labeled norepinephrine and normetanephrine can be isolated in the urine. It has been very difficult to detect the conversion of tyramine to norepinephrine in sympathetically innervated tissue. However, the conversion of tyramine to dopamine has been demonstrated in liver microsomes. Therefore, it is possible that this conversion of tyramine to norepinephrine observed in vivo is a reflection of a metabolic reaction taking place in the liver rather than in the sympathetic neuron.

Storage

A great conceptual advance made in the study of catecholamines more than 30 years ago was the recognition that in almost all tissues a large percentage of the norepinephrine present is located within highly specialized subcellular particles (colloquially referred to as "granules") in sympathetic nerve endings and chromaffin cells. Much of the norepinephrine in the CNS is also presumably located within similar particles. These granules contain adenosine triphosphate (ATP) in a molar ratio of catecholamine to ATP of about 4 : 1. Because of this perhaps fortuitous ratio, it is generally supposed that the anionic phosphate groups of ATP form a salt link with norepinephrine, which exists as a cation at physiological pH, and thereby serve as a means to bind the amines within the vesicles. Some such complex of the amines with ATP, with ATP associated with protein, or directly with protein, is probable since the intravesicular concentration of amines, at least in the adrenal chromaffin granules and probably also in the splenic nerve granules (0.3–1.1 M), would be hypertonic if present in free solution and might be expected to lead to osmotic lysis of the vesicles.

Transporters localized to the membranes of synaptic vesicles and chromaffin granules have recently been cloned. These vesicular monoamine transporters (VMATs) define another new family of proteins that display 12 hydrophobic, putative transmembrane domains and move transmitters into acidic intracellular compartments such as neurotransmitter vesicles. These transporters display no significant sequence homology with the plasma membrane Na^+/Cl^--coupled family, but they do resemble bacterial drug-resistance transporters. VMATs can also be distinguished from plasma membrane transporters that utilize transmembrane Na^+ gradients for active removal of transmitters by the fact that VMATs have adapted an "H^+ economy" to drive transmitter accumulation in the vesicles. All the amine storage vesicles studied in brain and adrenal chromaffin cells contain a vascular-type H^+ pumping ATPase similar to that found in lysosomes and Golgi membranes.

The catecholamine storage vesicles in adrenal chromaffin and splenic nerve appear to have a number of general properties:

1. They possess an outer limiting membrane with a transporter protein.
2. With appropriate fixation they possess an electron-dense core when viewed in the electron microscope.
3. They contain the enzyme dopamine-β-hydroxylase.
4. They have a high concentration of catecholamine and ATP in a 4 : 1 ratio in adrenal and 6–8 : 1 in splenic nerve.

5. Chromaffin and splenic nerve granules contain a characteristic soluble protein (chromogranin) also suggested to be involved in the storage process.

The possible functions that have been proposed for the catecholamine storage vesicles are as follows:

1. They bind and store norepinephrine thereby retarding its diffusion out of the neuron and protect it from being destroyed by monoamine oxidase (MAO), which is considered to be intraneuronal.

2. They serve as a depot of transmitter that may be released upon the appropriate physiological stimulus.

3. They oxidize dopamine to norepinephrine.

4. They take up dopamine from the cytoplasm, protecting it from oxidation by monoamine oxidase. The removal of interneuronal molecules into a storage system also effectively lowers the concentration gradient across neuronal membrane and thus acts as an amplification stage for the overall process of monoamine uptake by the plasma membrane transporter (see below).

Vesicular transport has been observed for several classic transmitters in addition to the monoamines, including acetylcholine, glutamate, GABA, and glycine. Of the vesicular transporters, the VMATs have been the ones most intensively studied and are the ones for which the most molecular information has been obtained. The availability of an excellent experimental model to study VMATs, the adrenal chromaffin granules of the adrenal medulla, and potent specific inhibitors (reserpine and tetrabenazine), have greatly facilitated these studies.

Identification of genes coding for VMATs has provided new tools, new approaches, and stimulated a renewed interest in the field. Studies on the regulation of synthesis and activity of the vesicular transporters are now feasible, and it is possible to examine the influence of acute and chronic drug treatments as well as assess transporter protein levels in various pathological states.

Release

Much of the current knowledge regarding the release of cate-cholamines derives from the study of both adrenal medullary tissue and adrenergically innervated peripheral organs. Here it can be directly demonstrated that norepinephrine is released from the nerve terminals during periods of nerve stimulation and that adrenergic fibers can sustain this output of transmitter during prolonged periods of stimulation if synthesis and reuptake of the transmitter are not impaired. In the periphery, it is possible to isolate the innervated end-organs, to collect a perfusate from the vascular system during nerve stimulation, and to analyze this for the presence and quantity of a given putative transmitter. However, we really know little about the mechanism by which axonal nerve impulses arriving at the terminal cause the release of norepinephrine (excitation-secretion coupling). Our best appreciation of the events comes from the possibly analogous release of catecholamines from the adrenal medulla, which has been extensively studied. For a comprehensive review of this area the reader is referred to Winkler.

The mechanism of catecholamine release by the adrenal medulla is thought to take place as follows: With activity in the preganglionic fibers, acetylcholine is released and thought to combine with the plasma membrane of the chromaffin cells. This produces a change in membrane protein conformation, altering the permeability of this membrane to Ca^{2+} and other ions, which then move inward. The influx of Ca^{2+} is believed to be the main stimulus responsible for the mobilization of the catecholamines and for their secretion. The current view is that the catecholamines are released from the chromaffin cell by a process of exocytosis, along with chromogranin, ATP, and a little dopamine-β-hydroxylase. Whether these cellular phenomena are applicable to the sympathetic nerve endings in general remains to be determined.

Teleologically, it appears rather unlikely that transmitter release from sympathetic nerve terminals does occur exclusively by a process of exocytosis. Exocytosis requires that the entire content of the granular vesicle be released (i.e., catecholamine, ATP, and soluble protein). Since the nerve terminal region, as far as we know now, cannot sustain any sort of protein biosynthesis, high rates of axonal flow from the nerve cell body would be required to replenish the protein lost during the process of exocytosis. Alternatively, one might propose a "protein reuptake" mechanism in order to recapture that protein released during the process of synaptic transmission. In peripheral nerve, release of norepinephrine has been shown to be frequency dependent within a physiological range of frequencies. This release of norepinephrine, similar to the release of ACh, is Ca^{2+} dependent. Some evidence has also been presented that indicates that newly synthesized norepinephrine may be released preferentially. This preferential release is additional evidence to support the contention that norepinephrine exists in more than one pool within the sympathetic neuron.

It has been a great deal more difficult to demonstrate release of a putative transmitter from a given type of nerve ending in the central nervous system. Thus it is impossible to perfuse specific localized brain areas through their vascular system. In addition, it is also difficult, but not necessarily impossible, to stimulate selectively a well-defined neuronal pathway within the brain. (See below.) However, with the independent development of the push–pull cannula, microdialysis techniques, and electrochemical detectors, it has become possible to collect catecholamines from certain deep nuclear masses of the CNS or to measure release directly by in vivo voltammetry. In addition, many less-sophisticated techniques such as cortical cups, ventricular perfusions, brain slices, and isolated spinal cords, have been used to demonstrate release of putative transmitters, including norepinephrine and dopamine from CNS tissue. All of these physiological techniques for studying release have some disadvantages and limitations, and many are somewhat gross and indirect. Nevertheless, they do represent increasing sophistication in research on release occurring in the CNS.

Regulation of Release

Much evidence is accumulating to suggest that endogenous humoral factors may regulate release by a direct local action on catecholamine nerve terminals. Not only can the local synaptic concentration of catecholamines modulate their own release by interacting with presynaptic autoreceptors, but also prostaglandins, vasoactive amines, polypeptides such as angiotensin II, and acetylcholine, have all been implicated in regulation of catecholamine release. The most convincing evidence, however, has been generated in support of a role for presynaptic receptors in the modulation of impulse-induced catecholamine release in both central and peripheral catecholamine neurons. In most catecholamine-containing systems tested, administration of catecholamine agonists appears to attenuate stimulus-induced release while administration of catecholamine receptor blockers augments release. These pharmacological studies have established the concept that presynaptic receptors modulate release by responding to the concentration of catecholamine in the synapse (high concentrations inhibiting release and low concentrations augmenting release). Most recently presynaptic autoreceptors have also been implicated in the regulation of dopamine synthesis. (See *Autoreceptors*, below.)

Several presynaptic receptors are involved in the inhibition of transmitter release from adrenergic nerves. These include α_2-adrenergic autoreceptors as well as muscarinic, opiate, and dopamine receptors. Other presynaptic receptors are linked to facilitation rather than inhibition of norepinephrine release. These receptor subtypes include β_2-adrenergic adrenoceptors and nicotinic and angiotensin II receptors. The precise mechanism by which autoreceptors influence neurotransmitter release from adrenergic neurons is unclear although it has been suggested that the activation of α_2-adrenoceptors may inhibit norepinephrine release by several mechanisms including: (1) an attenuation of the rate of Ca^{2+} entry through inhibition of voltage-sensitive Ca^{2+} channels; (2) a blockade of the spread of the action potential along the terminal varicosity; (3) an opening of potassium channels leading to hyperpolarization of the neuron terminals; and, (4) an inhibition of adenylate cyclase resulting in a decrease of intracellular cAMP and Ca^{2+} concentration.

Presynaptic β-adrenergic receptors are also present in some sympathetic tissues. These presynaptic receptors are usually believed to be of the $\beta 2$ subtype. The $\beta 2$-adrenoceptor–mediated facilitation of norepinephrine release may be due to stimulation of adenylate cyclase, leading to an increase in cAMP and a subsequent increase in intracellular Ca^{2+} concentrations. Alternatively, it has been suggested that the facilitation may be due to a transsynaptic signal involving a local renin–angiotensin response. This hypothesis posits that β-adrenoceptor agonists activate postsynaptic $\beta 2$-adrenoceptors in the vascular wall. This activation results in the synthesis of angiotensin II, which diffuses across the synaptic cleft to activate presynaptic angiotensin II receptors that facilitate norepinephrine release.

The presence of presynaptic facilitatory and inhibitory adrenergic receptors on the same nerve terminals may provide for a fine-tuning control of stimulus-evoked release of norepinephrine. Since the affinity of epinephrine for $\beta 2$-adrenoceptors is much greater than that of norepinephrine, this subset of adrenoceptors is more likely to be activated by endogenous epinephrine than norepinephrine. Thus, low synaptic concentrations of epinephrine could activate presynaptic $\beta 2$-adrenoceptors preferentially, leading to an increase in norepinephrine release. At higher concentrations of epinephrine or norepinephrine, activation of $\alpha 2$-adrenoceptors would predominate, and norepinephrine release would be diminished rather than enhanced.

Prostaglandins of the E series are also potent inhibitors of neurally induced release of norepinephrine in a great number of tissues, and their action appears to be dissociated from any interaction with presynaptic receptors. These substances are released from sympathetically innervated tissues, and most evidence indicates that inhibition of local prostaglandin production is associated with an increase in the release of norepinephrine and subsequent effector responses induced by neuronal activity. The control of norepinephrine release by this prostaglandin-mediated feedback mechanism appears to operate through restriction of calcium availability for the norepinephrine release process and seems to be most efficient within the physiological frequency range of nerve impulses.

Metabolism

The metabolism of exogenously administered or endogenous catecholamines differs markedly from that of acetylcholine in that the
speed of degradation of the amines is considerably slower than that
of the ACh–ACh-esterase system. The major mammalian enzymes
of importance in the metabolic degradation of catecholamines are
monoamine oxidase and catechol-O-methyltransferase (COMT)
(Fig. 8-5). Monoamine oxidase is an enzyme that converts catecholamines to their corresponding aldehydes. This aldehyde intermediate is rapidly metabolized, usually by oxidation by the enzyme
aldehyde dehydrogenase to the corresponding acid. In some circumstances, the aldehyde is reduced to the alcohol or glycol by aldehyde
reductase. In the case of brain norepinephrine, reduction of the
aldehyde metabolite appears to be the favored route of metabolism.
Neither of these latter enzymes has been extensively studied in neuronal tissue. Monoamine oxidase is a particle-bound protein, localized largely in the outer membrane of mitochondria, although a partial microsomal localization cannot be excluded. There is also some
evidence for a riboflavin-like material in monoamine oxidase isolated from liver mitochondria. Monoamine oxidase is usually considered to be an intraneuronal enzyme, but it occurs in abundance
extraneuronally. In fact, most experiments indicate that chronic
denervation of a sympathetic end-organ leads only to a relatively
small reduction in monoamine oxidase, suggesting that the greater
proportion of this enzyme is, in fact, extraneuronal. However, it is
the intraneuronal enzyme that seems to be important in catecholamine metabolism. Monoamine oxidase present in human and
rat brain exists in at least two different forms, designated type A and
type B based on substrate specificity and sensitivity to inhibition by
selected inhibitors. Clorgyline is a specific inhibitor of the A-type
enzyme, which has a substrate preference for norepinephrine and
serotonin. Deprenyl is a selective inhibitor of the B-type enzyme,
which has a substrate preference for β-phenylethylamine and benzy-
lamine as substrates. Dopamine, tyramine, and tryptamine, appear to
be equally good substrates for both forms of the enzyme. The role

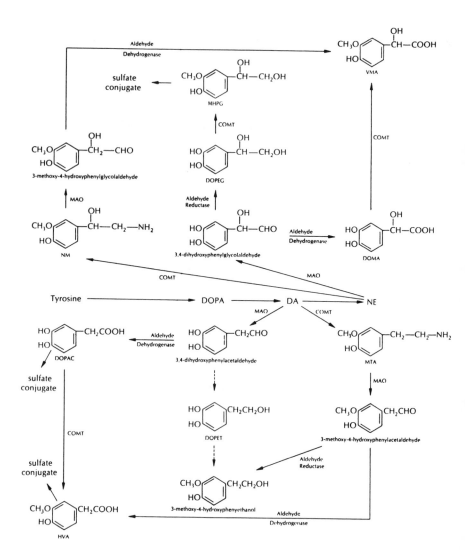

FIGURE 8-5. Dopamine and norepinephrine metabolism. The following abbreviations are used: DOPA, dihydroxyphenylalanine; DA, dopamine; NE, norepinephrine; DOMA, 3,4-dihydroxymandelic acid; DOPAC, 3,4-dihydroxyphenylacetic acid; DOPEG, 3,4-dihydroxyphenylglycol; DOPET, 3,4-dihydroxyphenylethanol; MOPET, 3-methoxy-4-hydroxyphenylethanol; MHPG, 3-methoxy-4-hydroxy-phenylglycol; HVA, homovanillic acid; VMA, 3-methoxy-4-hydroxy-mandelic acid; NM, normetanephrine; MTA, 3-methoxytyramine; MAO, monoamine oxidase; COMT, catechol-O-methyl transferase; Dashed arrows indicate steps that have not been firmly established.

and relative importance of these two types of monoamine oxidase in physiological and pathological states is currently unknown but this is an important area for further research. It has been speculated that under certain circumstances monoamine oxidase could serve to regulate norepinephrine biosynthesis by controlling the amount of substrate dopamine available to the enzyme dopamine-β-hydroxylase. Monoamine oxidase is not an exclusive catabolic enzyme for the catecholamines since it also oxidatively deaminates other biogenic amines such as 5-hydroxytryptamine, tryptamine, and tyramine. The intraneuronal localization of monoamine oxidase in mitochondria or other structures suggests that this would limit its action to amines that are present in a free (unbound) form in the axoplasm. Here, monoamine oxidase can act on amines that have been taken up by the axon before they are granule bound or it can even act on amines that are released from the granules before they pass out through the axonal membrane. Interestingly, the latter possibility seems of minor physiological importance, since monoamine oxidase inhibition does not potentiate the effects of peripheral sympathetic nerve stimulation.

The second enzyme of importance in the catabolism of catecholamines is COMT discovered by Axelrod in 1957. This enzyme is a relatively nonspecific enzyme that catalyzes the transfer of methyl groups from S-adenosyl methionine to the m-hydroxyl group of catecholamines and various other catechol compounds. COMT is found in the cytoplasm of most animal tissue, being particularly abundant in kidney and liver. A substantial amount of this enzyme is also found in the CNS and in various sympathetically innervated organs. The precise cellular localization of COMT has not been determined although it has been suggested (with little foundation) that it functions extraneuronally. The purified enzyme requires S-adenosyl methionine and Mg^{2+} ions for activity. As with monoamine oxidase, inhibition of COMT activity does not markedly potentiate the effects of sympathetic nerve stimulation, although in some tissue it tends to prolong the duration of the response to stimulation. Therefore, neither monoamine oxidase nor COMT would seem to be the primary mechanism for terminating the action of norepinephrine liberated at sympathetic nerve terminals. It may be, however, that these enzymes play a more important role in terminating transmitter action and regulating catecholamine function in the CNS.

Uptake

When sympathetic postganglionic nerves are stimulated at frequencies low enough to be comparable to those encountered physiologically, very little intact norepinephrine overflows into the circulation suggesting that local inactivation is very efficient. This local inactivation is not significantly blocked when COMT or monoamine oxidase or both are inhibited, and it is believed to involve mainly reuptake of the transmitter by sympathetic neurons.

Much attention has been focused on the role of tissue uptake mechanisms in the physiological inactivation of catecholamines. But only in recent years has this concept received direct experimental support although a number of earlier findings had suggested that catecholamines might be inactivated by some sort of nonmetabolic process. For example, more than 40 years ago, Burn suggested the possibility that exogenous norepinephrine might be taken up in storage sites in peripheral tissue. About 30 years ago it was demonstrated that an increase in the catecholamine content of cat and dog heart occurred after administration of norepinephrine and epinephrine in vivo. However, in view of the very high doses (mg) employed in these experiments, the significance of these findings remained questionable. It was really not until labeled catecholamines with high specific activity became available that similar experiments using doses of norepinephrine comparable to those likely to be encountered under physiological circumstances could be performed. Uptake studies of this nature carried out in vivo have indicated that approximately 40–60% of a relatively small intravenous dose of norepinephrine is metabolized enzymatically by COMT and monoamine oxidase, while the remainder is inactivated by uptake into various tissues. This uptake appears to be extremely rapid and in most cases the magnitude of the uptake is related to the density of sympathetic innervation and the proportion of cardiac output received by the tissue in question. Thus, after administration of 3H-norepinephrine in vivo, the greatest uptake and binding occur in tissues such as spleen and heart. The brain is also capable of taking up norepinephrine but since the uptake from the circulation is prevented by the blood–brain barrier, efficient uptake can only be observed when brain slices, minces, or brain synaptosomes are exposed to norepinephrine or when the labeled amines are administered intraventricularly or intracisternally.

A large amount of data now suggests that the major site of this uptake (and subsequently binding) is actually in sympathetic nerves. This evidence can be summarized as follows:

1. In most cases norepinephrine uptake correlates directly with the density of sympathetic innervation (or the endogenous content of norepinephrine) providing that sufficient allowance is made for differences in regional blood flow to various tissues.

2. In tissues without a normal sympathetic nerve supply, the ability to take up exogenous catecholamines is severely impaired. This can be demonstrated by surgical sympathectomy, chemical sympathectomy, or immunosympathectomy.

3. Labeled norepinephrine taken up by heart, spleen, artery, vein, or other tissues can be subsequently released by sympathetic nerve stimulation.

4. Autoradiographic studies at the electron microscope level have directly localized labeled catecholamines within the neuronal elements of the tissue.

5. Fluorescence microscopy has also provided equally direct evidence for the localization of norepinephrine uptake to sympathetic nerves.

A number of studies demonstrate that the uptake of catecholamines is active as it proceeds against a concentration gradient. For example, slices or minces of heart, spleen, and brain can concentrate norepinephrine to levels five to eight times those in the incubation medium. In intact tissues, even greater concentration ratios between tissue and medium may be obtained. In isolated rat heart perfused with Krebs solution containing 10 ng/ml of norepinephrine, Iversen found concentrations of labeled norepinephrine rise 30–40 times above that present in the perfusion medium. If we assume that norepinephrine uptake occurs almost exclusively into cardiac sympathetic neurons, the uptake process clearly has an exceptionally high affinity for norepinephrine. In fact, the actual concentration ratios between exogenous norepinephrine accumulated within the sympathetic neurons and that present in the medium could approach 10,000:1.

This uptake process is a saturable membrane transport process dependent on temperature and requiring energy. The stereochemically preferred substrate is L-norepinephrine; furthermore, norepinephrine is taken up more efficiently than its N-substituted derivatives. The uptake process is sodium dependent and can be blocked by inhibition of Na^+, K^+-activated ATPase. In fact, most evidence suggests that catecholamine uptake is mediated by some sort of active membrane transport mechanism located in the presynaptic terminal membrane of postganglionic sympathetic neurons.

A complementary DNA clone encoding a human norepinephrine transporter has recently been isolated. The isolated cDNA sequence predicts a protein of 617 amino acids with 12 highly hydrophobic regions compatible with membrane-spanning domains. Expression of the cDNA clone in transfected HeLa cells indicates that the norepinephrine transport activity is sodium dependent and sensitive to norepinephrine transport inhibitors. Striking sequence homology is notable between the norepinephrine transporter and the rat and human GABA transporters, suggesting a new transporter gene family.

The plasma membrane transporters have been intensively studied at the biochemical, pharmacological, and molecular levels. It has become clear that these Na^+ and Cl^--coupled transporters represent a group of integral membrane proteins encoded by a closely related family of genes that include the transporters of monoamines, GABA, glycine, and choline. A different class of plasma membrane transporters is represented by the glutamate transporter (cf. Chapter 6). Plasma membrane transporters can be classified into families and subfamilies based on ion dependence, topology, and sequence relatedness. The norepinephrine transporters (NET) are members of the

Na$^+$/Cl$^-$-dependent neurotransmitter transporter family that includes the dopamine transporters (DAT) and the serotonin transporter (SERT). Expression of NET, SERT, and DAT in nonneuronal cells has established model systems for analysis of the structural basis of transporter specificity for transmitters and transporter-specific antagonists. The availability of pure transporter proteins has facilitated the development of transporter-specific antibodies and nucleic acid probes and stimulated renewed interest in the endogenous control mechanisms acutely regulating monoamine transport in vivo. Also, the availability of human cDNA encoding the norepinephrine transporter offers the opportunity to determine whether alterations in transporter genes could have important etiological implications for major depressive or affective disorders.

Since many aspects of norepinephrine neurotransmission, including synthesis and release, are tightly regulated, it would not be unexpected to find that the NET are also subject to regulatory control. In fact, more than 30 years ago, Schneider and Gillis noted that, following stimulation of the sympathetic input to the heart, a rapid increase in the retention of norepinephrine occurred in the cat atrium. A number of similar observations have been made over the past three decades, suggesting that reuptake of norepinephrine increases in parallel with an increased rate of sympathetic discharge and norepinephrine release. The mechanism responsible for the increase in NET activity remains to be determined. Clearly the presence of serine/threonine phosphorylation sites on human NET raises the possibility that this effect might be mediated by protein phosphorylation. The role of protein phosphorylation or other post-translational modifications in regulating transporter function is only beginning to be evaluated. However, the fact that NET proteins can now be visualized suggests that this should be a fertile area for future investigations. Little is currently known about promotor regions and other regulatory elements involved in transcription regulation of the NET and SERT genes. However, recent studies have indicated that rats treated chronically with antidepressant drugs exhibit a decrease in the levels of SERT mRNA and a reduction in NET using [3H]nisoxetine autoradiography.

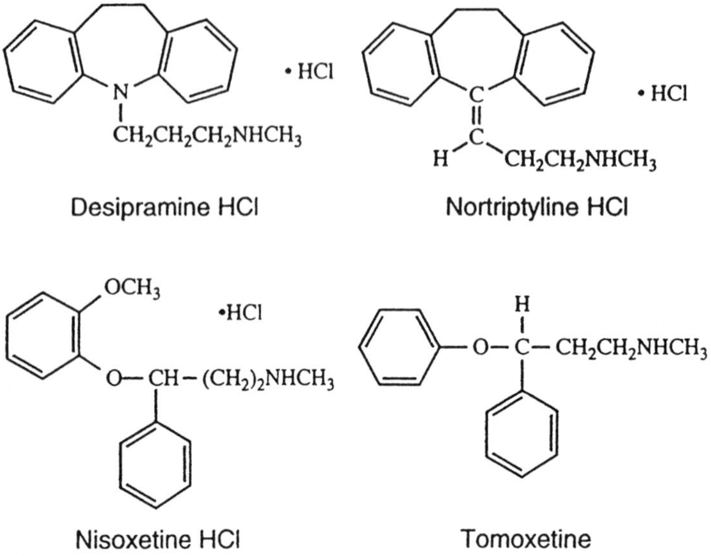
• HCl
N
CH2CH2CH2NHCH3
Desipramine HCl
C
C
H
CH2CH2NHCH3
• HCl
Nortriptyline HCl
OCH3
•HCl
O— CH – (CH2)2NHCH3
Nisoxetine HCl
H
O– C– CH2CH2NHCH3
Tomoxetine

FIGURE 8-6. Representative compounds that selectively inhibit norepinephrine reuptake.

The availability of cDNAs encoding transporter proteins may facilitate the development of sensitive screening techniques to help develop new and selective agents that target specific neurotransmitter transporters. Although this screening technology has not yet led to the development of uniquely selective inhibitors of the monoamine transporters, a number of selective inhibitors of norepinephrine uptake are available, several of which are used clinically in the treatment of affective disorders. Several of these norepinephrine uptake blockers are illustrated in Figure 8-6.

Adrenergic Receptors

When catecholamines are released from either noradrenergic nerve terminals or the adrenal medulla, they are recognized by and bind to specific receptor molecules on the plasma membrane of the neuroeffector cells that transduce catecholamine interactions with the cell into a physiological response. Depending on the nature of the receptor, this interaction sets off a cascade of membrane and intracellular events that cause the cell to carry out its specialized function. Classically, peripheral adrenergic receptors have been divided into two distinct classes called α- and β-receptors. The development of synthetic compounds active at adrenergic receptors has allowed further differentiation of adrenergic receptors into α1, α2, β1, and β2 subtypes (Fig. 8-7). Genes for these subtypes of adrenergic receptors have been cloned; this has shown that these receptors are members of a larger family of hormone receptors that mediate their activity through an interaction with one of a series of guanine nucleotide–binding regulatory proteins (G proteins). In addition to traditional classification based on their pharmacological profile, adrenergic receptors can be divided into three major classes by their differential coupling to G proteins (see Table 8-1). The β-adrenergic receptors activate G_s to stimulate adenyl cyclase. The α2-adrenergic receptors decrease adenyl cyclase activity through coupling to G_i. The α1-adrenergic receptor stimulates phospholipase C action through a still ill-defined G_x. As schematically illustrated in Figure 8-7, β-adrenoceptor signals are transmitted through the effector cell membrane via the adenylate cyclase system. The occupation of the β-adrenoceptor by the catecholamines stimulates the adenylate cyclase to generate cyclic adenosine-3′5′-monophosphate (cAMP) by a series of intramembrane events (see Chapter 5). Initially, the receptor interacts with a guanosine triphosphate (GTP)–dependent protein, G_s. The GTP-dependent G_s is linked to a catalytic unit (C) that generates cAMP upon activation by G_s; the catalytic unit then converts AMP to cAMP. The latter compound triggers a series of intracellular events involving protein kinases. The protein kinases activate further unknown biochemical changes to generate the final biological response to the transmitter.

The order of potency for stimulation of β1-adrenergic receptors by catecholamines is isoproterenol > norepinephrine ≥ epinephrine. For the β2-adrenoceptor the order of potency is isoproterenol > epinephrine > norepinephrine. Adrenergic blocking agents such as propranolol and alprenolol prevent the activation of the β-recep-

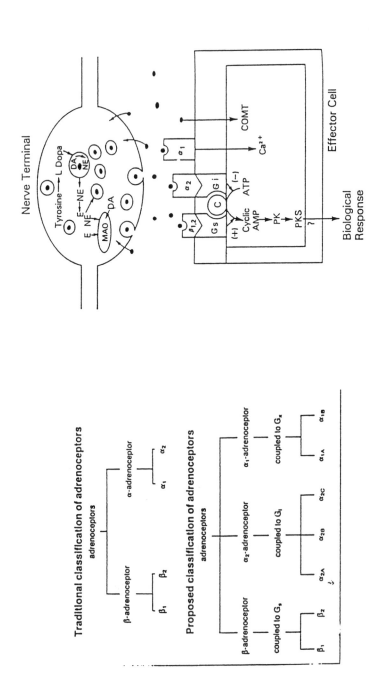

FIGURE 8-7. Classification and coupling of adrenoceptors. See text for details. Abbreviations: G_i, inhibitory G protein; G_s, stimulatory G protein; PK, protein kinase; PKS, protein kinase substrate; β1, β2, α1, α2: adrenergic receptor subtypes.

TABLE 8-1. Characterization of Adrenergic Receptor Subtypes

| Receptor Subtype | Receptor-Specific Agents | | Tissue Distribution | Major G Protein | Effector System |
	Agonist	Antagonist			
$\alpha_1 A$	Cirazoline	Corynanthin	Vas deferens, brain	G_q	↑ Phospholipase C
$\alpha_1 B$	Methoxamine	e	Liver, brain		↑ Ca^{2+} channel
$\alpha_1 C$	Phenylephrine	Indoramin	Olfactory bulb		↑ Phospholipase A_2
$\alpha_1 D$		Prazosin	Vas deferens, brain		↑ Phospholipase D
					→ Adenylyl cyclase
$\alpha_2 A$	Guanabenz	RX 821002	Aorta, brain	G_i	↑ $K^+ \downarrow Ca^{2+}$
$\alpha_2 B$	p-Aminoclonidine	Yohimbine	Liver, kidney		↑ Na^+-H^+ exchange
$\alpha_2 C$	BHT-920	SKF 86426	Brain		↑ Phospholipase C, A_1
β_1	Isoproterenol	Alprenolol	Heart, brain, pineal	G_s	↑ Adenyl cyclase
β_2		Propranolol	Lung, prostate		
β_3		Pindolol	Adipose tissue		↑ Ca^{2+} channel

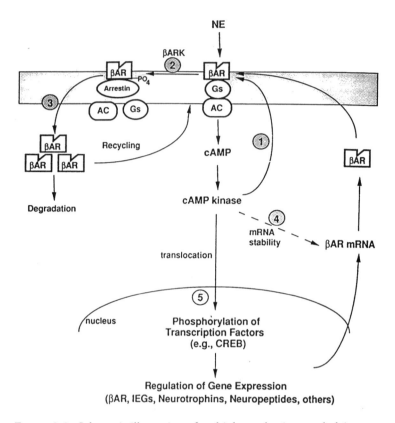

FIGURE 8-8. Schematic illustration of multiple mechanisms underlying regulation of the β-adrenergic receptor (βAR). (1) βAR stimulation of the cAMP system results in phosphorylation of the receptor by cAMP-dependent protein kinase, which leads to uncoupling of the receptor from G_s. The activated protein kinase would also phosphorylate many other proteins not shown in the figure, which would then mediate many of the actions of βAR activation. (2) Prolonged activation of the receptor leads to phosphorylation by another kinase, βAR kinase (βARK), which only phosphorylates the agonist-activated form of the receptor. This results in binding of the receptor to β-arrestin, which competes with G_s and thereby inhibits βAR-stimulation of adenylyl cyclase (AC). (3) Loss of βARs from the membrane occurs when receptors are internalized and sequestered into intracellular vesicles. This pool of receptors is then available for either recycling back to the membrane or degradation. Such sequestration, internalization, degradation, and membrane re-insertion may be mediated via receptor phosphory-

tors. The interaction of adrenergic agents with their receptors is saturable, stereospecific, and reversible. Prolonged exposure of β-adrenoceptors to endogenous or exogenous agonists often reduces the responsiveness of these receptors (desensitization). Desensitization can also be caused by uncoupling of β-receptors from the adenylate cyclase and by a decrease in the number of receptors (down-regulation). (See Fig. 8-8.) Depriving the β-adrenoceptors of catecholamine (by chemical or surgical denervation) increases the responsiveness of these receptors.

The recently identified β3-adrenoceptor shows lower affinity for agonists with a rank order of potency norepinephrine > isoproterenol > epinephrine and a lower affinity for known β-adrenoceptor antagonists. All three β-adrenoceptor subtypes appear to be linked to adenylyl cyclase activation through a stimulatory G protein, and at present there is no evidence for subtype-related differences in receptor-mediated cyclase interactions. However, the β3-adrenoceptor is insensitive to blockade by most β-adrenoceptor antagonists. No selective β3-adrenoceptor antagonist has yet been

---

lation and dephosphorylation involving the cAMP and/or βARK pathways. Another mechanism by which receptor activation leads to down-regulation of the βAR is via regulation of receptor mRNA levels, which may occur by two primary mechanisms. (4) The level of receptor mRNA is regulated by the stability or half-life of the mRNA. Although the mechanisms responsible for regulation of mRNA stability have not been identified, they may also involve cAMP-dependent protein kinase. (5) The level of receptor mRNA is also regulated via changes in βAR gene transcription. This effect is mediated by the cAMP pathway and appears to involve the translocation of cAMP-dependent protein kinase catalytic subunit into the nucleus and the phosphorylation of constitutively expressed transcription factors (e.g., CREB). It might also depend upon the subsequent induction of other transcription factors (e.g., immediate early gene [IEG] products such as c-Fos). In addition to regulation of receptor gene transcription, such regulation of transcription factors would mediate the effects of βAR activation on the expression of many other genes—for example, those for G proteins, cAMP-dependent protein kinase, neurotrophins, and neuropeptides. This, in turn, would mediate many of the more long-term effects of βAR activation on brain function. (From Duman and Nestler, 1995)

identified. In contrast to the β_1- and β_2-adrenoceptors, the β_3-adrenoceptor is resistant both to agonist-mediated short-term desensitization and, on a prolonged time scale, to agonist-induced receptor down-regulation. The biological importance of this differential response to the regulation by agonists is uncertain.

The α_1-adrenoceptor appears to be associated with calcium mobilization (Fig. 8-7), possibly via stimulation of phospholipase C and phospholipase. The α_1-receptor is distinguished from the α_2-adrenoceptor by its ability to be inhibited by the selective α_1-blocking agent prazosin. The α_2-adrenergic receptor is negatively linked to the adenylate cyclase complex via an inhibitory G protein (G_i). Like G_s, G_i is activated by guanosine triphosphate but in this case it inhibits the generation of AMP by the catalytic unit (C).

In the peripheral sympathetic nervous system, the α_2-receptor is mainly localized on presynaptic nerve terminals where it functions to modulate norepinephrine release. Stimulation of this receptor by catecholamines inhibits impulse-dependent release of transmitter; blockade of this receptor facilitates release.

Although it is clear that activation of α_2-adrenoceptors inhibits adenylyl cyclase activity (Fig. 8-7) mediated through an inhibitory G protein, this may not in all cases represent the signal transduction mechanism responsible for the effects associated with receptor activation. For example, the α_2-mediated inhibition of neurotransmitter release is generally insensitive to inactivation of inhibitory G proteins by pertussis toxin, suggesting the involvement of another signal transduction mechanism.

Dynamics of Adrenergic Receptors

Adrenergic receptors do not seem to be static entities but change in number and affinity in response to altered synaptic activity. In both brain and the peripheral sympathetic nervous system, destruction of adrenergic neurons is associated with functional supersensitivity of postsynaptic sites. Conversely, increasing synaptic norepinephrine by administering uptake blockers (tricyclic antidepressants) or inhibitors of monoamine oxidase, leads to functional subsensitivity.

These changes appear to be a compensatory response to altered levels of synaptic transmitter. The number of α_1- and α_2-receptors also increases after noradrenergic neurons in the brain and/or sympathetically innervated tissue have been destroyed by administration of 6-hydroxy-dopamine. It is interesting that after lesions in norepinephrine-containing neurons are made in the cerebral cortex, the number of β_1-receptors increases markedly, but no change occurs in the number of β_2-receptors. This may be because β_2-receptors have a low affinity for norepinephrine and the concentration of epinephrine in the cerebral cortex is relatively low. Likewise, the chronic administration of tricyclic antidepressants leads to a selective decrease in the density of β_1-adrenergic receptors in the cerebral cortex. This finding implies that the β_1-adrenergic receptors in the cortex are functionally innervated by adrenergic neurons.

Desensitization

The waning of a stimulated response in the face of continuous agonist exposure has been termed desensitization. Desensitization has been demonstrated in many hormone- and neurotransmitter-receptor systems, including those that activate different G proteins and effector systems. Desensitization of the cAMP response elicited by β-adrenergic agonists is a useful model system for studying this phenomenon. Desensitization has been viewed historically as two separate processes: heterologous, and homologous desensitization. Heterologous desensitization occurs when exposure of cells to a desensitizing agent leads to diminished responsiveness to a number of different stimuli. In contrast, homologous desensitization is much more specific and involves only the loss of responsiveness to the specific desensitizing agent. A number of molecular mechanisms operating at the receptor level have been implicated in the process of desensitization. The heterologous pattern of desensitization is thought to be mediated by phosphorylation of receptors coupled to G_s by cAMP-dependent PKA, whereas the homologous pattern of desensitization is thought to involve phosphorylation of receptors by a novel, cAMP-independent kinase, such as β-adrenergic receptor kinase or βARK. Recent studies on the β-adrenergic receptor have suggested that the principal mechanisms underlying rapid, agonist-induced desensitization of β-adrenergic receptor function in intact mammalian cells are a combination of these two processes and involve phosphorylation of the receptor by both types of kinases. Phosphorylation of the receptor is thought to lead to uncoupling of the β-adrenergic receptor from the stimulatory G protein (G_s). It is not yet clear exactly how phosphorylation leads to uncoupling of β-adrenergic receptor from G_s; for example, phosphorylation of one specific residue may directly impair interaction of the receptor with G_s, whereas phosphorylation of another residue may allow cytosolic factors such as β-arrestin to bind to the phorphorylated β-adrenergic receptor and decrease or prevent the coupling of the β-adrenergic receptor with G_s, leading to desensitization.

PHARMACOLOGY OF NORADRENERGIC NEURONS

In this section we will focus on the pharmacology of peripheral noradrenergic neurons and discuss possible sites of drug involvement in the life cycle of the catecholamines. Figure 8-9 depicts a schematic model of a noradrenergic nerve varicosity.

Site I—Precursor transport. As noted earlier, there appears to be an active uptake or transport of tyrosine as well as other aromatic amino acids into the CNS. In addition, it is also probable that tyrosine is actively taken up by adrenergic neurons both in the periphery and in the CNS. At present, we have no pharmacologic agent that specifically antagonizes the uptake of tyrosine into the brain or into the catecholamine-containing neuron. However, it is known that various aromatic amino acids will compete with each other for transport into the CNS. This competition might become important under a pathological situation in which blood aromatic amino acids are elevated. Thus, in phenylketonuria, when plasma levels of phenylalanine are elevated to about 10^-3 M, it might be expected that tyrosine and tryptophan uptake into the brain might be diminished.

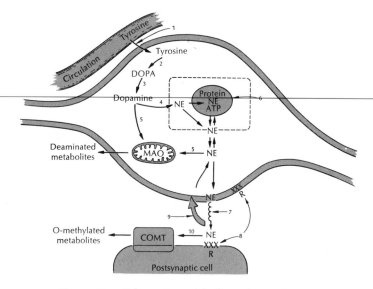

FIGURE 8-9. Schematic model of noradrenergic neuron.

Site 2—Tyrosine hydroxylase. The conversion of tyrosine to DOPA is the rate-limiting step in this biosynthetic pathway and thus is the step most susceptible to pharmacological manipulation. The various classes of tyrosine hydroxylase inhibitors were considered earlier in this chapter. An effective inhibitor of tyrosine hydroxylase both in vivo and in vitro, α-methyl-ρ-tyrosine is one of the few tyrosine hydroxylase inhibitors that have been employed clinically and found to be effective in the treatment of inoperable pheochromocytomas; it is of little or no value in the treatment of essential hypertension.

Sites 3 and 4—DOPA-decarboxylase and dopamine-β-hydroxylase. Potent inhibitors of DOPA-decarboxylase and dopamine-β-hydroxylase do exist but they are generally not effective in vivo in reducing tissue concentrations of catecholamines.

Due to the nonspecificity of these two enzymes (see above) as well as the relatively nonspecific vesicular storage sites, certain drugs or even amines or amine precursors in foodstuff can be synthesized to phenylethylamine derivatives with a catechol or a β-hydroxyl group and taken up and stored in the storage vesicles. These compounds can subsequently be released by sympathetic nerves, in part replacing some of the norepinephrine normally released by sympathetic impulses. Since these "false neurotransmitters" are usually less potent in their action, replacement of some of the endogenous, releasable pool of norepinephrine with these agents results in a diminished sympathetic response. Thus, a drug such as α-methyl-dihydrox-phenylalanine which is converted in the sympathetic neuron to α-methyl-norepinephrine, replaces a portion of the endogenous norepinephrine and itself can be subsequently released as a "false transmitter." The false transmitter theory also has been used to explain the hypotensive action of various monoamine oxidase inhibitors. Tyramine, an endogenous metabolite that is normally disposed of by monoamine oxidase in the liver and other tissues, is no longer rapidly deaminated in the presence of monoamine oxidase inhibitors and thus can be converted to the β-hydroxylated derivative octopamine, which then accumulates in sympathetic neurons. Upon activation of these nerves, octopamine is released and acts as a false neurotransmitter, producing a resultant decrease in sympathetic tone (observed with monoamine oxidase inhibitors). "False neurotransmitters" can also be formed in the brain but at the present time the action of these agents in the CNS is unclear.

Site 5—Metabolic degradation by monoamine oxidase. The monoamine oxidase inhibitors belong to quite a heterogeneous group of drugs that have in common the ability to block the oxidative deamination of various biogenic amines. These agents all produce marked increases in tissue amine concentrations. These compounds are in most cases employed clinically for the treatment of depression. However, some of these agents are used for the treatment of hypertension and angina pectoris. In the case of monoamine oxidase inhibitors, it should be immediately pointed out that while these drugs inhibit monoamine oxidase, there is no firmly established correlation between monoamine oxidase inhibition and therapeutic effect. In fact, it is generally well appreciated that monoamine oxidase inhibitors inhibit not only enzymes involved in oxidative deamination of monoamines but also many other enzymes unrelated to monoamine oxidase such as succinic dehydrogenase, dopamine-β-hydroxylase, 5-hydroxytryptophan decarboxylase, choline dehydrogenase, and diamine oxidase.

Site 6—Storage. A number of drugs, most notably the Rauwolfia alkaloids, interfere with the storage of monoamines by blocking the uptake of amine into the storage granules or by disrupting the binding of the amines. Thus drugs interfering with the storage mechanism will cause amines to be released intraneuronally. Amines released intraneuronally appear to be preferentially metabolized by monoamine oxidase.

Site 7—Release. The release of catecholamines is dependent on Ca^{2+} ions. Thus, in an in vitro system it is possible to block release by lowering the Ca^{2+} concentration of the medium. Although drugs such as bretylium act to inhibit the release of norepinephrine in the peripheral nervous system, this drug does not easily penetrate the brain and therefore has little or no effect on central monoaminergic neurons.

Site 8—Interaction with receptors. In the peripheral nervous system, α- and β-adrenergic blocking agents are effective. These agents block both the effects of sympathetic nerve stimulation and the action of exogenously administered norepinephrine. Some of these agents also have central activity. Evidence has indicated that presynaptic α-receptors, termed α2-receptors, are involved in the regulation of norepinephrine release. A blockade of these receptors facilitates release, while stimulation leads to an inhibition of release. Presynaptic β-receptors have also been identified in certain tissues. In contrast to α2-receptors, activation of these presynaptic β-receptors facilitates rather than attenuates norepinephrine release.

Site 9—Reuptake. Drugs such as cocaine, desipramine, amitryptyline, and other related tricyclic antidepressants, are effective inhibitors of norepinephrine uptake at both peripheral and central sites. Pharmacological intervention at these sites makes more norepinephrine available to interact with postsynaptic receptors and thus tends to potentiate adrenergic transmission.

Site 10—Catechol-O-methyl transferase. Catechol-O-methyl transferase can be inhibited by pyrogallol or various tropolone derivatives. Inhibition of this enzyme in most sympathetically innervated tissue such as heart, does not significantly potentiate the effects of nerve stimulation. In vascular tissue, however, inhibition of this enzyme does lead to a significant prolongation of the response to nerve stimulation. It appears possible that in this tissue catechol-*O*-methyl transferase may play some role in terminating transmitter action.

CNS Catecholamine Neurons

The cellular organization of the brain and spinal cord has been studied for many decades by classic histological and silver impregnation techniques. With the development in the last decade of completely different histological techniques based on the presence of a given type of transmitter substance or on specific synthetic enzymes involved in the formation of a given transmitter, it became possible to map chemically defined neuronal systems in the CNS of many species. With the formaldehyde fluorescence histochemical method of Falck and Hillarp, it has been possible to delineate norepinephrine-, dopamine-, and 5-hydroxytryptamine-containing systems in mammalian brain. By the time such techniques had been employed for several years, it became clear that the distribution of these chemically defined neuronal systems did not necessarily correspond to that of systems described earlier with the classic techniques.

It was really not until the mid-1960s that the histochemical fluorescent technique was applied to brain tissue and the anatomy of the monoamine-containing neuronal systems was described. By a variety of pharmacological and chemical methods, it has subsequently been possible to discriminate between norepinephrine-, epinephrine-, and dopamine-containing neurons and to describe in detail the distribution of these catecholamine-containing neurons in the mammalian CNS. Several recent and thorough surveys of these systems are available.

Figure 8-10 provides a schematic model of a central monoamine-containing neuron as observed by fluorescence and electron microscopy. The cell bodies contain relatively low concentrations of

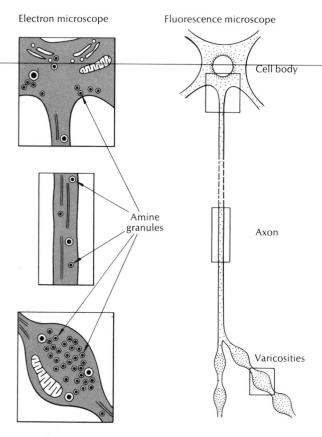

Schematic diagram of Monoamine Neuron

Electron microscope   Fluorescence microscope

Cell body

Amine
granules

Axon

Varicosities

FIGURE 8-10. Schematic illustration of a central monoamine-containing neuron. The right side depicts the general appearance and intraneuronal distribution of monoamines based upon fluorescence microscopy. The cell bodies contain a relatively low concentration of catecholamine (about 10–100 μg/gm) while the terminal varicosities contain a very high concentration 1000–3000 μg/gm). The preterminal axons contain very low concentration of amine. At the electron microscopic level (left side), dense core granules can be observed in the cell bodies and axons but appear to be highly concentrated in the terminal varicosities. (Illustration adapted after K. Fuxe and T. Hökfelt, *Triangle* 30: 73, 1971)

amine (about 10–100 μg/gm), while the terminal varicosities contain a very high concentration (about 1000–3000 μg/gm). The axons, on the other hand, consist of highly branched, largely unmyelinated fibers that have such a low concentration of amine that they are barely visible in untreated adult animals. With the electron microscope, depending on the type of fixation, it is possible to observe the presence of small granular vesicles that are thought to represent the subcellular storage sites containing the catecholamines. These granular vesicles are concentrated in the terminal varicosities of the central noradrenergic neuron just as they are observed in the peripheral sympathetic neuron.

Systems of Catecholamine Pathways in the CNS

Detailed analysis of catecholamine pathways in the CNS has been greatly accelerated by recent improvements in the application of fluorescence histochemistry such as the use of glyoxylic acid as the fluorogen and by the application of numerous auxiliary mapping methods such as anterograde and retrograde markers, immunocytochemical localization of catecholamine-synthesizing enzymes, and direct methods for observing noradrenergic terminals with the electron microscope. Extensive progress in the functional analysis of these systems has also been made possible by evaluation with microelectrodes of the effects produced by selective electrical stimulation of the catecholamine (especially norepinephrine) cell-body groups. From such studies it is clear that the systems can be characterized in simple terms only by ignoring large amounts of detailed cytological and functional data and that the earlier catecholamine wiring diagrams are no longer tenable. Furthermore, anatomic details for monoamine systems in rodents seem to bear only rudimentary homologies to their detailed selective anatomic configurations in human and nonhuman primates.

There are two major clusterings of norepinephrine cell bodies from which axonal systems arise to innervate targets throughout the entire neuraxis:

Locus Ceruleus

This compact cell group within the caudal pontine central gray is named for the pigment the cells bear in humans and in some higher primates; in the rat the nucleus contains about 1500 neurons on each side of the brain (Fig. 8-11). In humans, the locus ceruleus is composed of about 12,000 large neurons on each side of the brain. The axons of these neurons form extensive collateral branches that

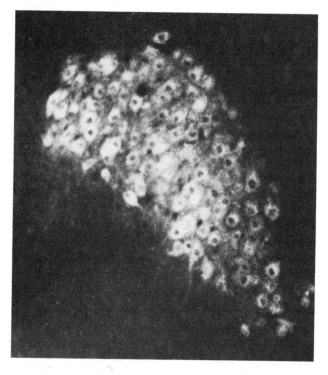

FIGURE 8-11. Formaldehyde-induced fluorescence of the rat nucleus locus ceruleus. In this frontal section through the principal portion of the nucleus, intensely fluorescent neurons can be seen clustered closely together. Within the neurons, the cell nucleus, which is not fluorescent after this treatment, appears dark except for the nucleolus. (Bloom, unpublished) (× 650)

project widely along well-defined tracts. At the electron microscope level, terminals of these axons exhibit—under appropriate fixation methods—the same type of small granular vesicles seen in the peripheral sympathetic nerves (Fig. 8-2).

Fibers from the locus ceruleus form five major noradrenergic tracts (Fig. 8-12): the central tegmental tract (or dorsal bundle described by Ungerstedt), a central gray dorsal longitudinal facsiculus tract, and a ventral tegmental–medial forebrain bundle tract. These tracts remain largely ipsilateral, although there is a crossing over in some species of up to 25% of the fibers. These three ascending tracts then follow other major vascular and fascicular routes to innervate all cortices, specific thalamic and hypothalamic nuclei, and the olfactory bulb. Another major fascicle ascends in the superior cerebellar peduncle to innervate the cerebellar cortex. The fifth

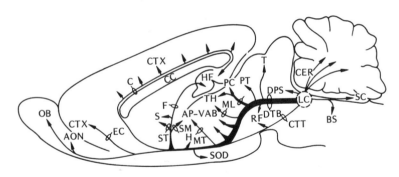

FIGURE 8-12. Diagram of the projections of the locus ceruleus viewed in the sagittal plane. See text for description. Abbreviations: AON, anterior olfactory nucleus; AP–VAB, ansa peduncularis–ventral amygdaloid bundle system; BS, brainstem nuclei; C, cingulum; CC, corpus callosum; CER, cerebellum; CTT, central tegmental tract; CTX, cerebral neocortex; DPS, dorsal periventricular system; DTB, dorsal catecholamine bundle; EC, external capsule; F, fornix; H, hypothalamus; HF, hippocampal formation; LC, locus ceruleus; ML, medial lemniscus; MT, mammillothalamic tract; OB, olfactory bulb; PC, posterior commissure; PT, pretectal area; RF, reticular formation; S, septal area; SC, spinal cord; SM, stria medullaris; SOD, supraoptic decussations; ST, stria terminalis; T, tectum; TH, thalamus. (Diagram compiled by R. Y. Moore, from the observations of Lindvall and Björklund, 1974b; Jones and Moore, 1977)

major tract is that which descends into the mesencephalon and spinal cord, where the fibers course in the ventral–lateral column. At their terminals, the locus ceruleus fibers form a plexiform network in which the incoming fibers gain access to a cortical region by passing through the major myelinated tracts, turning vertically toward the outer cortical surface, and then forming characteristic T-shaped branches that run parallel to the surface in the molecular layer; this pattern is seen in the cerebellar, hippocampal, and cerebral cortices.

Virtually all of the noradrenergic pathways that have been studied physiologically so far are efferent pathways of the locus ceruleus neurons; in cerebellum, hippocampus, and cerebral cortex, the major effect of activating this pathway is to produce an inhibition of spontaneous discharge. This effect has been associated with the slow type of synaptic transaction in which the hyperpolarizing response of the target cell is accompanied by increased membrane resistance. The mechanism of this action has been experimentally related to the second messenger scheme in which the noradrenergic receptor elicits its characteristic action on the target cells by activating the synthesis of cAMP in or on the postsynaptic membrane. Pharmacologically and cytochemically, target cells responding to norepinephrine or the locus ceruleus projection in these cortical areas exhibit β-adrenergic receptors.

The Lateral Tegmental Noradrenergic Neurons

A large number of noradrenergic neurons lie outside of the locus ceruleus where they are more loosely scattered throughout the lateral ventral tegmental fields. In general, the fibers from these neurons intermingle with those arising from the locus ceruleus, those from the more posterior tegmental levels contributing mainly descending fibers within the mesencephalon and spinal cord, and those from the more anterior tegmental levels contributing to the innervation of the forebrain and diencephalon. Because of the complex intermingling of the fibers from the various noradrenergic cell body sources, the physiological and pharmacological analysis of the effects of brain lesions becomes extremely difficult. The neurons of the lateral tegmental system may be the primary source of the noradrenergic fibers observed in the basal forebrain, such as amygdala and septum. No specific analysis of the physiology of these projections has yet been reported; therefore it remains to be established whether the β-receptive cAMP mechanism associated with the cortical projections of the locus ceruleus group also apply to the synapses of the lateral tegmental neurons.

EPINEPHRINE NEURONS

In the sympathetic nervous system and the adrenal medulla, epinephrine shares with norepinephrine the role of final neurotransmitter, the proportion of this sharing being a species-dependent, hormonally modified arrangement. Until method refinement, however, little evidence could be gathered to document the existence of epinephrine in the CNS because the chemical methods for analyzing epinephrine levels or for detecting activity attributable to the synthesizing enzyme phenylethanolamine-N-methyl transferase, were unable to provide unequivocal data. With the development of sensitive immunoassays for phenylethanolamine-N-methyl transferase and their application in immunohistochemistry, and with the application of gas chromatography–mass fragmentography (GCMF) and high-performance liquid chromatography (HPLC) with electrochemical detection to brain neurotransmitter measurements, the necessary data were rapidly acquired and the existence of epinephrine-containing neurons in the CNS confirmed. By immunohistochemistry, epinephrine-containing neurons are defined as those that are positively stained (in serial sections) with antibodies to tyrosine hydroxylase, dopamine-β-hydroxylase, and with antibodies to phenylethanolamine-N-methyl transferase. These cells are found largely in two groups: one group—called C-1 by Hökfelt—is intermingled with noradrenergic cells of the lateral tegmental system; the other group—called C-2—is found in the regions in which the noradrenergic cells of the dorsal medulla are also found. The axons of these two epinephrine systems ascend to the hypothalamus with the central tegmental tract, then via the ventral periventricular system into the hypothalamus. A third group of cells (C-3) in the midline (medial longitudinal fascicle) has also been described (Fig. 8-13). Within the mesencephalon, the epinephrine-containing fibers innervate the nuclei of visceral efferent and afferent systems, especially the dorsal motor nucleus of the vagus nerve. In addition, epinephrine fibers also innervate the locus ceruleus, the intermediolateral cell columns of the spinal cord, and the periventricular regions of the fourth ventricle.

Although there are considerably fewer epinephrine than dopamine and norepinephrine neurons in brain, they do appear to have a discrete anatomical distribution and are believed to subserve physiological functions discrete from other catecholamines. Selective plasma membrane transporters have been described for

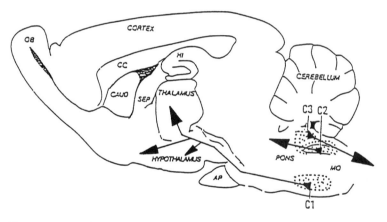

FIGURE 8-13. Schematic illustration of the distribution of the main central neuronal pathways containing epinephrine. The C-1 group represents a rostral continuation of the noradrenergic cell group in the ventrolateral medulla oblongata. The dorsal C-2 group of cells is mainly located in the dorsal vagal complex with the vast majority of cell bodies in the solitary tract nucleus. A C-3 group of cells in the midline (medial longitudinal fascicle) has also been described. The ventral group of epinephrine-containing neurons gives rise to both ascending and descending pathways, innervating largely periventricular regions such as the periaqueductal central gray and various hypothalamic nuclei and the lateral sympathetic column in the spinal cord. (Data from Hökfelt et al., 1984)

dopamine and norepinephrine, and these transporters have been extensively studied both in vivo and in vitro. However, little attention has been given to the hypothetical epinephrine transporter, and at the present time a specific transporter for epinephrine has not been isolated nor have drugs that selectively block epinephrine reuptake been described. All the pharmacological agents that have been shown to block the norepinephrine transporter (see Fig. 8-7) selectively are also effective inhibitors of epinephrine uptake. Thus, it is uncertain whether a selective transporter for epinephrine actually exists in the mammalian CNS or peripheral nervous system. If an epinephrine transporter could be cloned and expressed, this might permit a search for selective inhibitors that, if discovered, could provide valuable pharmacological tools for elucidating the functional role of central epinephrine-containing neurons.

Except for tests of epinephrine in the locus ceruleus (where it inhibits firing), no other tests of the cellular physiology of this system have thus far been reported. Our understanding concerning the function of epinephrine-containing neurons in brain is still very limited, but, based on the distribution of epinephrine in specific brain regions, attention has been directed to their possible role in neuroendocrine mechanisms and blood pressure control.

Coexistence of Classic Transmitters and Peptides

The findings of numerous peptides in the CNS has raised the question of the relationship of the neurons containing these substances to neurons containing classic neurotransmitters. It was originally believed that the peptide-containing nerves represented separate identifiable systems. However, when studies in the peripheral nervous system clearly indicated the coexistence of peptides with the sympathetic and parasympathetic transmitters norepinephrine and ACh, respectively, this initiated a series of systematic studies in the CNS to determine if classic neurotransmitter in the brain coexisted with one or more peptides. These studies revealed that coexistence of classic transmitter and peptides may represent a rule rather than an exception. The functional significance of the occurrence of a classic transmitter and a peptide in the same neuron and their possible release from the same nerve ending is still unclear in the CNS. However, some insight may be obtained from studies in the periphery where two distinct types of interactions have been noted.

In the salivary gland, the parasympathetic cholinergic neurons contain a VIP-like peptide while the sympathetic norepinephrine neurons contain a neuropeptide Y–like peptide. In this system, the peptides seem to augment the action of the classic transmitters. Thus, VIP induces vasodilation and enhances the secretory effects of ACh while neuropeptide Y causes vasoconstriction like norepinephrine. In contrast, in the vas deferens, where the norepinephrine neurons innervating this tissue also contain neuropeptide Y, the peptide seems to inhibit the release of norepinephrine via a presynaptic action. There is also some indication of preferential storage and release of norepinephrine and neuropeptide Y with release of the peptide occurring preferentially at higher frequencies or during burst firing.

The physiological significance of multiple messengers at the synapse in the CNS is still unclear, but an appreciation of coexistence phenomena may influence our view on interneuronal communication and in a larger perspective may be of importance in our efforts to treat or prevent various disease states or abnormalities of the nervous system.

Catecholamine Metabolism

Catecholamine metabolism was discussed earlier, and only a few aspects pertinent to central catecholamine metabolism will be covered here.

In the peripheral sympathetic nervous system, the aldehyde intermediate produced by the action of monoamine oxidase on norepinephrine and normetanephrine can be oxidized to the corresponding acid or reduced to the corresponding glycol. Oxidation usually exceeds reduction, and quantitatively, vanilylmandelic acid (VMA) is the major metabolite of norepinephrine and is readily detectable in the urine. In fact, the urinary levels of VMA are routinely measured in clinical laboratories in order to provide an index of peripheral sympathetic nerve function as well as to diagnose the presence of catecholamine-secreting tumors such as pheochromocytomas and neuroblastomas. In the CNS, however, reduction of the intermediate aldehyde formed by the action of monoamine oxidase on catecholamines or catecholamine metabolites predominates, and a major metabolite of norepinephrine found in brain is the glycol derivative 3-methoxy-4-hydroxy-phenethyleneglycol (MHPG). Very little, if any, VMA is found in brain.

In many species a large fraction of the MHPG formed in the brain is sulfate conjugated. However, in primates, MHPG exists primarily in an unconjugated "free" form in the brain. Some normetanephrine is also found in the brain and spinal cord. Destruction of noradrenergic neurons in the brain or spinal cord causes a reduction in the endogenous levels of these metabolites although not as marked as the corresponding reduction in endogenous norepinephrine. Recently, it has been demonstrated that direct electrical stimulation of the locus ceruleus or severe stress produces an increase in the turnover of norepinephrine as well as an increase in the accumulation of the sulfate conjugate of MHPG in the rat cerebral cortex. These effects are completely abolished by ablation of the locus ceruleus or by transection of the dorsal pathway, which suggests that the accumulation of MHPG-sulfate in brain may reflect the functional activity of central noradrenergic neurons. Since MHPG-sulfate readily diffuses from the brain into the CSF or general circulation, estimates of its concentration in the CSF or in urine are thought to provide a possible reflection of activity of noradrenergic neurons in the brain. Even though MHPG is proportionately a minor metabolite or norepinephrine in the peripheral sympathetic nervous system, a fairly large portion of the MHPG in the urine still derives from the periphery. In the rodent, it has been estimated that the brain provides a minor contribution (25%–30%) to urinary MHPG while in primates, the brain contribution is quite large

(60%). Thus, at least in the rodent, it is quite probable that relatively large changes in the formation of MHPG by the brain are necessary to produce detectable changes in urinary MHPG. For example, in the rat, destruction of the majority of norepinephrine-containing neurons in the brain by treatment with 6-hydroxy-dopamine leads to only about a 25% decrease in urinary MHPG levels. Nevertheless, measurement of urinary changes in MHPG is still a reasonable strategy for obtaining some insight into possible alterations of brain norepinephrine metabolism in patients with psychiatric illnesses. Measurement of CSF levels of MHPG also provides another possible approach to assessing central adrenergic function in human subjects. More recent studies have suggested that plasma levels of MHPG might provide a reflection of central noradrenergic activity. In these studies, stimulation of the locus ceruleus in the rat was shown to result in a significant increase in the levels of plasma MHPG; and administration of drugs that are known to alter noradrenergic activity in rodents, have predictably changed plasma levels and venous–arterial differences in MHPG in nonhuman primates. Although changes in urinary, plasma, and CSF levels of MHPG and their relationship to central noradrenergic function have to be interpreted with considerable caution, these measurements do provide a starting point for clinical investigation.

BIOCHEMICAL ORGANIZATION

In general, noradrenergic neurons in the CNS appear to behave in a fashion very similar to the postganglionic sympathetic noradrenergic neurons in the periphery and have the same presynaptic regulatory controls. Electrical stimulation of the noradrenergic neurons in the locus ceruleus causes both an activation of tyrosine hydroxylase and an increase in the turnover of norepinephrine as well as a frequency-dependent accumulation of MHPG-sulfate (a major metabolite of norepinephrine in the rat CNS) in the cortex. Interruption of impulse flow in these neurons decreases turnover of norepinephrine but has little or no effect on the steady-state level of transmitter. Similar effects are also obtained in the serotonin-containing neurons of the midbrain raphe (stimulation increases synthesis and turnover, and interruption of impulse flow slows turnover). It is well known that acute severe stress increases the turnover of norepinephrine in the CNS, apparently as a result of increase in impulse flow in noradrenergic neurons. Recent experiments have confirmed this speculation. If impulse flow in noradrenergic neurons projecting to the cortex is acutely interrupted by destruction of the locus ceruleus, the increase in norepinephrine turnover and accumulation of MHPG-sulfate induced by stress is completely blocked. Also, microdialysis experiments have demonstrated stress-induced increases in norepinephrine release in hipocampus and frontal cortex.

PHARMACOLOGY OF CENTRAL CATECHOLAMINE-CONTAINING NEURONS

The psychotropic drugs (drugs that alter behavior, mood, and perception in humans and behavior in animals) can be divided into two main categories: the psychotherapeutic drugs; and, the psychotomimetic agents. Psychotherapeutic drugs can be further divided, on the basis of their activity in humans, into at least four general classes: antipsychotics; antianxiety drugs; antidepressants; and, stimulants.

In recent years a great deal more information has become available concerning the anatomy, biochemistry, and functional organization of catecholamine systems. This has fostered knowledge concerning the mechanisms and sites of action of many psychotropic drugs. It is now appreciated that drugs can influence the output of monoamine systems by interacting at several distinct sites. For example, drugs can act to influence the output of catecholamine systems by (1) acting presynaptically to alter the life cycle of the transmitter (i.e., synthesis, storage, and release, etc. [see Figs. 8-14 and 9-8]); (2) acting postsynaptically to mimic or block the action of the transmitter at the level of postsynaptic receptors; and (3) acting at the level of cell body autoreceptors to influence the physiological activity of catecholamine neurons. The activity of catecholamine neurons can also be influenced by postsynaptic receptors via negative neuronal feedback loops. In the latter two actions, drugs appear to exert their influence by interacting with catecholamine receptors. Figure 8-15 illustrates the various neuronal pathways and local mechanisms regulating synaptic homeostasis in central noradrenergic neurons. In general, catecholamine receptors can be subdivided into two broad categories: those that are localized directly on catecholamine neurons (often referred to as presynaptic receptors); and, those on other cell types that are often termed simply postsynaptic receptors since they are postsynaptic to the catecholamine neurons, the source of the endogenous ligand. When it became appreciated that catecholamine neurons, in addition to possessing receptors on the nerve terminals, appear to have receptors distributed over all parts of the neurons (i.e., soma, dendrites, and preterminal axons), the term presynaptic receptors really became inappropriate as a description for all these receptors. Carlsson, in 1975, suggested that "autoreceptor" was a more appropriate term to describe these receptors since the sensitivity of these catecholamine receptors to the neurons' own transmitter seemed more significant than their location relative to the synapse. The term autoreceptor achieved rapid acceptance and pharmacological research in the succeeding years resulted in the detection and description of autoreceptors on neurons in almost all chemically defined neuronal systems.

The presence of autoreceptors on some neurons may turn out to be only of pharmacological interest since they may never encounter effective concentrations of the appropriate endogenous agonist in vivo. Others, however, in addition to their pharmacological responsiveness, may play a very important physiological role in the maintenance of presynaptic events. This certainly appears to be the case for catecholamine autoreceptors (see Chapter 9).

The pharmacology of central and peripheral norepinephrine neurons is quite similar (Figs. 8-9 and 8-14). The main differences seem to be related primarily to drug delivery and the greater complexity of the neuronal pathways regulating synaptic homeostasis of the central norepinephrine systems (Fig. 8-15).

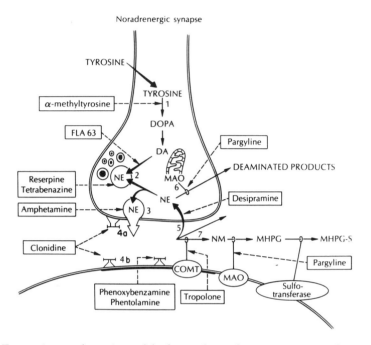

FIGURE 8-14. schematic model of central noradrenergic neuron indicating possible sites of drug action.

Site 1: *Enzymatic synthesis:*

    a. Tyrosine hydroxylase reaction blocked by the competitive inhibitor, α-methyltyrosine.

    b. Dopamine-β-hydroxylase reaction blocked by a dithiocarbamate derivative, Fla-63 bis-(1-methyl-4-homopiperazinyl-thiocarbonyl)-disulfide.

Site 2: *Storage:* Reserpine and tetrabenazine interfere with the uptake—storage mechanism of the amine granules. The depletion of norepinephrine produced by reserpine is long-lasting and the storage granules are irreversibly damaged. Tetrabenazine also interferes with the uptake—storage mechanism of the granules, except the effects of this drug are of a shorter duration and do not appear to be irreversible.

Site 3: *Release:* Amphetamine appears to cause an increase in the net release of norepinephrine. Probably the primary mechanism by which amphetamine causes release is by its ability to block effectively the reuptake mechanism.

Site 4: *Receptor interaction:*

    a. Presynaptic α2-autoreceptors

    b. Postsynaptic α-receptors

Effect of Drugs on the Electrophysiological Activity of Noradrenergic Neurons

A number of drugs have been found to influence the activity of the noradrenergic neurons in the locus ceruleus. Table 8-2 lists the drugs that have been studied and their influence on locus cell firing. Amphetamine inhibits locus cell firing, and it appears to do so by activating a neuronal feedback loop. The effects of amphetamine on locus firing are partially blocked by chlorpromazine. L-amphetamine is equipotent with D-amphetamine in its inhibitory effects. α-Receptor agonists and antagonists also influence locus cell firing. This is due in part to the interaction of these adrenergic agents with autoreceptors on the norepinephrine cell bodies or dendrites in the locus. α2-Agonists such as clonidine and guanfacine cause a suppression of locus cell firing, and these inhibitory effects are reversed

---

Clonidine appears to be a very potent autoreceptor-stimulating drug. At higher doses clonidine will also stimulate postsynaptic receptors. Phenoxybenzamine and phenotolamine are effective α-receptor blocking agents. Recent experiments indicate that these drugs may also have some presynaptic α2 receptor blocking action. However, yohimbine and piperoxane are more selective as α2-receptor-blocking agents.

Site 5: *Reuptake:* Norepinephrine has its action terminated by being taken up into the presynaptic terminal. The tricyclic drug desipramine is an example of a potent inhibitor of this uptake mechanism.

Site 6: *Monoamine Oxidase (MAO):* Norepinephrine or dopamine present in a free state within the presynaptic terminal can be degraded by the enzyme MAO, which appears to be located in the outer membrane of mitochondria. Pargyline is an effective inhibitor of MAO.

Site 7: *Catchol-O-methyl transferase (COMT):* Norepinephrine can be inactivated by the enzyme COMT, which is believed to be localized outside the presynaptic neuron. Tropolone is an inhibitor of COMT. The normetanephrine (NM) formed by the action of COMT on NE can be further metabolized by MAO and aldehyde reductase to 3-methoxy-4-hydroxyphenylglycol (MHPG). The MHPG formed can be further metabolized to MHPG-sulfate by the action of a sulfotransferase found in the brain. Although MHPG-sulfate is the predominant form of this metabolite found in rodent brain, the free unconjugated MHPG is the major form found in primate brain.

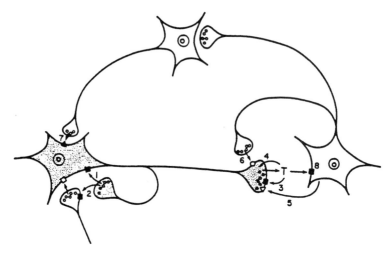

FIGURE 8-15. Neural pathways regulating synaptic homeostasis of locus ceruleus neurons. Influence of locus ceruleus neuron (shaded) on its target cells can be regulated both by modulation of transmitter release (1-7) and by amplification of signal provided by that release (8). This involves a large variety of cell surface receptors including receptors that respond to norepinephrine itself (■) as well as receptors responding to other chemical signals (○,□,▲,△). Principal pathways for regulating norepinephrine release: 1, direct action of recurrent collaterals onto soma; 2, indirect action of recurrent colaterals, mediated via influence on presynaptic afferents; 3, direct action of norepinephrine on presynaptic terminal; 4, alterations in rate of norepinephrine reuptake; 5, humoral signals generated by target; 6, neural signals providing short-loop negative feedback from target; 7, neural signals providing long-loop negative feedback from target; and 8, extent to which signal is amplified can be modulated by short-term modification of the sensitivity of the target, by long-term changes in number of receptors, and by other means such as release of cotransmitter. (Modified from Stricker and Zigmond, 1986)

by α2-antagonists. Piperoxane, yohimbine, and idazoxane, α2-antagonists, cause a marked increase in single cell activity. Morphine and morphine-like peptides such as enkephalins and endorphins have been shown to exert an inhibitory influence on locus cell firing. The inhibitory effects of morphine or enkephalin are reversed by

TABLE 8-2. Correlation Between Brain Levels of MHPG and Electrophysiological Activity of Noradrenergic Neurons in the Locus Ceruleus

Drug or Experimental Condition	Change in Locus Ceruleus Unit Activity	Change in Brain MHPG
Alpha$_2$-agonists	Decrease	Decrease
Clonidine		
Guanfacine		
Alpha$_2$-antagonists	Increase	Increase
Yohimbine		
Piperoxane		
Idazoxane		
Tricyclic antidepressants	Decrease	Decrease
Desipramine		
Imipramine		
Amitriptyline		
Amphetamine	Decrease	Decrease, increase (dose dependent)
Morphine	Decrease	Increase
Methylxanthines	Increase	Increase
Isobutylmethylxanthine		
Naloxone	No change	No change
Naloxone-precipitated morphine withdrawal	Increase	Increase
Stress	Increase	Increase
Electrical stimulation LC	Increase	Increase
Transection of dorsal NE bundle	Decrease	Decrease

the opiate antagonist naloxone but are uninfluenced by α2-antagonists. Methylxanthines such as isobutyl-methylxanthines (IBMX), which induce a quasi-opiate withdrawal syndrome, cause an increase in locus cell firing; this increase is antagonized by α2-agonists such as clonidine. Locus ceruleus cells in chronic morphine-treated rats dramatically increase locus cell firing during naloxone-induced withdrawal, and this increase in firing is suppressed by clonidine. These animal studies, viewed in conjunction with the clinical data demonstrating the effectiveness of treating opiate withdrawal with clonidine, have suggested that norepinephrine hyperactivity may be an important component of the opiate withdrawal syndrome in humans. These studies have thus provided new insight into the rational design of a new class of drugs (α_2-agonists) to treat opiate withdrawal, although the role of the locus ceruleus in this treatment effect has not actually been established.

Numerous pharmacological studies, mostly conducted in rodents, have demonstrated that there is a good correlation between drug-induced changes in the firing rate of locus ceruleus neurons monitored by extracellular single-unit recording and alterations in brain levels of MHPG. For example, drug-induced suppression of central noradrenergic activity produced by administration of clonidine or tricyclic antidepressants is accompanied by a reduction in MHPG. The α_2-antagonists (e.g., idazoxane, piperoxane, or yohimbine) or experimental conditions (e.g., stress or naloxone-precipitated withdrawal) that cause an increase in noradrenergic activity produce an increase in the brain levels of MHPG. The magnitude of the increase in MHPG produced by naloxone-precipitated morphine withdrawal exceeds that produced by administration of supramaximal doses of α_2-antagonists such as piperoxane or yohimbine. The biochemical observation is consistent with electrophysiological studies indicating that the increase in locus ceruleus cell activity produced during naloxone-precipitated withdrawal exceeds that achieved by administration of α_2-antagonists such as piperoxane and yohimbine. Since MHPG levels in brain measured under controlled experimental conditions provide an index of physiological activity in the locus ceruleus, measures of this metabolite in accessible body fluids may be useful for the assessment of changes in central norepinephrine function in intact animals or patients. Many clinical researchers are currently monitoring this metabolite in CSF, plasma, and urine in the hope that this measure will provide insight concerning alterations in central noradrenergic function.

Physiological Functions of Central Noradrenergic Neurons

Many functions have been proposed for the central norepinephrine neurons and their several sets of synaptic connections. Among the hypotheses that have the most supportive data are ideas concerning their role in affective psychoses (described below), in learning and memory, in reinforcement, in sleep–wake cycle regulation, and in the anxiety–nociception hypothesis. It has also been suggested that a major function of central noradrenergic neurons is not on neuronal activity and related behavioral phenomena at all, but rather a more general role in cerebral blood flow and metabolism. However, data available at present fit better into a more general proposal: The main function of the locus ceruleus and its projections is to determine the brain's global orientation concerning events in the external world and within the viscera. Such a hypothesis of central noradrenergic neuron function has been generated by observations of locus ceruleus unit discharge in untreated, awake, behaviorally responsive, rats and monkeys. These studies reveal the locus ceruleus units to be highly responsive to a variety of nonnoxious sensory stimuli and also reveal that the responsiveness of these units varies as a function of the animal's level of behavioral vigilance. Increased neuronal activity in the locus ceruleus is associated with unexpected sensory events in the subject's external environment while decreased noradrenergic activity is associated with behaviors that mediate tonic vegetative behaviors. Such a global-orienting function can also incorporate other aspects of presumptive function expressed by earlier data, but none of those more discrete functions can be documented as necessary or sufficient explanations of locus ceruleus function.

Pharmacology of Adrenergic Neurons

Limited experiments have suggested that the pharmacology of central adrenergic neurons is similar to that of central noradrenergic neurons. Agents that block tyrosine hydroxylase, DOPA-decarboxylase, and dopamine β-hydroxylase, lead to a reduction of both norepinephrine and epinephrine in the brain. Depleting agents such as reserpine, which cause release of norepinephrine and dopamine, also are effective in releasing epinephrine. Monoamine oxidase inhibitors cause an elevation of norepinephrine, dopamine, and epinephrine; but inhibitors of the A form of monoamine oxidase are much more effective in elevation of epinephrine. In fact, most of the pharmacological data are consistent with the hypothesis that, at least in the rat hypothalamus, oxidative deamination is an important metabolic process by which epinephrine is degraded and that type A monoamine oxidase is predominantly involved in this degradation.

Similar to observations made in central noradrenergic neurons, α2-agonists such as clonidine, decrease epinephrine formation and α2-antagonists increase epinephrine turnover. These data are also consistent with the possibility that α2-receptors are involved in the regulation of synthesis and release of epinephrine and perhaps also in the control of the functional activity of adrenergic neurons.

SELECTED REFERENCES

Amara, S. G. (1995). Monoamine transporters: Basic biology with clinical implications. *Neurosci.* 1(5):259–268.

Barker, E. L. and R. D. Blakely (1995). Norepinephrine and serotonin transporters: Molecular targets of antidepressant drugs. In *Psychopharmacology: The Fourth Generation of Progress* (F. E. Bloom and D. J. Kupfer, eds.), pp. 321–334, Raven Press, New York.

Bönisch, H. and M. Brüss (1993). The noradrenaline transporter of the neuronal plasma membrane. *Ann. N.Y. Acad. Sci.* 193–202.

Bylund, D. B., D. C. Eikenberg, J. P. Hieble, S. Z. Langer, R. J. Lefkowitz, K. P. Minneman, P. B. Molinoff, R. R. Ruffolo, Jr., and U. Trendelenburg (1994). IV. International Union of Pharmacology Nomenclature of Adrenoceptors. *Pharmacol. Rev.* 46(2), 121–136.

Charney, D. S., J. D. Bremner, and D. E. Redmond, Jr. (1995). Noradrenergic neuronal substrates for anxiety and fear: Clinical associations based on preclinical research. In *Psychopharmacology: The Fourth Generation of Progress* (F. E. Bloom and D. J. Kupfer, eds.), pp. 387–396, Raven Press, New York.

Chumpradit, S., M.-P. Kung, C. Panyachotipun, V. Prapansiri, C. Foulon, B. P. Brooks, S. A. Szabo, S. Tejani-Butt, A. Frazer, and H. F. Kung (1992). Iodinated tomoxetine derivatives as selective ligands for serotonin and norepinephrine uptake sites. *J. Med. Chem.* 35(23), 4492–4497.

Dalhström, A. and A. Carlsson (1986). Making visible the invisible. In *Discoveries in Pharmacology*, Vol 3, *Pharmacology Methods, Receptors and Chemotherapy* (M. J. Parnham and J. Bruinvels, eds.), pp. 97–125, Elsevier Science Publishers, B. V, Amsterdam.

Duman, R. S. and E. J. Nestler (1995). Signal transduction pathways for catecholamine receptors. In *Psychopharmacology: The Fourth Generation of Progress* (F. E. Bloom and D. J. Kupfer, eds.), pp. 303–320, Raven Press, New York.

Euler, U. S. von (1956). *Noradrenaline.* Charles C. Thomas, Springfield, IL.

Foote, S. L. and G. S. Aston-Jones (1995). Pharmacology and physiology of central noradrenergic systems. In *Psychopharmacology: The Fourth Generation of Progress* (F. E. Bloom and D. J. Kupfer, eds.), pp. 335–246, Raven Press, New York.

Fuller, R. W. (1982) Pharmacology of brain epinephrine neurons. *Annu. Rev. Pharmacol. Toxicol.* 22, 31–55.

Gibson, C. J. (1985). Control of monoamine synthesis by precursor availability. In *Handbook of Neurochemistry*, 2nd ed. (A. Lajtha, ed.), Vol 8, p. 309, Plenum, New York.

Goldstein, M. (1995). Long- and short-term regulation of tyrosine hydroxylase. In *Psychopharmacology: The Fourth Generation of Progress* (F. E. Bloom and D. J. Kupfer, eds.), pp. 189–196, Raven Press, New York.

Hökfelt, T., O. Johansson, and M. Goldstein (1984). Central catecholamine neurons as revealed by immunohistochemistry with special reference to adrenaline neurons. In *Handbook of Chemical Neuroanatomy*, Vol. 2, part I, *Classical Transmitters in the CNS* (A. Bjorklund, and T. Hökfeld, eds.), Elsevier, Amsterdam.

Holmes, P. V. and J. N. Crawley (1995). Coexisting neurotransmitters in central noradrenergic neurons. In *Psychopharmacology: The Fourth Generation of Progress* (F. E. Bloom and D. J. Kupfer, eds.), pp. 347–354, Raven Press, New York.

Lefkowitz, R. J., S. Cotecchia, M. A. Kjelsberg, J. Pitcher, W. J. Koch, J. Inglese, and M. G. Cargon (1993). Adrenergic receptors: Recent insights into their mechanism of activation and desensitization. In *Advances in Second Messenger and Phosphoprotein Research*, Vol 28 (B. L. Brown and P.R.M. Dobson, eds.), pp. 1–9, Raven Press, New York.

Lindvall, O. and A. Björklund (1983). Dopamine and norepinephrine containing neuron systems: Their anatomy in the rat brain. In *Chemical Neuroanatomy* (P. C. Emson, ed.), Raven Press, New York.

Maas, J. W. (ed.) (1983). *MHPG: Basic Mechanisms and Psychopathology.* Academic Press, New York.

Mefford, I. N. (1988). Epinephrine in mammalian brain. *Prog. Neuro-Psychopharmacol. Biol. Psychiatry* 12, 365–388.

Melikian, H. E., J. K. McDonald, H. Gu, G. Rudnick, K. R. Moore, and R. D. Blakely (1994). Human norepinephrine transporter: Biosynthetic studies using a site-directed polyclonal antibody. *J. Biol. Chem.* 269(16), 12290–12297.

Milligan, G., P. Svoboda, C. M. Brown (1994). Why are there so many adrenoceptor subtypes? *Biochem. Pharmacol.* 48(6): 1059–1071.

Moore, R. Y. and F. E. Bloom (1979). Central catecholamine neuron systems: Anatomy and physiology of the norepinephrine and epinephrine systems. *Annu. Rev. Neurosci.* 2, 113.

Pacholczyk, T., R. D. Blakely, and S. G. Amara (1991). Expression cloning of a cocaine- and antidepressant-sensitive human noradrenaline transporter. *Nature* 250, 350–353.

Redmond, D. E., Jr. (1987). Studies of the nucleus locus coeruleus in monkeys and hypothesis for neuropsychopharmacology. In *Psychopharmacology: The Third Generation of Progress* (H. Y. Meltzer, ed), pp. 967–975, Raven Press, New York.

Robbins, T. W. and B. J. Everitt (1995). Central norepinephrine neurons and behavior. In *Psychopharmacology: The Fourth Generation of Progress* (F. E. Bloom and D. J. Kupfer, eds.), pp. 363–372, Raven Press, New York.

Schuldiner, S., A. Shirvan, and M. Linial (1995). Vesicular neurotransmitter transporters: From bacteria to humans. *Physiol. Rev.* 75(2), 369–392.

Valentino, R. J. and G. S. Aston-Jones (1995). Physiological and anatomical determinants of locus coeruleus discharge: Behavioral and clinical implications. In *Psychopharmacology: The Fourth Generation of Progress* (F. E. Bloom and D. J. Kupfer, eds.), pp. 373–386, Raven Press, New York.

Zigmond, R. E., M. A. Schwarzschild, and A. R. Rittenhouse (1989). Acute regulation of tyrosine hydroxylase by nerve activity and by neurotransmitters via phosphorylation. *Annu. Rev. Neurosci.* 12, 415–461.

9 | Dopamine

Dopamine is a relative newcomer to the field of monoamine transmitters in mammalian brain. Until the mid-1950s dopamine was exclusively considered to be an intermediate in the biosynthesis of the
catecholamines norepinephrine and epinephrine. Significant tissue
levels of dopamine were first demonstrated in peripheral organs of
ruminant species. A short time later, Montagu, Carlsson, and
coworkers found that dopamine was also present in the brain in
about equal concentrations to those of norepinephrine. The very
marked differences in regional distribution of the two catecholamines, dopamine and norepinephrine, both within the CNS
and in bovine peripheral tissue, led Swedish investigators to propose
a biological role for dopamine independent of its function as a precursor for norepinephrine biosynthesis. Studies demonstrating that
most brain dopamine is confined to the basal ganglia led to the hypothesis that it might be involved in motor control and that decreased striatal dopamine could be the cause of extrapyramidal
symptoms in Parkinson's disease. The discovery of profound depletions of dopamine in the striatum of parkinsonian patients and the
demonstration that L-DOPA has beneficial effects in these patients
substantiated the clinical relevance of this theory. These and other
largely pharmacological studies were the impetus for the almost explosive developments in dopamine research during the past three
decades. With the development of a fluorescent histochemical
method for the visualization of catecholamines in freeze-dried tissue
sections and the application of this technique to brain tissue, the
anatomy of brain dopamine systems could also be described.

Dopaminergic Systems

The central dopamine-containing systems are considerably more
complex in their organization than the noradrenergic systems. Not
only are there many more dopamine cells (the number of mesen-

cephalic dopamine cells has been estimated at about 15,000–20,000 on each side, while the number of noradrenergic neurons in the entire brainstem is reported to be about 5000 on each side), there are also several major dopamine-containing nuclei as well as specialized dopamine neurons that make extremely localized connections within the retina and olfactory bulb. When dopamine systems were first visualized in the CNS, it was thought that all dopamine cells within the zona compacta of the substantia nigra projected to the caudate putamen nucleus. Dopamine cells in the ventral tegmental area surrounding the interpeduncular nucleus were believed to project exclusively to parts of the limbic system. This beautiful simplicity lasted a relatively short time. Soon, dopamine inputs to the cortex were discovered and within a few years, primarily through the use of retrograde tracing techniques, it became apparent that dopamine cells within the A8, A9, and A10 area, form an anatomically heterogeneous population in terms of their projection areas (Björklund and Lindvall, 1984). Retrograde tracing techniques also led to another important discovery—that dopamine cells project topographically to the areas that they innervate. Thus, although there is some overlap, those dopamine cells nearest to each other in a given area are more likely to innervate a common region than are dopamine cells distant from each other.

Based on this progress in detailing the anatomy of the dopamine systems (Fig. 9-1), it is convenient to consider these systems under three major categories in terms of the length of the efferent dopamine fibers.

1. *Ultrashort Systems.* Among the ultrashort systems are the *interplexiform amacrine-like neurons*, which link inner and outer plexiform layers of the retina, and the *periglomerular dopamine cells* of the olfactory bulb, which link together mitral cell dendrites in separated adjacent glomeruli.

2. *Intermediate-Length Systems.* The intermediate-length systems include (a) the *tuberohypophysial dopamine cells*, which project from arcuate and periventricular nuclei into the intermediate lobe of the pituitary and into the median emi-

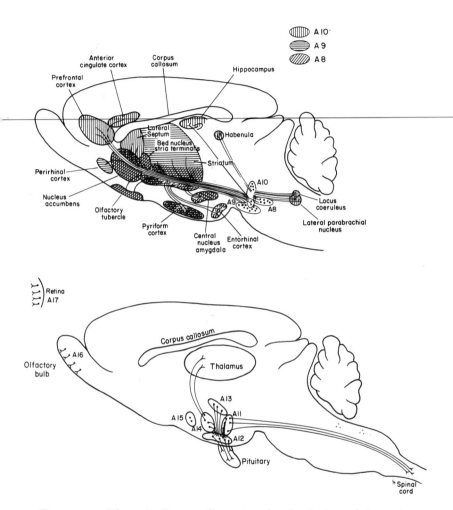

FIGURE 9-1. Schematic diagram illustrating the distribution of the main central neuronal pathways containing dopamine. The stippled regions indicate the major nerve terminal areas and their cell groups of origin. The cell groups in this figure are named according to the nomenclature of Dahlström and Fuxe (1965).

nence (often referred to as the tuberoinfundibular system); (b) the *incertohypothalamic neurons*, which link the dorsal and posterior hypothalamus with the dorsal anterior hypothalamus and lateral septal nuclei; and (c) the *medullary periventricular* group, which includes those dopamine cells in the perimeter of the dorsal motor nucleus of the vagus nerve, the nucleus tractus solitarius, and the cells scattered in the tegmental radiation of the periaqueductal grey matter.

3. *Long-Length Systems.* The long-length systems are the long projections linking the ventral tegmental and substantia nigra dopamine cells with three principal sets of targets: the *neostriatum* (principally the caudate and putamen); the *limbic cortex* (medial prefrontal, cingulate, and entorhinal areas); and *other limbic structures* (the regions of the *septum, olfactory tubercle, nucleus accumbens septi, amygdaloid complex,* and *piriform cortex*). These latter two groups have frequently been termed the *mesocortical* and *mesolimbic dopamine* projections, respectively. Under certain conditions these limbic target systems, when compared to the nigrostriatal system, exhibit some unique pharmacological properties, which are discussed in detail in the latter portion of this chapter.

With regard to cellular analysis of function, almost all reported data are from studies of the nigrostriatal dopamine projection. Here most studies using the iontophoretic administration of dopamine indicate that the predominant qualitative response is inhibition. This effect, like that of apomorphine and cAMP, is potentiated by phosphodiesterase inhibitors, providing the physiological counterpart to the second messenger hypothesis suggested by biochemical studies. On the other hand, the effects on the properties of caudate neurons when electrical stimulation is applied to the ventral tegmentum and substantia nigra are considerably less homogeneous, ranging from excitation to inhibition with considerable variations in latency and sensitivity to dopamine antagonism. In one study neither the excitations nor the inhibitions elicited by nigral stimulation were prevented by 6-hydroxy-dopamine–induced destruction of the dopamine cell bodies. Unambiguous pharmacological and electrophysiological analyses of the dopamine pathway thus remain to be accomplished. Although at present the bulk of the evidence favors an inhibitory role for dopamine, inquisitive students will want to examine the arguments raised by both sides and make their own evaluations of the effects reported. Important basic information is needed to rule out the spread of current to nearby nondopamine tracts and to determine accurately the latency of conduction reported for these extremely fine unmyelinated fibers.

In the last decade, it has become apparent that midbrain dopamine neurons are quite heterogeneous in terms of their biochemistry, physiology, pharmacology, and regulatory properties when compared to the prototypic nigrostriatal system in which most of the earlier studies were performed. While midbrain dopamine neurons differ in a number of important ways, their functional organization generally reflects features of transmitter dynamics that are shared by all dopamine neurons. These features have been most thoroughly studied in the nigrostriatal pathway and are summarized below (see Fig. 9-2).

DOPAMINE SYNTHESIS

Dopamine synthesis, like that of all catecholamines in the nervous system, originates from the amino acid precursor tyrosine, which must be transported across the blood–brain barrier into the dopamine neuron. A number of conditions can affect tyrosine transport, diminishing its availability and thus altering dopamine formation. The rate-limiting step in dopamine synthesis once tyrosine gains entry into the neuron is the conversion of L-tyrosine to L-dihydroxyphenylalanine (L-DOPA) by the enzyme tyrosine hydroxylase. DOPA is subsequently converted to dopamine by L-aromatic amino acid decarboxylase. This latter enzyme turns over so rapidly that DOPA levels in the brain are negligible under normal conditions. Because of the high activity of this enzyme and the low en-

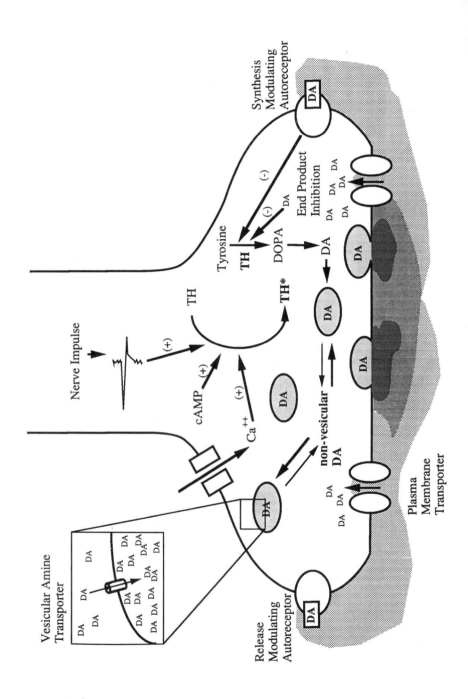

FIGURE 9-2. Schematic model of a prototypic dopaminergic nerve terminal illustrating the life cycle of dopamine and the mechanisms that modulate dopamine synthesis, release, and storage. Invasion of the terminal by a nerve impulse results in the Ca^{2+}-dependent release of dopamine. This release process is attenuated by release-modulating autoreceptors. Increased impulse flow also stimulates tyrosine hydroxylation. This appears to involve the phosphorylation of tyrosine hydroxylase (TH), resulting in the conversion to an activated form with greater affinity for tetrahydrobiopterin cofactor and reduced affinity for the end-product inhibitor dopamine. The rate of tyrosine hydroxylation can be attenuated by (a) activation of synthesis-modulating autoreceptors, which may function by reversing the kinetic activation of TH; and (b) end-product inhibition by intraneuronal dopamine that competes with cofactor for a binding site on TH. Release- and synthesis-modulating autoreceptors may represent distinct receptor sites. Alternatively, one site may regulate both functions through distinct transduction mechanisms. The plasma membrane dopamine transporter is a unique component of the dopamine terminal that serves an important physiological role in the inactivation and recycling of dopamine release into the synaptic cleft. The vascular amine transporter (VAT) transports cytoplasmic dopamine into storage vesicle, decreasing cytoplasmic concentration of dopamine and preventing metabolism by monoamine oxidase. VAT modulates the concentration of free dopamine in the nerve terminals.

dogenous levels of DOPA normally present in the brain, it is possible to enhance dramatically the formation of dopamine by providing this enzyme with increased amounts of substrate. Since the levels of tyrosine in the brain are relatively high and are above the K_m for tyrosine hydroxylase, under normal conditions it is not feasible to augment dopamine synthesis significantly by increasing brain levels of this amino acid. Since tyrosine hydroxylase is the rate-limiting enzyme in the biosynthesis of dopamine, this enzyme sets the pace for the formation of dopamine and is particularly susceptible to physiological regulation and pharmacological manipulation. Endogenous mechanisms for regulating the rate of dopamine synthesis in dopamine neurons primarily involve modulation of tyrosine hydroxylase activity. The activity of this key enzyme is subject to four major regulatory influences:

1. Dopamine and other catecholamines function as end-product inhibitors of tyrosine hydroxylase (TH) by competing with a tetrahydrobiopterin (BH-4) cofactor for a binding site on the enzyme.
2. The availability of BH-4 may also play a role in regulating TH activity. Tyrosine hydroxylase can exist in two kinetic forms that exhibit different affinities for BH-4. The conversion from low- to high-affinity forms is thought to involve direct phosphorylation of the enzyme, and the proportion of TH molecules in the high-affinity state appears to be a function of the state of neuronal firing.
3. Presynaptic dopamine receptors also modulate the rate of tyrosine hydroxylation. These receptors are activated by dopamine released from the nerve terminal, resulting in feedback inhibition of dopamine synthesis. Autoreceptors can modulate both synthesis and release of dopamine and represent important sites for the pharmacological manipulation of dopaminergic function by dopamine agonists and antagonists.
4. Dopamine synthesis also depends on the rate of impulse flow in the nigrostriatal pathway. During periods of in-

creased impulse flow, the rate of tyrosine hydroxylation is increased primarily through a kinetic activation of tyrosine hydroxylase that increases its affinity for BH-4 and decreases its affinity for the normal end-product inhibitor, dopamine. Under conditions of increased impulse flow, tyrosine hydroxylation is also more susceptible to precursor regulation by tyrosine availability.

Dopamine Release

Calcium-dependent release of dopamine from the nerve terminal is thought to occur in response to invasion of the terminal by an action potential. The extent of dopamine release appears to be a function of the rate and pattern of firing. The burst-firing mode leads to an enhanced release of dopamine. Dopamine release is also modulated by presynaptic release-modulating autoreceptors. In general, dopamine agonists inhibit while dopamine antagonists enhance the evoked release of dopamine.

Dopamine Uptake and the Dopamine Transporter

Dopamine-nerve terminals possess high-affinity dopamine uptake sites that are important in terminating transmitter action and in maintaining transmitter homeostasis. Uptake is accomplished by a membrane carrier that is capable of transporting dopamine in either direction, depending on the existing concentration gradient. The dopamine transporter is a unique component of the functioning dopamine nerve terminal, which plays an important physiological role in the inactivation and recycling of dopamine released into the synaptic cleft by actively pumping extracellular dopamine back into the nerve terminal. Substantial progress has been made over the past decade in studying the dopamine transporter. An important advance in dopamine uptake research was the development of a new class of very potent and selective dopamine uptake inhibitors, the GBR series. Figure 9-3 illustrates the chemical structure of some of the dopamine uptake inhibitors.

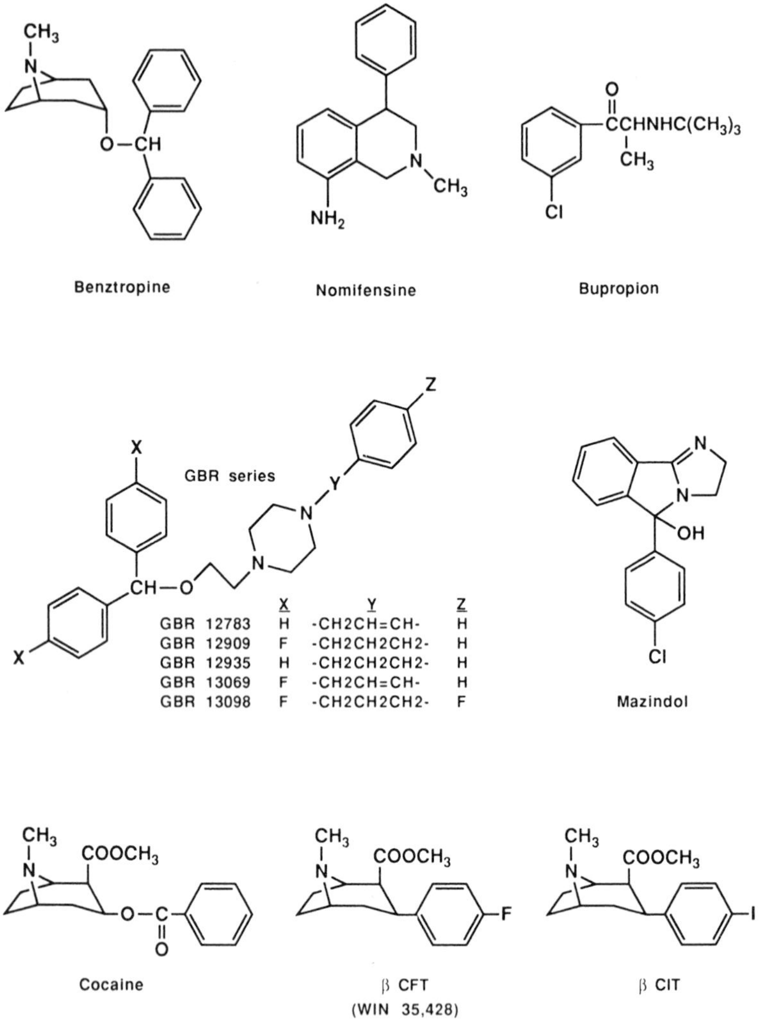

FIGURE 9-3. Chemical structures of some dopamine uptake inhibitors.

With recent developments in the isolation and structural elucidation of a variety of membrane-bound receptors and the availability of several potent and selective radioligands for the dopamine uptake site, the stage was set for the isolation and molecular characterization of the dopamine transporter. This was successfully achieved in 1991 by several groups.

A complementary DNA encoding a rat dopamine transporter (DAT) has been isolated that exhibits high sequence similarity with the previously cloned norepinephrine and γ-aminobutyric acid transporters. The DAT is a 619 amino acid protein with 12 hydrophobic putative membrane spanning domains and is a member of the family of Na^+/Cl^--dependent plasma membrane transporters. Using the energy provided by the Na^+ gradient generated by the Na^+/K^+-transporting ATPase, the DAT recaptures dopamine soon after its release, thereby modulating its concentration in the synapse and its time-dependent interaction with both pre- and postsynaptic receptors. The molecular characterization and cloning of the rat, bovine, and human transporter for dopamine has been achieved, and these proteins are highly conserved between species. A schematic representation of the dopamine transporter showing proposed orientation in the plasma membrane and potential sites of glycosylation and phosphorylation is given in Fig. 9-4.

A number of studies have suggested that DAT is a useful phenotypic marker for DA neurons and their nerve terminals and perhaps even better in some cases than tyrosine hydroxylase, which is present in all catecholaminergic neurons and highly regulated. Nevertheless, caution should be exercised in the use of this marker, since it is now clear that DAT expression varies significantly among DA cell groups and that during development embryonic midbrain DA neurons express dopamine and tyrosine hydroxylase well before the expression of DAT. It is noteworthy that the extensive heterogeneity in several classes of neurotransmitter receptors does not appear to be evident in the monoamine transporters, particularly DATs.

Both dopamine and norepinephrine transporters (DAT and NET) are members of the same family of substrate-specific, high-affinity, Na^+- and Cl^--dependent membrane transporters (see

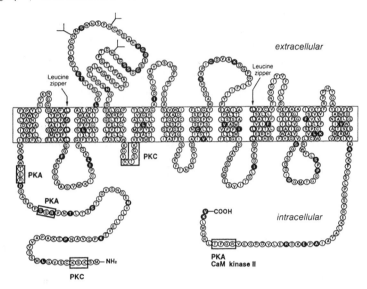

FIGURE 9-4. Schematic diagram of the primary amino acid sequence of the human dopamine transporter. Amino acid residues in *black* are those that differ between rat and human sequences. Y-shaped symbols represent potential sites for *N*-glycosylation. Boxed residues represent consensus phosphorylation site for protein kinase A (PKA), protein kinase C (PKC) or Ca^{2+}-calmodulin-dependent kinase II (CaM kinase II). Disposition of the transporter protein with respect to the plasma membrane is putative. Two leucine zipper motifs are indicated. (From Giros and Caron, 1993).

Chapter 8). Although these catecholamine transporters have highly similar sequences, they exhibit important differences in their selectivity for the catecholamines and for neurotoxins like 1-methyl-4-phenylpyridininium (MPP^+) and very distinct pharmacologies (cf. Figs. 9-3 and 8-4).

Mesolimbic dopamine neurons are implicated in the reinforcing properties of a variety of drugs of abuse, including psychomotor stimulants such as cocaine and amphetamine. Cocaine and related drugs bind to DAT and prevent dopamine transport in a fashion that correlates well with their behavioral reinforcing and psychomotor-stimulating properties. In fact, DATs have often been referred to as the brain's principal "cocaine receptors."

Receptor-binding studies have demonstrated that compounds that bind to the DAT also inhibit dopamine uptake with a rank order of potency proportional to the affinity demonstrated in binding studies. This relationship of uptake inhibitory potency and binding potency suggests that the two processes may be intimately linked and that any compound that binds to DAT will also block dopamine transport. However, point mutation studies of the cloned DAT indicate that reuptake inhibition and binding potency may be distinct processes that, under certain conditions are separable, not inextricably linked. These preliminary mutagenesis data suggest that it may be possible to develop agents that can prevent binding of stimulants like cocaine to the DAT while still allowing normal DA transport to ensue, thus supporting the feasibility of developing cocaine antagonists for the treatment of drug overdose, withdrawal, or addiction.

DAT has also assumed importance in the study of MPTP-induced and idiopathic Parkinson's disease. The selectivity of dopamine neurotoxins like MPTP seems to depend on their high affinity for the DAT. In primates, MPTP toxicity can be prevented by pretreatment with DAT inhibitors, but, once transported into the neuron, the toxin destroys the dopamine neurons, ultimately producing parkinsonism (see Chapter 13). It is noteworthy that expression of the cloned DAT in COS cells confers sensitivity to MPP^+ toxicity while expression of a vesicular transporter clone confers resistance to MPP^+ in sensitive CHO cells. Thus, the levels of vesicular transporter and DAT expression in combination could conceivably dictate the response to exogenous or endogenously generated neurotoxins. In this regard it is interesting that regional differences in the levels of DAT expression appear to correlate with the extent of DA cell loss after MPTP treatment or in Parkinson's disease.

Dopamine Metabolism

Released dopamine is converted to dihydroxyphenylacetic acid (DOPAC) by intraneuronal monoamine oxidase (MAO) after reuptake by the nerve terminal. Released dopamine is also converted to homovanillic acid (HVA), probably at an extraneuronal site through the sequential action of catechol-O-methyltransferase (COMT) and MAO. In rat brain, DOPAC is the major metabolite and considerable amounts of DOPAC and HVA are present in sulfate-conjugated as well as free forms. In humans and other primates, however, the major brain metabolite is HVA, and only a small amount is found in the conjugated form.

The primary metabolites of dopamine in the CNS are HVA, DOPAC, and a small amount of 3-methoxytyramine (3-MT). In dopamine systems, in contrast to norepinephrine systems (see Chapter 8), the acidic rather than the neutral metabolites appear to predominate. Accumulation of HVA in the brain or CSF has often been used as an index of the functional activity of dopaminergic neurons in the brain. 3-MT is also a useful index provided precautions are taken to minimize postmortem increases in this metabolite. Antipsychotic drugs, which increase the turnover of dopamine (in part because of their ability to increase the activity of dopaminergic neurons and to augment dopamine release), also increase the amount of HVA and 3-MT in the brain and CSF. In addition, electrical stimulation of the nigrostriatal pathway increases brain levels of HVA and 3-MT as well as the release of HVA into ventricular perfusates. In disease states such as parkinsonism, where there are degenerative changes in the substantia nigra and partial destruction of the dopamine neurons, a reduction of HVA is observed in the CSF. Similar changes are observed in MPTP-induced parkinsonism in humans and nonhuman primates.

In rat brain, it has been demonstrated that short-term accumulation of DOPAC in the striatum may provide an accurate reflection of activity in dopaminergic neurons of the nigrostriatal pathway. A cessation of impulse flow after the placement of acute lesions in the nigrostriatal pathway leads to a rapid decrease in striatal DOPAC. Conversely, the electrical stimulation of the nigrostriatal pathway results in a frequency-dependent increase in DOPAC within the striatum. Drugs that increase impulse flow in the nigrostriatal pathway, such as the antipsychotic phenothiazines, and butyrophenones, the anesthetics, and hypnotics, also increase striatal DOPAC. Drugs that block or decrease impulse flow, such as γ-hydroxybutyrate, 3-amino-1-hydroxypyrrolid-2-one (HA-966), apomorphine, and amphetamine, reduce DOPAC levels. Thus, there appears to be an excellent correlation between changes in impulse flow in dopaminergic neurons, which are induced either pharmacologically or mechanically, and changes in the steady-state levels of DOPAC. Unfortunately, in primates, DOPAC is a minor brain metabolite. Not only is it difficult to measure DOPAC in CSF, but in primate brain this metabolite is unresponsive to drug treatments that cause large changes in dopamine metabolism (i.e., HVA).

By means of the sensitive and specific technique of gas chromatography–mass fragmentography, it has been possible to measure accurately DOPAC, HVA, and their conjugates in rat and human plasma. Studies in rats have demonstrated that stimulation of the nigrostriatal pathway and administration of antipsychotic drugs produce an increase in plasma levels of DOPAC and HVA. Several studies in humans have also indicated that dopaminergic drugs can influence plasma levels of HVA, although the effect is quite modest in comparison to that observed in rodents. Thus, plasma levels of these metabolites may prove to be useful indices of central dopamine metabolism in several species but probably not in primates.

BIOCHEMICAL ORGANIZATION

The synthesis and release of dopamine is influenced by the activity of dopaminergic neurons, but in the CNS these neurons do not behave in a fashion completely analogous to peripheral or central noradrenergic neurons. Nor do all dopamine systems respond identically to neuronal activity. Increased impulse flow in the nigrostriatal or mesolimbic dopamine system does lead to both an increase in dopamine synthesis and turnover and a frequency-dependent increase in the accumulation of dopamine metabolites in the striatum and olfactory tubercle. This parallels other monoamine systems where an increase in impulse flow causes an increase in the synthesis and turnover of transmitter.

Short-term stimulation of central dopaminergic neurons increases tyrosine hydroxylation, the rate-limiting step in transmitter synthesis that is mediated in part by kinetic alterations in tyrosine hydroxylase. The tyrosine hydroxylase prepared from tissue containing the terminals of the stimulated dopamine neurons has an increased affinity for pteridine cofactor and a decreased affinity for the natural end-product inhibitor dopamine. As in central noradrenergic systems, it seems that a finite period of time is necessary for this activation to occur, and, once activated, the enzyme remains in this altered physical state for a short period after the simulation ends. The activation appears to involve phosphorylation of tyrosine hydroxylase.

However, if impulse flow is interrupted in the nigrostriatal or mesolimbic dopamine system, either mechanically or pharmacologically by treatment with γ-hydroxybutyrate, the neurons respond in a rather peculiar fashion by rapidly increasing the concentration of dopamine in the nerve terminals of the respective dopamine systems. Not only do the terminals accumulate dopamine by reducing release, but there is also a dramatic increase in the rate of dopamine synthesis. This increase occurs despite the steadily increasing concentration of endogenous dopamine within the nerve terminal.

The actual mechanisms whereby a cessation of impulse flow initiates changes in the properties of tyrosine hydroxylase are unclear, although many experiments have suggested that they may involve a decrease in the availability of intracellular calcium. Similar changes in the activity or properties of tyrosine hydroxylase are not observed in central noradrenergic neurons or in dopamine neurons (such as the mesoprefrontal dopamine neurons) lacking synthesis-modulating autoreceptors. At present, the physiological significance of this paradoxical response of certain dopaminergic neurons to a cessation of impulse flow is unclear. However, it is conceivable that these neurons may achieve some operational advantage by being able to increase their supply of transmitter rapidly during periods of quiescence.

POTENTIAL SITES OF DRUG ACTION ON DOPAMINERGIC NEURONS

There are many sites at which drugs can influence the function of dopamine neurons. The potential sites for modulation are illustrated in Fig. 9-5 and summarized in Table 9-1. For the purpose of this discussion, drug effects can be divided into three broad categories:

1. Nonreceptor-mediated effects on presynaptic function (Fig. 9-5).
2. Dopamine receptor-mediated effects.
3. Effects mediated indirectly as a result of drug interaction with other neurotransmitter systems that interact with dopamine neurons.

The relative importance of each of these potential sites of drug action will vary among different dopamine systems, depending on factors such as the presence or absence of autoreceptors, the efficiency of postsynaptic receptor-mediated neuronal feedback pathways, and the nature of the afferent inputs impinging on the dopamine neurons in question.

Nonreceptor-Mediated Effects

There are several stages in the life cycle of dopamine where drugs can influence transmitter dynamics, as illustrated in Fig. 9-5. There are many useful pharmacological tools for modifying dopaminergic activity and manipulating dopaminergic function, but most of these agents are not very selective for dopaminergic synapses and will interact with other catecholamine (norepinephrine and epinephrine) systems—and in some cases with other monoamine (5-HT) systems as well. For example, amphetamine, cocaine, benztropine, and nomifensine interact with the presynaptic transporter that normally

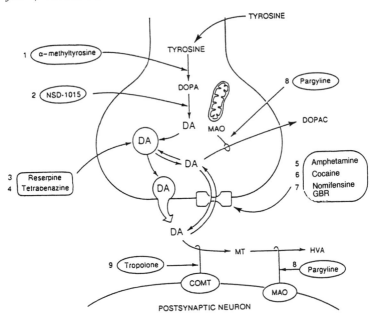

FIGURE 9-5. Schematic model of striatal dopaminergic nerve terminal. Drugs which alter dopamine (DA) life cycle include (1) α-methyltyrosine: competitive inhibitor of tyrosine hydroxylase; (2) NSD-1015: inhibitor of DOPA decarboxylase; (3) reserpine: irreversibly damages DA uptake-storage mechanisms and produces long-lasting depletion of DA; (4) tetrabenazine: also interferes with DA uptake-storage, but the effects are of shorter duration than reserpine and do not appear to be irreversible; (5) amphetamine: increases synaptic DA through a number of mechanisms, including induction of release of DA and blockade of DA reuptake; (6) cocaine: also blocks DA reuptake and induces DA release; (7) nomifensine and GBR: also block DA reuptake but lack DA-releasing ability; (8) pargyline: inhibitor of monoamine oxidase (MAO); (9) tropolone: inhibitor of catechol-O-methyltransferase (COMT). HVA, homovanillic acid; 3-MT, 3-methoxytyramine.

functions in the reuptake of released dopamine. However, these drugs also have an appreciable affinity for noradrenergic (and in some cases serotonergic) uptake sites.

The major exception to this generality is that there are drugs that target the specific monoamine transporters. Recently, uptake block-

TABLE 9-1. Potential Sites for Modulating Dopaminergic Function

Site	Consequences
Modulatory effects at dopamine (DA) receptors	
1. Stimulate postsynaptic DA receptors	A. Enhance dopaminergic transmission
	B. Enhance function of neuronal feedback loops
Block postsynaptic DA receptors	A. Block dopaminergic transmission
	B. Interfere with function of neuronal feedback loops
2. Stimulate presynaptic DA autoreceptors	A. Decrease DA synthesis and release
	A. Increase DA synthesis and release
Block presynaptic DA autoreceptors	A. Decrease firing rate and diminish DA output from nerve terminal
	B. Decrease somatodendritic DA turnover(?)
3. Stimulate somatodendritic DA autoreceptors	A. Interfere with feedback regulation of firing rate and DA output from terminal
Block somatodendritic DA autoreceptors	B. Interfere with feedback regulation of somatodendritic DA turnover
Modulatory effects at non-DA receptors on DA neurons	
4. Modify afferent input to cell body (i.e., block or mimic effects of transmitter released by afferent terminal)	A. Alter firing rate of DA cell and thus alter DA output from nerve terminal
	B. Alter somatodendritic turnover of DA (and co-localized peptides?)
5. Modify afferent input to nerve terminal	A. Alter release from nerve terminal of DA (and co-localized peptides?)

DA AUTORECEPTOR
POSTSYNAPTIC DA RECEPTOR

ers in the GBR series have been characterized that appear to be highly selective for the dopamine transport complex and should prove to be valuable experimental tools, perhaps leading to the development of therapeutic agents that can selectively augment dopamine function or be useful as diagnostic aids for visualization of the integrity of dopamine systems in vivo. In fact, striking results have been obtained with several new cocaine derivatives such as 3β-(4-fluorophenyl)tropane-2β-carboxylate (CFT) and 3β-(4-iodophenyl)tropane-2β-carboxylate (β-CIT), which exhibit high affinity for the dopamine transporter (see Fig. 9-3). These agents have been used in autoradiographic experiments and in PET and SPECT studies to image the striatal dopamine transporter in both normal and parkinsonian monkeys and humans. These studies have demonstrated (1) loss of striatal dopamine transporters in both experimental and idiopathic Parkinson's disease and (2) the restoration of dopamine transporter density and improvement of behavioral functions following nigral grafts in the caudate of transplanted MPTP monkeys. DAT ligands used for in vivo imaging in the future may be routinely employed in the diagnosis of certain diseases like parkinsonism and following the progression of the disease and the response to treatment. Imaging of DAT may also prove very useful for monitoring the viability of dopamine grafts and their outgrowth once transplanted into parkinsonian recipients.

Dopamine Receptor-Mediated Effects

Originally, it was thought that all drugs that affect dopamine activity, including the neuroleptics, worked through nonreceptor-mediated mechanisms such as those described above. However, it is now clear that many therapeutically important drugs interact with dopamine receptors. Drugs that affect dopamine receptors can be classified into two groups (Table 9-2):

1. Receptors on nondopamine cell types, which are usually referred to as postsynaptic receptors since they are postsynaptic to a dopamine-releasing cell.

2. Receptors on dopamine cells, which are termed autoreceptors to indicate their sensitivity to the neuron's own transmitter.

POSTSYNAPTIC DOPAMINE RECEPTORS

Postsynaptic dopamine receptors can be classified as either D_1 or D_2 receptor subtypes, based on functional and pharmacological criteria described below. Both types of receptors have been found in the projection areas of midbrain dopamine neurons, although it is unclear whether they are located on distinct subsets of dopamine-receptive cells in various projection fields. In the striatum, postsynaptic dopamine receptors regulate the activity of neuronal feedback pathways by which striatal neurons can communicate with dopamine cell bodies in the SN. This enables dopamine-innervated cells in the striatum to modulate the physiological activity of nigrostriatal dopamine neurons. In general, increased postsynaptic receptor stimulation results in decreased nigrostriatal dopamine activity.

Following chronic exposure to dopamine agonists or antagonists, postsynaptic dopamine receptors exhibit adaptive changes. For example, chronic exposure to dopamine antagonists or chemical denervation with 6-hydroxydopamine (6-OHDA) produces an increase in the number of dopamine-binding sites measured in receptor-binding assays. This may be related to the behavioral supersensitivity to dopamine agonists that also develops as a result of chronic antagonist administration or denervation. Conversely, repeated administration of dopamine agonists decreases the number of dopamine-binding sites and produces subsensitivity to subsequent administration of dopamine agonists in behavioral as well as biochemical and electrophysiological models. Changes such as these may be relevant to understanding the state of dopamine receptors in diseases believed to involve chronic dopaminergic hyper- or hypoactivity.

AUTORECEPTORS

Autoreceptors can exist on most portions of dopamine cells, including the soma, dendrites, and nerve terminals. Stimulation of

TABLE 9-2. Biochemistry, Physiology and Pharmacology of D_1 and D_2 Dopamine (DA) Receptors

D_1 Receptors Biochemical Manifestations Induced by Receptor Stimulation: - increase in cAMP formation -phosphorylation of DARPP-32		D_2 Receptors Biochemical Manifestations Induced by Receptor Stimulation: - decrease in cAMP formation or no change	
Location	Function	Location	Function
CNS: postsynaptic to DA neuron terminals dendrites (striatum, Nuc. Acc., Olf. tub., SN, etc.)	Enabling effect on behavioral and electrophysiological effect elicited by stim. of D_2 receptors. Function uncertain	Striatum and nuc. acc., DA nerve terminals	Autoreceptors inhibit DA synthesis and release and modulate turnover
		Retina	Mediate light-adaptive response of photoreceptors (↑ blink)
		SN and VTA: soma dendrites	Inhibits DA cell firing
Bovine parathyroid gland	Increases parathyroid hormone release	Striatum: cholinergic interneurons	Inhibits acetylcholine release
Vascular smooth muscle (canine renal and mesentery bed most used model system)	Vascular relaxation		

Vertebrate retina: in teleost, localized specifically to horizontal cells	Mediate responses of horizontal cells	Pituitary gland: anterior lobe	Inhibits cAMP and prolactin release. May also regulate Ca^{2+} channels
		Pituitary gland: intermediate lobe melanotrophs	Inhib. cAMP and αMSH release
		Chemosensitive trigger zone	Emesis
		Carotid body	Depresses spontaneous chemosensory discharge
		Sympathetic nerve terminals (numerous tissues)	Inhib. norepinephrine release
Pharmacology Selective agonists	SKF 38393 (partial agonist); dihydrexidine (full agonist); SKF 82526 (fenoldopam)	LY 171555 (quinpirole); RU 24926; [+] PHNO EMD-23-448 (autoreceptor selective)	
Selective antagonists	SCH 23390; SKF 83566; SCH 39166	(−)-Sulpiride; YM 09151-2; domperidone; raclopride	

dopamine autoreceptors in the somatodendritic region slows the firing rate of dopamine neurons, while stimulation of autoreceptors on the nerve terminals inhibits dopamine synthesis and release. Thus, somatodendritic and nerve terminal autoreceptors work in concept to exert feedback on dopaminergic transmission. Both somatodendritic and nerve terminal autoreceptors can be classified as D_2 receptors and exhibit similar pharmacological properties. Like postsynaptic receptors, somatodendritic and nerve terminal autoreceptors develop supersensitivity after chronic antagonist treatment or prolonged decreases in dopamine release and they desensitize in response to repeated administration of dopamine agonists. Interestingly, some evidence indicates that the autoreceptors are more readily desensitized than postsynaptic dopamine receptors. This has been suggested to play a role in the "on–off" effects observed during chronic L-DOPA therapy in Parkinson's disease.

Dopamine autoreceptors can be defined functionally in terms of the events they regulate, e.g., synthesis-modulating, release-modulating, impulse-modulating autoreceptors, and are therefore divided into three categories. However, it is not yet known whether distinct receptor proteins modulate each of these functions or whether the same receptor protein is coupled to each function through distinct transduction mechanisms. It is clear, however, that autoreceptor-mediated pathways for the regulation of dopamine release from the nerve terminal are distinct from autoreceptor-mediated pathways for the regulation of dopamine synthesis, since dopamine terminals in the prefrontal and cingulate cortices possess autoreceptors that regulate release but lack functional synthesis-modulating autoreceptors.

AUTORECEPTORS VERSUS POSTSYNAPTIC RECEPTORS: PHARMACOLOGICAL AND FUNCTIONAL CONSIDERATIONS

Autoreceptors and postsynaptic dopamine receptors differ in several ways. The most clear-cut difference is that autoreceptors are 5–10 times more sensitive to the effects of dopamine and apomorphine than postsynaptic dopamine receptors in behavioral, biochemical, and electrophysiological models. In the low-dose range, therefore, autoreceptor-mediated effects of dopamine agonists will predominate, resulting in diminished dopaminergic function, while higher doses will also stimulate postsynaptic receptors, leading to enhanced dopaminergic function. Autoreceptors also differ from postsynaptic receptors in their pharmacological profile. Dopamine agonists have recently been synthesized that are relatively selective for autoreceptors. As would be predicted, autoreceptor-selective agonists inhibit dopamine release, synthesis, and impulse flow in dopamine neurons, and elicit behavioral responses associated with diminished dopaminergic function. These agonists are very useful experimental probes for studying dopamine receptor function and may prove useful in diseases thought to involve excessive dopaminergic activity. Most recently, dopamine antagonists that selectively block autoreceptors have also been synthesized. By blocking dopamine autoreceptors they enhance dopamine function. Some of these agents appear to have a built-in ceiling on their response since, as the dose is increased, they exert an antagonistic action on postsynaptic dopamine receptors. The structures of several of these autoreceptor-selective agents are illustrated in Fig. 9-6.

Dopamine agonists and antagonists may act on several types of dopamine receptors to elicit biochemical changes in dopamine metabolism and alter the functional output of dopaminergic systems (Table 9-2). A drug's net effect on dopaminergic activity will depend on both its pre- and postsynaptic effects and the selectivity with which it acts at these different sites.

DRUG INTERACTIONS AT D_1 AND D_2 RECEPTORS

In the preceding discussion, dopamine receptors were broadly divided into presynaptic and postsynaptic categories. A second popular dopamine receptor classification that received increasing attention in the 1980s is based on the presence or absence of a positive coupling between the receptor and adenylate cyclase activity. On the basis of biochemical, physiological, and pharmacological studies, it is now well established that dopamine can act on at least two types

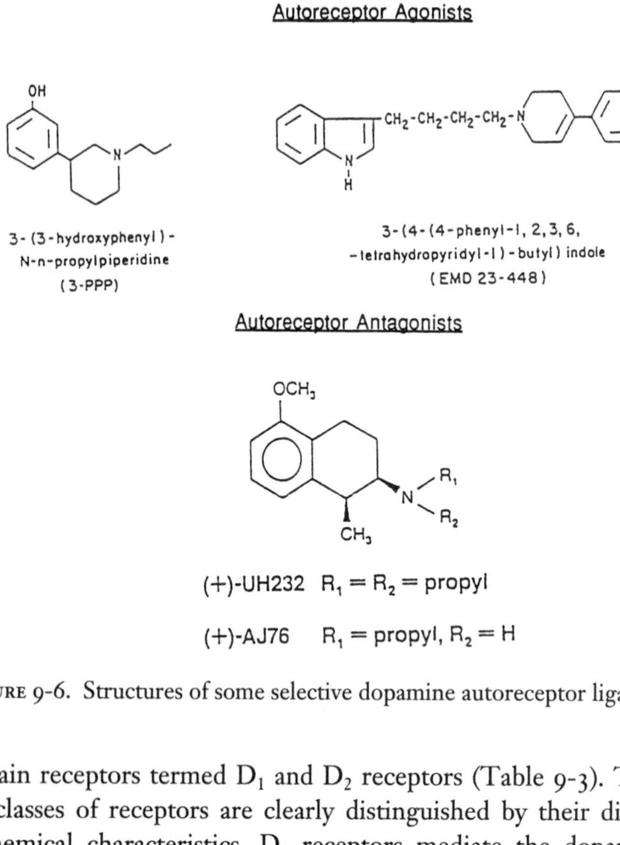
Autoreceptor Agonists
OH
N
CH2-CH2-CH2-CH2-N
N
H
3-(3-hydroxyphenyl)-
N-n-propylpiperidine
(3-PPP)
3-(4-(4-phenyl-1, 2, 3, 6,
-tetrahydropyridyl-1)-butyl) indole
(EMD 23-448)
Autoreceptor Antagonists
OCH3
R1
N
R2
CH3
(+)-UH232 R1 = R2 = propyl
(+)-AJ76 R1 = propyl, R2 = H

FIGURE 9-6. Structures of some selective dopamine autoreceptor ligands.

of brain receptors termed D_1 and D_2 receptors (Table 9-3). These two classes of receptors are clearly distinguished by their distinct biochemical characteristics. D_1 receptors mediate the dopamine-stimulated increase of adenylate cyclase activity. D_2 receptors are thought to mediate effects that are independent of D_1-mediated effects and also to exert an opposing influence on adenylate cyclase activity. The D_2 site is further characterized by picomolar affinity for antagonist, while the D_1 site is characterized by millimolar affinity for antagonist. The arrangements of D_1 and D_2 receptors and the nerve terminal autoreceptor are diagrammed in Fig. 9-7.

With the development of D_1 and D_2 selective agonists and antagonists, however, it has become common to rely on pharmacological characteristics when attempting to determine whether an effect is

TABLE 9-3. Signal Transduction Mechanisms Associated with
Dopamine Receptors

D_2 Receptors	D_1 Receptors
1. Inhibition of adenylate cyclase	Stimulation of adenylate cyclase
2. Inhibition of Ca^{2+} entry through voltage-sensitive Ca^{2+} channels	Stimulation of phosphoinositide turnover
3. Enhancement of K^+ conductances	
4. Modulation of phosphoinositide metabolism	

mediated by D_1 or D_2 receptors. The distinction between pharmacological and functional definitions is important because it is becoming clear that dopamine receptors with the same pharmacological characteristics do not necessarily have the same functional characteristics. For example, dopamine receptors with D_2 pharmacology are present in both the striatum and nucleus accumbens, but are coupled to an inhibition of adenylate cyclase only in the striatum. Regional differences in coupling between dopamine receptors and GTP-binding proteins have also been reported. Furthermore, many recent studies indicate that dopamine receptors can influence cellular function through mechanisms other than stimulating or inhibiting adenylate cyclase (Table 9-3). These may include direct effects on potassium and calcium channels, as well as the modulation of inositol phosphate production. It therefore seems unrealistic to expect equivalency between pharmacological and biochemical classifications of dopamine receptor subtypes.

Although D_1 and D_2 receptors can have opposite effects on adenylate cyclase activity, it is apparent that the physiological significance of their interaction is more complex. While D_1 and D_2 agonists can have opposite effects on oral movements, they also produce

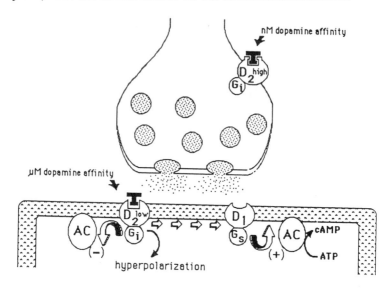

FIGURE 9-7. Schematic diagram depicting the anatomical arrangement of D_1 and D_2 receptors and the nerve terminal autoreceptor. The D_1 site is positively coupled with adenylate cyclase via a G_s protein. The D_2 postsynaptic site is negatively coupled with adenylate cyclase via a G_i protein, stimulation of which leads to hyperpolarization. The autoreceptors appear to exert their DA synthesis and release-modulating effects via a G_i regulatory protein. DA autoreceptors are sensitive to very low (nanomolar) concentrations of DA. Although it is not known whether the D_1 and D_2 receptors are situated on the same neuron, they appear to be functionally linked, as depicted by the arrows. Innervated postjunctional dopamine receptors operate in the low-affinity state (D_1 low and D_2 low) in contrast to the D_2 autoreceptor, which operates in the high-affinity state (D_2 high).

a synergistic increase in locomotor activity and behavioral stereotypes under certain circumstances. Electrophysiological experiments have suggested that D_1 receptor activation is required for full postsynaptic expression of D_2 effects. The interaction of the D_1 receptor with other neurotransmitter systems is also being explored. Most dopamine-receptive cells contain a protein, DARRP-32, whose phosphorylation is enhanced by the elevated levels of cAMP promoted through a D_1 mechanism. DARRP regulates the activity of a phosphatase, an enzyme that reverses the action of protein kinases. Through its effect on DARRP, the D_1 receptor may be able to influence the phosphorylation state of effector proteins that have been activated through calcium-calmodulin or protein kinase C mechanisms.

DYNAMICS OF DOPAMINE RECEPTORS

Destruction of the nigrostriatal dopamine systems has clear and reproducible behavioral consequences. Unilateral lesions of this system produce rotational behavior. Behavioral studies in lesioned rats indicate that dopamine receptors in the denervated striatum are supersensitive. Administration of dopamine agonists (such as apomorphine) that selectively stimulate dopamine receptors produces rotational behavior in rats with unilateral lesions of the nigrostriatal dopamine neurons. The degree of receptor sensitivity can be quantified by measuring the amount of rotational behavior. The number of dopamine receptors in the striatum ipsilateral to the lesion increases markedly, and this increase appears to correlate with the extent of the behavioral supersensitivity reflected by the rotational behavior. Thus, an increase in dopamine receptor density appears to be related to the behavioral supersensitivity observed following unilateral destruction of the nigrostriatal dopamine system.

Changes in the number of dopamine receptors are also observed in the striatum following chronic administration of dopamine antagonists. This had led to the speculation that serious side effects such as tardive dyskinesia following chronic treatment with a neuroleptic drug might be due to supersensitivity of dopamine receptors that have been chronically blocked.

Dopamine receptors have been observed to change in disease states. In schizophrenia, the density of the dopamine transporter and the D_1 dopamine receptor is normal. However, the D_2 receptor density is consistently elevated in postmortem studies of brain regions such as caudate and putamen, even in tissue obtained from neuroleptic-free individuals. Some preliminary evidence indicating abnormal D_2 structure as well as a reduced linkage between D_1 and D_2 receptors is available, warranting a detailed study of the genes for these two receptors in schizophrenia. The loss of midbrain dopamine in Parkinson's disease is accompanied by a matching loss of the dopamine transporter and a rise in both D_1 and D_2 receptor density. These alterations are found in the caudate nucleus and putamen tissues from unmedicated patients. Long-term treatment with L-DOPA appears to revert the receptor densities back toward normal levels. D_1 and D_2 dopamine receptors are decreased in the striatum of patients with Huntington's chorea, and there also appears to be a reduced or absent linkage between D_1 and D_2 receptors.

MOLECULAR BIOLOGY OF DOPAMINE RECEPTORS

Available evidence indicates that these two classes of dopamine receptors (D_1 and D_2) are distinct molecular entities, utilize different transducing units (Table 9-3), and have a different distribution in the brain (Table 9-2).

Recent developments in molecular biology, including cloning of the cDNA and/or genes for several members of the large family of G-protein-coupled receptors, have revealed that heterogeneity in the biochemical characteristics or pharmacology of individual receptors often indicates the presence of previously unsuspected molecular subtypes (see Chapter 11). For dopamine systems, even though the D_1/D_2 dopamine receptor classification is widely accepted, biochemical, pharmacological, and behavioral approaches over the past few years have produced data that are increasingly difficult to reconcile with the existence of only two dopamine receptor subtypes and suggest the presence of several novel subtypes of both D_1 and D_2 dopamine receptors. In fact, recent cloning studies have already identified four subtypes of D_2 receptors and two subtypes of D_1 receptors (Table 9-4). Two forms of the D_2 dopamine receptor—the D_2 short and D_2 long—were identified by gene cloning and shown to be derived from alternative splicing of a common gene. These two subtypes appear to have an identical pharmacology. A third sub-type of the D_2 receptor, termed a D_3 receptor, was isolated by the screening of rat brain cDNA and genomic libraries and a combination of reverse transcription and use of the polymerase chain reaction. This new D_3 dopamine receptor exhibits several novel characteristics. It has a different anatomical distribution, with the highest levels found in limbic brain structures, and its pharmacological profile, although similar to the D_2S and D_2L forms, shows some distinct differences; the D_3 receptor exhibits about a 100-fold increase in affinity for the dopamine agonist quinpirol.

The fourth subtype of the D_2 receptor recently cloned, the D_4 receptor gene, has high homology to the human D_2 and D_3 receptor genes. The pharmacological profile of this receptor resembles that of the D_2 and D_3 receptors, but its affinity for the atypical antipsychotic drug clozapine is an order of magnitude higher. The D_4 messenger RNA has an interesting regional distribution in monkey brain with high levels observed in frontal cortex, midbrain, amygdala, and medulla and lower levels detected in basal ganglia. The function of these D_2 receptor subtypes is presently unknown. All known varieties of the D_2 receptor have seven membrane-spanning domains similar to the structure originally proposed for β-adrenergic receptors. Differences in ligand binding and transduction mechanisms are presumably related to variations in the sequence of the receptor. The D_1 receptor of humans and rats has also been cloned, expressed, and characterized by several laboratories; this work in conjunction with other studies is consistent with the idea that other D_1 receptor subtypes may also exist. In fact, a gene encoding a 477 amino acid protein has recently been cloned that has a striking homology to the cloned D_1 receptor. This D_1 receptor subtype, called D_5, has a pharmacological profile similar to that of the cloned D_1 receptor but displays a 10-fold higher affinity for the endogenous agonist dopamine. Similar to the D_1 receptor, the D_5 receptor stimulates adenylate cyclase activity. This receptor is neuron specific and located primarily in the limbic areas of brain but is absent from the parathyroid, kidney, and adrenal gland.

TABLE 9-4. Comparison of Dopamine Receptor Subtypes

Receptor Isoforms	D_1	$D_{2(short)}$	$D_{2(long)}$	D_3	D_4	D_5
Chromosome	5q 35.1	11q 22–23		3q 13.3	11 p 15.5	4 p 15–16
Brain regions enriched	C/P	C/P	C/P	OT	FCX	TH
	OT	OT	OT	NA	Midbrain	Hi
	NA	NA	NA	IC	AMG	HYP
Posterior pituitary	Absent	Present	Present	Absent	?	Absent
Nigral dopamine cells	No	Yes	Yes	Yes	No	No
GTP regulation	Yes	Yes	Yes	Yes	Yes	No
Adenylyl cyclase	Stim.	Inhib.	Inhib.	Inhib.	Inhib.	Stim.
Affinity for dopamine	Micromolar	Micromolar	Micromolar	Nanomolar	Submicromolar	Submicromolar

Characteristic agonist	SKF-38393	Bromocriptine	Bromocriptine	7-OH-DPAT	?	SKF-38393
Characteristic antagonist	SCH-23390	Sulpiride	Sulpiride	UH 232	Clozapine	SCH-23390
Amino acids						
Rat	446	415	444	446	385	475
Human	446	414	443	400	387	477
Amino acid sequence homology in transmembrane						
Versus $D_{2(long)}$	44%	100%	100%	75%	53%	47%

C/P, caudate/putamen; OT, olfactory tubercle; NA, nucleus accumbens; IC, islands of Calleja; FCX, frontal cortex; AMG, amygdala; TH, thalamus; Hi, hippocampus; Hyp, hypothalamus.

DISTRIBUTION OF SUBTYPE-SPECIFIC DOPAMINE RECEPTOR MRNA
IN BRAIN

Recent advances in molecular biology have made it feasible to determine which specific cells are expressing a given gene, thus allowing the anatomical determination not only of what population of cells express a given gene but also of how these genes may be regulated in normal and pathological conditions. At least five genes encoding dopamine receptors have been discovered. Currently there are few if any pharmacological or immunological tools for measuring the distribution of the receptor proteins of the new dopamine receptors. Thus, our knowledge concerning the tissue distribution of these receptors, especially in cases where specific ligands or selective antibodies have yet to be developed, has come primarily from in situ hybridization experiments. The tissue distribution and characteristics of the mRNAs of the five different dopamine receptors are illustrated in Tables 9-4 and 9-5. These dopamine receptors appear to have overlapping as well as some unique anatomical distributions and in some cases distinct pharmacological profiles. In general, the distribution patterns found in rodents parallel those observed in primates, with several notable exceptions alluded to below. The D_1 and D_2 receptor mRNAs are present in all dopaminoceptive regions of rat brain. In brain regions such as the substantia nigra and ventral tegmental area, high levels of D_2 but not D_1 mRNA are detected. The absence of D_1 and D_5 receptor mRNA in the substantia nigra and ventral tegmental area argue against these receptors playing a role as autoreceptors. Receptor mRNAs for D_3, D_4, and D_5 receptors are largely present in tissues where D_1 and/or D_2 receptor mRNAs are also expressed. However, in most cases the relative abundance of mRNA for these receptors is one or several orders of magnitude lower than that found for D_1 and/or D_2 receptors. While the primate substantia nigra contains high levels of D_2 receptor mRNA, the ventral tegmental area in primate brain does not contain detectable levels of D_2 or D_3 receptor mRNA, suggesting that primate ventral tegmental area may not contain appreciable numbers of dopamine autoreceptors.

TABLE 9-5. Neuroanatomical Distribution of Dopamine Receptor mRNAs[a]

	D_1	D_2	D_3	D_4	D_5
Dopamine cell body regions (autoreceptors)					
Substantia nigra	o	+++	+	?+	o
Ventral tegmental area	o	+++	+	?+	o
Zona incerta	o	+++	o	o	o
Dopaminoceptive regions (postsynaptic receptors)					
Caudate putamen	+++	+++	+	o	o
Nucleus accumbens	+++	+++	+++	+	o
Septum	+	+	+	o	o
Olfactory tubercle	+++	+++	+	o	o
Amygdala	+++	+	+	+	o
Hippocampus	+	+	+	+	++
Cortex	+	+	+	+	o
Hypothalamus	+	+	+	+	+
Thalamus	+	+	+	+	++
Cerebellum	+	+	+	o	o

[a]Relative abundance of given mRNA in rat brain; +++, abundant expression; ++, moderate level of expression; +, low level of expression; o, no mRNA observed. (Data summary from Meador-Woodruff, 1994).

Anatomically, D_5 receptor mRNA has a rather discrete distribution in rat brain: it is found only in the hypothalamus, hippocampus, and the parafasicular nucleus of the thalamus. In the primate this distribution extends to other temporal lobe structures as well. The consensus of a number of studies is that the only brain region that expresses all five dopamine receptors is the hippocampus. Dopamine receptor mRNA has also been found outside the CNS. D_2 receptor mRNA is abundant in the pituitary and adrenal glands and in retina. Northern blot analyses have shown that neither D_1 nor D_3 receptor mRNA is detected outside the CNS, although, in kidney and heart,

D_1- and D_2-like activities have been described. Since low levels of D_5 receptor mRNA are expressed in the kidney, this could account for the D_1-like activity. The D_4 receptor mRNA recently found in rat heart may account for the D_2-like activity previously described in heart. The level of D_4 receptor mRNA found in rat brain is about 20-fold lower than the level in heart. In peripherally innervated tissue heart seems to be the exception, since no D_4 mRNA has been found in adrenal, kidney, or liver.

PHARMACOLOGY OF DOPAMINE RECEPTOR SUBTYPES

At present, no selective ligands have been developed that can distinguish between D_1 and D_5 receptors. Pharmacologically, the only characteristic that distinguishes D_1 from D_5 receptors is the increased affinity of the D_5 receptor for dopamine. Most neuroleptic drugs exhibit a higher affinity for D_2 dopamine receptors than for D_3 or D_4 receptors. The affinities of the five dopamine receptor subtypes for selected clinically relevant dopamine antagonists are summarized in Table 9-6.

The most interesting feature of the human D_4 receptor is its apparent high affinity for clozapine (an atypical neuroleptic) and its unique distribution in primate brain (frontal cortex > midbrain = amygdala > striatum), differing markedly from D_2 and D_3 receptor mRNA. This interesting pharmacology and unique distribution in brain generated a great deal of excitement, particularly from a clinical standpoint. The possibility that clozapine exerts its therapeutic effects via a D_4 receptor mechanism was seen immediately as offering a new and rational target for drug development. Seeman and colleagues provided tantalizing evidence that there may be an increase in the number of D_4 receptors in schizophrenic patients, further fueling the impetus to develop D_4 selective antagonists as potential antipsychotic drugs. But this provocative, albeit indirect, study needs to be replicated using more direct measures to assess D_4 receptor numbers in brains of normal and schizophrenic subjects. Also the significance of this finding, even if replicated, is still uncertain. The neuroleptics taken by patients throughout the course of

TABLE 9-6. Affinities of Clinically Relevant Antagonists for the
Five Dopamine Receptors

	K_1 values (nM)				
	D_1-like		D_2-like		
Ligand	D_1	D_5	D_2	D_3	D_4
Antagonists					
Chlorpromazine	90	130	3	4	35
Clozapine	170	330	230	170	21
Haloperidol	80	100	1.2	7	2.3
Nemonapride			0.06	0.3	0.15
Raclopride	18,000		1.8	3.5	2400
Remoxipride			300	1600	2800
Risperidone			5	6.7	7
SCH23390	0.2	0.3	1100	800	3000
Spiperone	350	3500	0.06	0.6	0.08
S-Sulpiride	45,000	77,000	15	13	1000

Dissociative constants (K_i) for ligands at the various dopamine receptors were derived from recent work by Seemen and Van T ol (1994).

their disease could modify dopamine receptor density, and the over-abundance of D_4 receptors observed in the autopsied brains of schizophrenic patients could be a result of drug treatment rather than a cause of the disease. Future clinical research efforts might profitably be directed to the use of in vivo imaging techniques (i.e., SPECT and PET) to evaluate dopamine receptor subtypes in schizophrenia when appropriate D_4- and D_3-selective ligands become available.

The identification of novel dopamine receptor subtypes has already had a dramatic impact on our understanding of dopaminergic systems. Recent focus on the human dopamine D_4 receptor indicates that the DNA sequence of this receptor is highly polymorphic at both the DNA and amino acid levels, exhibiting a least 25 alleles. A novel polymorphism of the D4 receptor was observed within the putative third cytoplasmic loop of the protein, suggesting that some polymorphic variants may display different pharmacological properties. This high frequency of variation in the coding region of a functional receptor protein is unprecedented and could confer differences in efficacy of drug treatment and/or predispose an individual to the development of dopamine-dependent neuropsychiatric disorders. The availability of receptor clones, receptor antibodies, and expressed receptor proteins has permitted gene mapping as well as in-depth studies of the circuitry of the dopaminergic systems and the mechanisms regulating them at both the genomic and cytoplasmic levels. It has also allowed the physical structure of the receptors to be ascertained and should permit the design and development of highly specific ligands. Hopefully, these new selective agents will be not only helpful in studying the function of dopamine systems in normal and pathological states but also useful in the therapeutic management of disorders associated with malfunction of specific dopaminergic systems. Strides toward this goal have already begun with the successful development of selective D_4 antagonists by several pharmaceutical companies. Perhaps before the next revision of this text we will know if these agents exhibit any therapeutic benefits in the treatment of schizophrenia.

PHARMACOLOGY OF DOPAMINERGIC SYSTEMS

Nigrostriatal and Mesolimbic Dopamine Systems

The nigrostriatal and mesolimbic dopamine neurons appear to respond in a similar manner to drug administration (Table 9-6). Acute administration of dopamine agonists (dopamine receptor stimulators) decrease dopamine cell activity, decrease dopamine turnover, and decrease dopamine catabolism. Acute administration of antipsychotic drugs (dopamine receptor blockers) increases dopaminergic cell activity, enhances dopamine turnover, increases dopamine catabolism, and accelerates dopamine biosynthesis. The increase in dopamine biosynthesis occurs at the tyrosine hydroxylase step and is explained in part as a result of the ability of antipsychotic drugs to block postsynaptic receptors and to increase dopaminergic activity via a neuronal feedback mechanism (Table 9-1). It is also apparent that some of the observed effects are enhanced as a result of interaction with nerve terminal autoreceptors. Blockade of nerve terminal autoreceptors increases both the synthesis and the release of dopamine. These systems respond to monoamine oxidase inhibitors (MAOI) with an increase in dopamine and a decrease in dopamine synthesis, as do the other dopamine systems discussed below.

Recently it has been shown that long-term treatment with antipsychotic drugs produces a different spectrum of effects on central dopaminergic neurons. For example, following long-term treatment with haloperidol, *nigrostriatal* dopamine neurons in the rat become quiescent, and dopamine metabolite levels and dopamine synthesis and turnover in the *striatum* return to normal limits. The kinetic activation of *striatal* tyrosine hydroxylase, shown to occur following an acute challenge dose of an antipsychotic drug, also subsides following long-term treatment. These results are usually interpreted as indicative of the development of tolerance in the nigrostriatal dopamine system. In contrast (see below), tolerance to the biochemical effects observed following acute administration of antipsychotic drugs does not appear to develop in the mesoprefrontal and mesocingulate cortical dopamine pathways after chronic administration.

Mesocortical Dopamine System

It has only become possible with the advent of more sensitive analytical techniques to measure changes in the low levels of dopamine and related metabolites in this system as well as in those discussed below. The response of the mesocortical dopamine systems to dopaminergic drugs is in most instances qualitatively similar to that of the nigrostriatal and mesolimbic systems (Table 9-7), although some notable exceptions have been recently observed. In fact, over the past few years the mesotelencephalic dopamine neurons, which were once believed to be three relatively simple and homogeneous

TABLE 9-7. Pharmacology of Central Dopaminergic Systems

Characteristics	Nigro-striatal	Meso-accumbal	Meso-prefrontal	Meso-piriform	Tubero-infundibular[a]	Tubero-hypophyseal[b]
Respond to DA antagonist (increase in synthesis, catabolism, and turnover)	Yes	Yes	Yes (small)	Yes	No	Yes (small)
Respond to DA agonists (decrease in synthesis, catabolism, and turnover)	Yes	Yes	Yes (small)	Yes	No	Yes
Respond to monoamine oxidase inhibitors (increase in DA, decrease in synthesis)	Yes	Yes	Yes	Yes	Yes	Yes
Presence of nerve terminal synthesis modulating autoreceptors	Yes	Yes	No	Yes	No	Yes

High-affinity DA transport	Yes	Yes	Yes	Yes	No	No
Respond to mild stress (increase in DA synthesis and catabolism, blocked by diazepam)	Yes	Yes? (small)	Yes	No	No	No
Respond to anxiogenic β-carbolines (increase in DA catabolism)	No	No	Yes	No	—	—

[a] Cell bodies of this group of neurons are located in the arcuate and periventricular nuclei, and their axons terminate in the external layer of the median eminence.

[b] Cell bodies of this group of neurons are located in the arcuate and periventricular nuclei and their axons terminate in the neurointermediate lobe of the pituitary (posterior pituitary).

systems, have been found to be an anatomically, biochemically, and electrophysiologically heterogeneous population of cells with differing pharmacological responsiveness. For example, although a great majority of midbrain dopamine neurons appear to possess autoreceptors on their cell bodies, dendrites, and nerve terminals, dopamine cells that project to the prefrontal and cingulate cortices appear to have either a greatly diminished number of these receptors or to lack them entirely. The absence (or insensitivity) of impulse-regulating somatodendritic as well as synthesis-modulating nerve terminal autoreceptors on the mesoprefrontal and mesocingulate cortical dopamine neurons may, in part, explain some of the unique biochemical, physiological, and pharmacological properties of these two subpopulations of midbrain dopamine neurons (Table 9-8). For example, the mesoprefrontal and mesocingulate dopamine neurons appear to have a faster firing rate and a more rapid turnover of transmitter than the nigrostriatal, mesolimbic, and *mesopiriform* dopamine neurons. Transmitter synthesis is also more readily influenced by altered availability of precursor tyrosine in midbrain dopamine neurons lacking autoreceptors (mesoprefrontal and mesocingulate) than in those midbrain dopamine neurons possessing autoreceptors. This may be related to the enhanced rate of physiological activity in this subpopulation of midbrain dopamine neurons, making them more responsive to precursor regulation. The mesoprefrontal and mesocingulate dopamine neurons also show a diminished biochemical and electrophysiological responsiveness to dopamine agonists and antagonists. Administration of low doses of apomorphine or autoreceptor selective dopamine agonists, in contrast to their inhibitory effect on other midbrain dopamine neurons, are ineffective in decreasing the activity or in lowering dopamine metabolite levels in these two cortical dopamine projections. Dopamine receptor-blocking drugs such as haloperidol produce large increases in synthesis and the accumulation of dopamine metabolites in the nigrostriatal, mesolimbic, and mesopiriform dopamine neurons, but have only a modest effect in the mesoprefrontal and mesocingulate dopamine neurons.

TABLE 9-8. Unique Characteristics of mesotelencephalic Dopamine Systems Lacking Autoreceptors (Mesoprefrontal and Mesocingulate) Compared to Those Possessing Autoreceptors (Mesopiriform, Mesolimbic, and Nigrostriatal)

1. A higher rate of physiological activity (firing) and a different pattern of activity (more bursting).
2. A higher turnover rate and metabolism of transmitter dopamine.
3. Greatly diminished biochemical and electrophysiological responsiveness to dopamine agonists and antagonists.
4. Lack of biochemical tolerance development following chronic antipsychotic drug administration.
5. Resistance to the development of depolarization-induced inactivation following chronic treatment with antipsychotic drugs.
6. Transmitter synthesis more readily influenced by altered availability of precursor tyrosine.

Heterogeneity among midbrain dopamine neurons is also found when one studies the effects of chronic antipsychotic drug administration. When classic antipsychotic drugs are administered repeatedly over time, the great majority of dopamine cells cease to fire due to the development of a state of depolarization inactivation. However, some midbrain dopamine cells appear to be unaffected by repeated antipsychotic drug administration. These dopamine cells turn out to be the neurons projecting to the prefrontal and cingulate cortices. Parallel findings are observed biochemically. Following chronic administration of antipsychotic drugs, tolerance develops to the metabolite-elevating effects of these agents in the midbrain dopamine systems that possess autoreceptors, but not in the systems that lack autoreceptors. When the atypical antipsychotic drug clozapine (a drug that possesses therapeutic efficacy but lacks Parkinson-like side effects and an ability to produce tardive dyskinesia) is administered repeatedly, dopamine neurons in the ventral tegmental area develop depolarization inactivation, but neurons in the substantia nigra do not. The reason for this differential effect is unknown at present. Foot shock, swim stress, or conditioned fear cause a selective (benzodiazepine-reversible) metabolic activation of the mesoprefrontal dopamine neurons without causing a marked or consistent effect on other midbrain dopamine neurons, including the mesocingulate dopamine neurons. Thus, this selective activation does not appear to be due solely to the absence of autoreceptors. The anxiogenic benzodiazepine receptor ligands such as the β-carbolines also produce a selective dose-dependent activation of mesoprefrontal dopamine neurons without increasing dopamine metabolism in other midbrain dopamine neurons.

In summary, certain mesotelencephalic dopamine systems, namely the mesoprefrontal and mesocingulate dopamine neurons, possess many unique characteristics compared to the nigrostriatal, mesolimbic, and mesopiriform dopamine systems (Table 9-8). Many of these unique characteristics may be the consequence of the lack of impulse-regulating somatodendrite and synthesis-modulating nerve terminal dopamine autoreceptors. However, some, such as the response to stress and anxiogenic β-carbolines, are clearly dependent upon other regulatory influences and not solely related to the absence of autoreceptors. These findings suggest that dopamine action at autoreceptors may be one of the important ways that dopamine cells possessing these receptors modulate their function. Since autoreceptors appear to play such an important role in regulation of the systems that possess them, future studies will need to address how midbrain dopamine systems that lack autoreceptors are regulated (role of afferent neuronal systems). In fact, some studies have suggested that a Substance P/Substance K innervation of the ventral tegmental area (A10) may influence the functional activity of mesocortical and mesolimbic dopamine neurons.

More recent studies have demonstrated the importance of NMDA receptors and of the glutamatergic input to the ventral tegmental area (VTA) in the regulatory control of the mesoprefrontal dopamine neurons. This input is believed to be at least partially responsible for converting pacemaker-like firing in dopamine cells into burst-firing patterns. NMDA receptors in the VTA appear to modulate differentially the dopamine projections to the prefrontal cortex and nucleus accumbens. The NMDA receptor is selectively activated by NMDA and regulated at several pharmacologically distinct sites including a high-affinity strychnine-insensitive glycine binding site (see Chapter 6). Competitive antagonists of this strychnine-insensitive glycine site, which cross the blood–brain barrier, have made possible the in vivo pharmacological modulation of the NMDA receptor via this site. In behavioral paradigms (restraint stress and conditioned fear) that are known to cause a metabolic activation of mesoprefrontal and mesaccumbens dopamine neurons, these agents (i.e. [+]-HA-966) selectively abolish the activation of the mesoprefrontal dopamine neurons. The stress-induced activation of the serotonin neurons in the prefrontal cortex and the dopaminergic activation of the nucleus accumbens are not altered by (+)-HA-966. These data indicate that under certain perturbed states the NMDA receptor complex and associated glycine modulatory site may play an important role in the afferent control of the dopamine neurons in the prefrontal cortex and may provide a potential target for pharmacological regulation of this important dopamine projection.

The observation that central dopamine systems are quite heterogeneous from both a biochemical and functional point of view holds promise that it will soon be possible to develop drugs targeted to modify or restore function to selective dopamine systems that are abnormal in various behavioral or pathological states. Some progress has already been achieved in developing agents that appear to have an action at selective dopamine receptor sites (Fig. 9-6). However, whether these agents will be useful in selectively modifying the function of subsets of midbrain dopamine neurons remains to be determined.

Tuberoinfundibular and Tuberohypophysial Dopamine Systems

The tuberoinfundibular dopamine system responds to pharmacological and endocrinological manipulations in a manner that is qualitatively different from the other three dopamine systems—nigrostriatal, mesolimbic, and mesocortical—described above (Table 9-7). The tuberoinfundibular neurons appear to be regulated in part by circulating levels of prolactin. Prolactin increases the activity of these neurons by acting within the medial basal hypothalamus, possibly acting directly on the tuberoinfundibular neurons. These neurons in turn release dopamine, which then inhibits prolactin release from the anterior pituitary. Haloperidol and other antipsychotic drugs have no effect on dopamine turnover in the tuberoinfundibular dopamine system until about 16 hours after drug administration, whereas the biochemical effects in other systems are observed within minutes and are maximal in several hours. The absence of an acute response to dopamine antagonists and agonists may be related to the lack of autoreceptors in this system. While less is known about the pharmacology of the tuberohypophyseal dopamine neurons, this system seems to respond to drugs in a manner qualitatively similar to the better-studied dopamine systems (Table 9-7).

SPECIFIC DRUG CLASSES

Antipsychotic Drugs (Major Tranquilizers)

Although there are at least seven major classes of antipsychotic drugs, all of which have in common the ability to ameliorate psychosis and evoke extrapyramidal reactions, the most widely used are the phenothiazines, thioxanthenes, and butyrophenones. As might be expected from the name given to chlorpromazine by the French—Largactil—the phenothiazines are notorious for their wide spectrum of pharmacological and biochemical effects. Fortunately, however, it turns out that many actions attributable to the antipsychotic phenothiazines are nonspecific properties of the phenothiazine moiety itself and do not appear to be correlated with the antipsychotic potency of the therapeutically active subgroup of phenothiazines. Along with many of their other actions, the antipsychotic phenothiazines have been shown to interact with both noradrenergic and dopaminergic neurons. However, the antipsychotic potency of these compounds appears to be best correlated with their ability to interact with dopamine-containing neurons. Much evidence has accumulated to indicate that antipsychotic drugs are effective blockers of dopamine receptors. In fact, it is believed that this reduction in dopamine activity expressed postsynaptically is responsible for the extrapyramidal side effects observed with these drugs. The ability of antipsychotic phenothiazines to block dopamine receptors on as yet chemically unidentified postsynaptic neurons may in some way also be related to the antipsychotic potency of these agents.

For many years it has been appreciated that antipsychotic drugs of both the phenothiazine and butyrophenone classes can increase the turnover of dopamine in the CNS. Since these drugs appear to have potent dopamine- receptor-blocking capabilities, it has been suggested that the increase observed in dopamine turnover results from blockade of both dopamine autoreceptors and postsynaptic dopamine receptors and a consequent feedback activation of the dopaminergic neurons, presumably by some sort of neuronal feedback loop. This speculation has been verified in part by direct extracellular recording techniques.

Antianxiety Drugs (Minor Tranquilizers)

The antianxiety drugs appear to interact with both noradrenergic and selected dopaminergic neurons. They are very effective in blocking and stress-induced increases in the turnover of both cortical norepinephrine and mesoprefrontal cortical dopamine. The actual mechanism by which these drugs exert their effects on the norepinephrine system is at the present time unknown, although in general they are believed to facilitate GABAergic transmission. Electrophysiological studies have demonstrated that many of these agents can suppress unit activity in the locus ceruleus. It has often been speculated that the relief of stress symptoms in humans is mediated by "turning off" the central noradrenergic neurons. (Noradrenergic systems were believed to be the primary monoamine system where antianxiety agents interact.) However, in view of the recent studies demonstrating an effect of stress on the mesoprefrontal dopamine system and reversal of the stress-induced activation by antianxiety drugs, it is clear that the mode of action of antianxiety drugs is an open question. Furthermore, in tests of anxiolytic action in rats, complete destruction of the telencephalic projections of the locus ceruleus fails to alter the effectiveness of benzodiazepines.

Antidepressants

There are several major classes of antidepressant drugs, the monoamine oxidase inhibitors (MAOI), the tricyclic drugs, and the atypical agents. The MAOIs interact with both noradrenergic and dopaminergic neurons. By definition, these agents inhibit the enzyme monoamine oxidase, which is involved in oxidatively deaminating catecholamines and thus cause an increase in the endogeneous levels of both dopamine and norepinephrine in the brain. At least from a biochemical standpoint, the tricyclic antidepressants appear to interact primarily with noradrenergic rather than dopaminergic neurons. The secondary amine derivatives such as desipramine are exceptionally potent inhibitors of norepinephrine uptake but have only a minimal effect on dopamine uptake. The tertiary amine derivatives are less effective in inhibiting norepinephrine uptake but are very effective in blocking serotonin uptake. The atypical antidepressants do not appear to have a selective interaction with monoamine systems when given by acute administration.

For many years it was postulated that the therapeutic action of antidepressant drugs was directly related to their uptake-blocking capability. However, since the therapeutic action of antidepressant drugs is usually delayed in onset for a week or more, the hypothesis that the rapid, acute actions of these drugs in blocking norepinephrine and serotonin reuptake are responsible for the long-term clinical antidepressant effects has been challenged. This hypothesis has been further discredited by recent investigations demonstrating that clinically effective antidepressant drugs such as iprindole and mianserin fail to inhibit neuronal uptake of 5-HT and norepinephrine significantly, while effective uptake inhibitors such as amphetamine, femoxetine, and cocaine are not effective in the treatment of depression. Current research has consequently shifted from the study of acute drug effects on amine metabolism to studies on the adaptive changes in norepinephrine and 5-HT receptor function induced by chronic antidepressant drug administration. In general, studies of this nature have revealed that chronic treatment with clinically effective antidepressant drugs causes a diminished responsiveness of central β-receptors and enhanced responsiveness of central serotonin and α_1-receptors. The time course of these receptor alterations more closely parallels the time course of the clinical antidepressant effects in humans. At present, however, the relationship between these receptor alterations and clinical antidepressant effects is unknown.

Stimulants

The therapeutic usage of this class of drugs is becoming less and less common as awareness of their abuse potential increases. At the present time, in fact, their use is largely restricted to the treatment of narcolepsy and hyperkinetic children and as general anorectic agents. The principal drugs in this category are the various analogues and isomers of amphetamine and methylphenidate.

For many years it has been known that ingestion of large amounts of amphetamine often leads to a state of paranoid psychosis that may be hard to distinguish from the paranoia associated with schizophrenia. It now appears that this paranoid state can be readily and reproducibly induced in human volunteers given large amounts of amphetamine so that the drug may provide a convenient "model psychosis" for experimental study. It is of interest in this regard that antipsychotic drugs such as chlorpromazine can readily reverse the amphetamine-induced psychosis.

On a biochemical level it was no surprise to learn that amphetamine and related compounds interact with catecholamine-containing neurons, since amphetamine is a close structural analogue of the

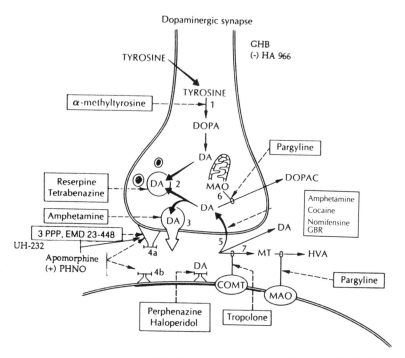

FIGURE 9-8. Schematic model of a central dopaminergic neuron indicating possible sites of drug action.

Site 1: *Enzymatic synthesis:* Tyrosine hydroxylase reaction blocked by the competitive inhibitor, α-methyltyrosine and other tyrosine hydroxylase inhibitors. (See Chapter 9).

Site 2: *Storage:* Reserpine and tetrabenazine interfere with the uptake-storage mechanism of the amine granules. The depletion of dopamine produced by reserpine is long-lasting and the storage granules appear to be irreversibly damaged. Tetrabenazine also interferes with the uptake—storage mechanism of the granules, except that the effects of this drug do not appear to be irreversible.

Site 3: *Release:* γ-Hydroxybutyrate and HA966 effectively blocks the release of dopamine by blocking impulse flow in dopaminergic neurons. Amphetamine administered in high doses releases dopamine, but most of the releasing ability of amphetamine appears to be related to its ability to effectively block dopamine reuptake.

Site 4: *Receptor interaction:* Apomorphine is an effective dopamine receptor-stimulating drug, with both pre- and postsynaptic sites of action. Both -3-PPP and EMD-23 (an indolebutylamine) are autoreceptor agonists. Perphenazine and haloperidol are effective dopamine receptor-blocking drugs.

catecholamines. However, there was no clear evidence that amphetamine produced its CNS effects through a catecholamine mechanism until it was demonstrated that α-methyl-—tyrosine (a potent inhibitor of tyrosine hydroxylase) prevented most of the behavioral effects of D-amphetamine. The question as to which catecholamine, norepinephrine or dopamine, is involved in the behavioral effects of amphetamine is still an open question. But it is generally believed that the so-called stereotypic behavior in animals (i.e., compulsive gnawing, sniffing) induced by amphetamine is associated with a dopaminergic mechanism and that the increase in locomotor activity involves a noradrenergic mechanism or both.

Almost all classes of psychotropic drugs interact in one way or another with catecholamine-containing neurons. Figure 9-8 outlines the life cycle of the transmitters of dopaminergic and noradrenergic neurons in the CNS and indicates possible sites at which drugs may intervene in this cycle. These schematic models also provide examples of drugs or chemical agents that interfere at these various sites within the life cycle of the transmitter substances. These numerous sorts of interactions ultimately result in an increase, decrease, or no change in the functional activity of the catecholamine neuron in question.

Only recently did it become clear exactly how various pharmacological agents alter activity in defined catecholamine neuronal systems in the brain. In most cases, the turnover of monoamines depends essentially upon impulse flow in the neuron. An increase in impulse flow usually causes an increase in turnover, and a decrease

---

Site 5: *Reuptake:* Dopamine has its action terminated by being taken up into the presynaptic terminal. Amphetamine as well as benztropine, the anticholinergic drug, are potent inhibitors of this reuptake mechanism.

Site 6: *Monoamine Oxidase (MAO):* Dopamine present in a free state within the presynaptic terminal can be degraded by the enzyme MAO, which appears to be located in the outer membrane of the mitochondria. Dihydroxyphenylacetic acid (DOPAC) is a product of the action of MAO and aldehyde oxidase on dopamine. Pargyline is an effective inhibitor of MAO. Some MAO is also present outside the dopaminergic neuron.

Site 7: *Catechol-O-methyl transferase (COMT):* Dopamine can be inactivated by the enzyme COMT, which is believed to be localized outside the presynaptic neuron. Tropolone is an inhibitor of COMT.

in impulse flow causes a reduction in turnover. As mentioned above, however, this is not always the case in the dopamine system, if synthesis is used as an index of turnover. Measurement of turnover has been employed to gain some insight into the activity of various types of monoamine-containing neurons during different behavioral states or after administration of different psychotropic drugs. As might be predicted, psychotropic drugs can have a variety of effects, and these effects can alter the turnover of a given transmitter substance without necessarily altering impulse flow in the neuronal system under study. For example, a drug can have a direct effect on the synthesis, degradation, uptake, or release of a given transmitter which will in turn alter the turnover of the transmitter in question but will not necessarily lead to an increase or decrease in the activity of the neuronal system that utilizes that substance as a transmitter. So it becomes clear that an alternation in turnover of a transmitter is not necessarily a clear indication that there has been a change in impulse flow in a given neuronal pathway. Therefore, the most direct way to determine if a drug alters impulse flow in a chemically defined neuronal system is to measure the activity of that system while the animal is under the influence of the drug.

It now appears that there are also a number of ways in which drugs can alter impulse flow. For example, a drug can act directly on the nerve cell body; it can act on other neurons, which then influence impulse flow in the neuron under study; or it can act at the postsynaptic receptor to cause stimulation or blockade, which then results in some sort of feedback influence on the presynaptic neuron. This feedback information could either be neuronal or perhaps transsynaptically mediated by release of some local chemical from the postsynaptic membrane. The combined histochemical–neurophysiological identification of dopaminergic and noradrenergic neurons has made possible the direct study of the effects of various drugs on the firing of these chemically defined neurons.

Effect of Drugs on the Activity of Dopaminergic Neurons

More than a decade ago, entirely on the basis of indirect biochemical evidence, Swedish pharmacologists speculated that antipsychotic drugs, as a result of their ability to block dopaminergic receptors, caused a feedback activation of dopaminergic neurons. Direct extracellular recording experiments have validated these earlier speculations. Antipsychotic drugs of both the phenothiazine and butyrophenone classes cause a dramatic increase in the firing rate of dopamine neurons in rat brain, an action presumably mediated by a neuronal feedback loop. Phenothiazines without antipsychotic potency are completely inactive in this regard. Of course, it is still a big jump to determine whether this interesting ability of antipsychotic drugs to block dopamine receptors is in any way related to their antipsychotic properties. In fact, we already know that there is a temporal dissociation of the two actions.

A naturally occurring metabolite of mammalian brain, γ-hydroxybutyrate, also has an interesting interaction with dopaminergic neurons. When γ-hydroxybutyrate is administered to rats in anesthetic doses, this compound causes a complete cessation of impulse flow in the nigrostriatal pathway. This inhibition does not appear to be mediated via activation of a neuronal feedback loop. On the other hand, the activity of other monoamine-containing neurons such as the norepinephrine-containing neurons in the locus ceruleus or the 5-HT–containing neurons of the midbrain raphe are relatively unaffected. Thus, this drug appears to be capable of producing a reversible lesion in the nigrostriatal and mesolimbic dopamine systems. The ability of this drug to block impulse flow in dopamine neurons most likely explains its unique biochemical properties as well as why it can effectively antagonize the ability of the antipsychotic drugs to increase the accumulation of dopamine metabolites in the striatum. Whether the drug will prove useful in the treatment of certain disease or drug-induced states thought to be related to dopaminergic hyperactivity remains to be determined.

There is also compelling biochemical and behavioral evidence that amphetamine, L-dihydroxyphenylalanine, and apomorphine interact with dopamine neuronal systems in the CNS. Until recently, however, there has been no direct evidence that these drugs affect neuronal activity. Since all three drugs presumably lead either directly or indirectly to a stimulation of dopamine receptors, it has been predicted that they might show the activity of dopaminergic neurons via some sort of neuronal feedback loop. Extracellular unit-recording experiments conducted on dopamine cells in the zona compacta and ventral tegmental area have borne out this speculation for all three drugs. In doses as low as 0.25 mg/kg, D-amphetamine inhibits firing of these dopamine neurons. It appears to act indirectly through the release of newly synthesized dopamine since α-methyl——tyrosine can prevent or reverse the depression of dopamine cell firing. L-amphetamine will also inhibit the firing of dopaminergic neurons, but is much less potent than D-amphetamine in producing this effect. L-dihydroxyphenylalanine also inhibits firing of these neurons and appears to exert its effect indirectly by being converted to dopamine. Pretreatment with a decarboxylase inhibitor will antagonize completely the inhibitory effect of dihydroxyphenylalanine of these cells. Likewise, apomorphine appears to inhibit effectively the firing of dopamine cells, but this drug seems to act directly on dopamine receptors because its actions are not altered by inhibition of dopamine synthesis with α-methyl——tyrosine. It is also noteworthy that antipsychotic drugs have the ability to block and reverse the depressant effects of D-amphetamine, L-dihydroxyphenylalanine, and apomorphine on the firing of dopamine cells. This observation suggests that all three drugs may exert their effects on firing rate through some sort of postsynaptic feedback pathway. Other experiments indicate that apomorphine also has a direct effect on dopaminergic neurons of the nigrostriatal and mesolimbic systems. This effect is believed to be mediated by an interaction with autoreceptors on dopamine cell bodies. As indicated earlier, apomorphine acts on dopamine autoreceptors in dosages that are ineffective on postsynaptic dopamine receptors. Administered in autoreceptor-selective doses, apomorphine is ineffective in influencing the activity of dopamine neurons that lack autoreceptors.

Parkinson's Disease

Parkinson's disease (PD) is a progressive neurodegenerative disorder of the basal ganglia that is characterized by tremor, muscular rigidity, difficulty in initiating motor activity, and loss of postural reflexes. It is observed in approximately 1% of the population over age 55. It has been known for over 70 years that PD is characterized pathologically by loss of pigmented cells in the substantia nigra, but only since 1960 has it been appreciated that there is a substantial loss of dopamine in the striatum. It is now clear that PD can be defined in biochemical terms as primarily a dopamine-deficiency state resulting from degeneration or injury to dopamine neurons. The most striking degenerative loss of dopamine neurons is observed in the nigrostriatal system. Even in patients with mild symptoms, a striatal dopamine loss of 70%–80% is observed, while severely impaired subjects have striatal dopamine depletions in excess of 90%. Since the dopamine transporter is heavily expressed in the terminals of dopamine neurons that are lost in PD, it is not surprising that striatal binding of agents that label this site (cocaine, nomifensine, GBR, and mazindol) is lost in the parkinsonian striatum. This alteration corresponds well with the loss of functional dopamine uptake visualized in vivo by position emission tomography using (^{18}F)-L-DOPA uptake or nomifensine. Although striatal dopamine loss represents the primary neurochemical abnormality in the PD brain, typical parkinsonism is accompanied by the loss of other dopamine systems and other monoamine neurons as well. Some degeneration of dopamine-containing neurons is also apparent in the mesolimbic, mesocortical, and hypothalamic systems, norepinephrine-containing neurons in the locus ceruleus, and serotonin neurons. Non-monoamine systems are also affected, with depletions observed in somatostatin, neurotensin, Substance P, enkephalin, and cholecystokinin-8. Since many of these nondopamine systems indirectly interact with mesotelencephalic dopamine systems, it is obvious that changes in some of them are bound to influence in a complex way the function of dopamine neurons.

The strategy for treating PD has been to restore the dopamine deficit in the brain by pharmacological means or more recently by neural grafting of dopamine-containing cells. There are a number of theoretical strategies for drug therapy in PD, including substrate supplementation direct and indirect dopamine agonists, metabolic inhibitors (MAOI), and uptake inhibitors. The most successful treatment has been the use of L-DOPA. Direct dopamine agonists also have some benefit in patients whose responsiveness to L-DOPA is greatly reduced or erratic. So far, the only direct-acting dopamine agonist that has found extensive use in PD is bromocriptine, primarily a D_2 agonist. Other agents belonging to this class will no doubt prove useful in the future as supplements or alternatives to L-DOPA.

In view of the behavioral and electrophysiological studies that suggest that D_1 receptor activation is necessary for the effects of D_2 receptor stimulation to be maximally expressed in normal animals as well as in animals with supersensitive dopamine receptors, the functional interaction between D_1 and D_2 receptors could have important implications in PD, where stimulation of postsynaptic dopamine receptors confers symptomatic benefit. Knowledge of the optimal ratio of relative drug activity at D_1 and D_2 receptors that is required to elicit effective stimulation of dopamine-mediated function may provide a basis for the design of new drugs. Also, as more knowledge accumulates concerning the distribution and function of various dopamine receptor subtypes, this should facilitate the development of new agents to treat dopamine deficiency states.

The MPTP-treated parkinsonian primate has provided a very useful animal model in which to examine therapeutic strategies for the treatment of PD (see Chapter 13). In fact, this model has already been successfully exploited to design, refine, and evaluate neural transplantation techniques and to test new pharmacological strategies for the therapeutic management of PD. Dihydrexidine, a D_1 full agonist, is capable of reversing the parkinsonian behavior in this primate model, while the partial D_1 agonist SKF-38393 is ineffective. Whether the therapeutic effects of dihydrexidine in this model are solely related to its D_1 agonist properties or its combined action on D_1 and D_2 receptors remains to be determined.

Dopamine Hypothesis of Schizophrenia

The etiology of schizophrenia has also been linked to defective dopamine neurotransmission. The growing conviction that antipsychotic agents act therapeutically by decreasing central dopaminergic transmission led to the formulation of the dopamine theory of schizophrenia. In its simplest form, this hypothesis states that schizophrenia may be related to a relative excess of mesolimbic dopaminergic activity (see Chapter 13).

SELECTED REFERENCES

Andersen, P. H., J. A. Gingrich, M. D. Bates, A. Dearry, P. Falardeau, S. E. Senogles, and M. G. Caron (1990). Dopamine receptor subtypes: Beyond the D_1/D_2 classification. *Trends Pharmacol. Sci. II*, 231–236.

Andreasen, N. C. (1988). Brain imaging: Applications in psychiatry. *Science 239*, 1381–1388.

Bannon, M. J. and R. H. Roth (1983). Pharmacology of mesocortical dopamine neurons. *Pharmacol. Rev. 35(1)*, 53.

Bannon, M. J., J. G. Granneman, and G. Kapatos (1995). The dopamine transporter: Potential involvement in neuropsychiatric disorders. In *Psychopharmacology: The Fourth Generation of Progress* (F. E. Bloom and D. J. Kupfer, eds.), pp. 179–188, Raven Press, New York.

Björklund, A. and O. Lindvall (1984). Dopamine-containing systems in the CNS. In *Handbook of Chemical Neuroanatomy*, Vol 2, *Classical Transmitters in the CNS, Part I* (A. Björklund and T. Hökfelt, eds.), Elsevier, Amsterdam.

Chio, C. L., R. F. Drong, D. T. Riley, G. S. Gill, J. L. Slightom, and R. M. Huff (1994a). D_4 dopamine receptor-mediated signaling events determined in transfected Chinese hamster ovary cells. *J. Biol. Chem. 269*, 11813–11819.

Chio, C. L., M. E. Lajiness, and R. M. Huff (1994b). Activation of heterologously expressed D_3 dopamine receptors: Comparison with D2 dopamine receptors. *Mol. Pharmacol. 45*, 51–60.

Chiodo, L. A., and A. S. Freeman (eds.) (1987). *Neurophysiology of Dopaminergic Systems: Current Status and Clinical Perspectives*. Lakeshore Publishing, Gross Pointe, MI.

Chiodo, L. A., A. S. Freeman, and B. S. Bunney (1995). Dopamine autoreceptor signal transduction and regulation. In *Psychopharmacology: The Fourth Generation of Progress* (F. E. Bloom and D. J. Kupfer, eds.), pp. 221–226, Raven Press, New York.

Civelli, O., J. R. Bunzow, and D. K. Grandy (1993). Molecular diversity of the dopamine receptors. *Annu. Rev. Pharmacol. Toxicol. 32*, 281–307.

Civelli, O. (1995). Molecular biology of the dopamine receptor subtypes. In *Psychopharmacology: The Fourth Generation of Progress* (F. E. Bloom and D. J. Kupfer, eds.), pp. 155–162, Raven Press, New York.

Elsworth, J. D., M. Al-Tikriti, J. R. Sladek Jr., J. R. Taylor, R. B. Innis, D. E. Redmond, Jr., and R. H. Roth (1994). Novel radioligands for the dopamine transporter demonstrate the presence of nigral grafts in caudate nucleus of MPTP treated monkey with improved behavioral function. *Exp. Neurol. 126*, 300–304.

Elsworth, J. D. and R. H. Roth (1995). Dopamine autoreceptor pharmacology and function: recent insights. In *The Dopamine Receptors* (K. A. Neve, ed.), Human Press, Totowa, New Jersey, in press.

Gingrich, J. A. and M. G. Caron (1993). Recent advances in the molecular biology of dopamine receptors. *Annu. Rev. Neurosci. 16*, 299–321.

Giros, B. and M. G. Caron (1993). Molecular characterization of the dopamine transporter. *TIPS 14(2)*, 43–49.

Horn, A. S. (1990). Dopamine uptake: A review of progress in the last decade. *Prog. Neurobiology 34*, 387–400.

Kebabian, J. W., M. Beaulieu, and Y. Itoh (1984). Pharmacological and biochemical evidence for the existence of two categories of dopamine receptor. *Can. J. Neurol. Sci. 11*, 114.

Kilty, J. E., D. Lorang, and S. G. Amara (1991). Cloning and expression of a cocaine-sensitive rat dopamine transporter. *Science, 254*, 578–579.

Lemoal, M. and H. Simon (1991). Mesocorticolimbic dopaminergic network: Functional and regulatory roles. *Physiol. Rev. 71*, 155.

Meador-Woodruff, J. H. (1994). Update on dopamine receptors. *Ann. Clin. Psychiatry 6(2)*, 79–90.

Meador-Woodruff, J. H., S. P. Damask, and S. J. Watson, Jr. (1994). Differential expression of autoreceptors in the ascending dopamine systems of the human brain. *Proc. Natl. Acad. Sci. U.S.A. 91*, 8297–8301.

Meador-Woodruff, J. H., A. Mansour, D. K. Grandy, S. P. Damask, O. Civelli, and S. J. Watson, Jr. (1992). Distribution of D5 dopamine receptor mRNA in rat brain. *Neurosci. Lett. 145*, 209–212.

Meador-Woodruff, J. H., D. K. Grandy, H.H.M. VanTol, S. P. Damask, B. S. Karley, Y. Little, O. Civelli, S. J. Watson, Jr. (1994). Dopamine receptor gene expression in the human medial temporal lobe. *Neuropsychopharmacology 10(4)*, 239–248.

Moore, K. E. and K. J. Lookingland (1995). Dopaminergic neuronal systems in the hypothalamus. In *Psychopharmacology: The Fourth Generation of Progress* (F. E. Bloom and D. J. Kupfer, eds.), pp. 245–256, Raven Press, New York.

O'Malley, K. L., S. Haron, L. Tang, and R. D. Todd (1992). The rat dopamine D4 receptor: Sequence, gene structure and demonstration of expression in the cardiovascular system. *New Biologist 4(2)*, 137–146.

Roth, R. H. and J. D. Elsworth (1995). Biochemical pharmacology of midbrain dopamine neurons. In *Psychopharmacology: The Fourth Genera-

tion of Progress (F. E. Bloom and D. J. Kupfer, eds.), pp. 227–244, Raven Press, New York.

Schwartz, J.-C., D., Levesque, M.-P. Martres, and P. Sokoloff (1993). Dopamine D₃ receptor: Basic and clinical aspects. *Clin. Neuropharmacol.* 16(4), 295–314.

Seeman, P. (1995). Dopamine receptors: Clinical correlates. In *Psychopharmacology: The Fourth Generation of Progress* (F. E. Bloom and D. J. Kupfer, eds.), pp. 295–302, Raven Press, New York.

Seeman, P. and H.H.M. VanTol (1994). Dopamine receptor pharmacology. *TIPS, 15,* 264–270.

Shimada, S., S. Kitayama, C.-L. Lin, A. Patel, E. Nanthakumar, P. Gregor, M. Kuhar, and G. Uhl (1991). Cloning and expression of a cocaine-sensitive dopamine transporter complementary DNA. *Science 254,* 576–577.

Sibley, D. R. and F. J. Monsma, Jr. (1992). Molecular biology of dopamine receptors. *TIPS 13(2),* 61–69.

Sunahara, R. K., H. C. Guan, B. F. O'Dowd, P. Seeman, L. G. Laurier, G. Ng, S. R. George, J. Torchia, H.H.M. VanTol, and H. B. Niznik (1991). Cloning of the gene for a human dopamine D₅ receptor with higher affinity for dopamine than D₁. *Nature 350,* 614–619.

Tang, L., R. D. Todd, A. Heller, and K. L. O'Malley (1994). Pharmacological and functional characterization of D₂, D₃ and D₄ dopamine receptors in fibroblast and dopaminergic cell lines. *J. Pharmacol. Exp. Ther. 268,* 495–502.

Taylor, J. R., M. S., Lawrence, D. E. Redmond, Jr., J. D. Elsworth, R. H. Roth, D. E. Nichols, and R. B. Mailman (1991). Dihydrexidine, a full dopamine D₁ agonist, reduces MPTP-induced parkinsonism in African green monkeys. *Eur. J. Pharmacol. 199,* 389–391.

VanTol, H.H.M., J. R. Bunsow, H. C. Guan, R. K. Sunahara, P. Seeman, H. B. Niznik, and O. Civelli (1991). Cloning of the gene for a human dopamine D₄ receptor with high affinity for the antipsychotic clozapine. *Nature 350,* 610–614.

Wolf, M. E., A. Y. Deutch and R. H. Roth (1987). Pharmacology of central dopaminergic neurons. In *Handbook of Schizophrenia, Vol 2, Neurochemistry and Neuropharmacology of Schizophrenia.* (F. A. Henn and L. E. DeLisi, eds.), pp. 101–147, Elsevier.

Wolf, M. E. and R. H. Roth (1990). Autoreceptor regulation of dopamine synthesis. *Ann. N.Y. Acad. Sci. 604,* 323–343.

Xu, M., X.-T. Hu, D. C. Cooper, R. Moratalla, A. M. Graybiel, F. J. White, and S. Tonegawa (1994). Elimination of cocaine-induced hyperactivity and dopamine-mediated neurophysiological effects in dopamine D₁ receptor mutant mice. *Cell 79,* 945–955.

10 | Serotonin (5-Hydroxytryptamine) and Histamine

SEROTONIN

Of all the neurotransmitters discussed in this book, serotonin remains historically the most intimately involved with neuropsychopharmacology. From the mid-nineteenth century, scientists had been aware that a substance found in serum caused powerful contraction of smooth muscle organs, but over a hundred years passed before scientists at the Cleveland Clinic succeeded in isolating this substance as a possible cause of high blood pressure. By this time, investigators in Italy were characterizing a substance found in high concentrations in chromaffin cells of the intestinal mucosa. This material also constricted smooth muscular elements, particularly those of the gut. The material isolated from the bloodstream was given the name "serotonin," while that from the intestinal tract was called "enteramine." Subsequently, both materials were purified, crystallized, and shown to be 5-hydroxytryptamine (5-HT), which could then be prepared synthetically and shown to possess all the biological features of the natural substance. The indole nature of this molecule bore many resemblances to the psychedelic drug LSD, with which it could be shown to interact on smooth muscle preparations in vitro. 5-HT is also structurally related to other psychotropic agents (Fig. 10-1).

When 5-HT was first found within the mammalian CNS, the theory arose that various forms of mental illness could be due to biochemical abnormalities in its synthesis. This line of thought was even further extended when the tranquilizing substance reserpine

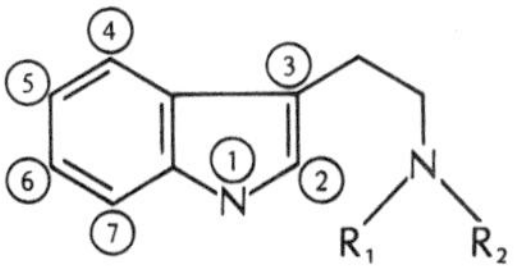

FIGURE 10-1. Structural relationships of the various indolealkyl amines.

Compound	Substitutions
Tryptamine	R_1 and R_2 = H
Serotonin	Tryptamine with 5 hydroxy
Melatonin	5 Methoxy, N-acetyl
DMT*	R_1 and R_2 = methyl
DET*	R_1 and R_2 = ethyl
Bufotenine*	5 Hydroxy, DMT
Szara psychotrope*	6 Hydroxy, DET
Psilocin*	4 Hydroxy, DMT
Harmaline*	6 Methoxy; R_1 forms isopropyl link to C_2
5-MT	5-Methoxytryptamine
5,6 DHT	5,6 Dihydroxytryptamine
5,7 DHT	5,7 Dihydroxytryptamine

*Psychotropic or behavioral effects.

was observed to deplete brain 5-HT; throughout the duration of the depletion, profound behavioral depression was observed. As we shall see, many of these ideas and theories are still maintained, although we now have much more ample evidence with which to evaluate them.

BIOSYNTHESIS AND METABOLISM OF SEROTONIN

Serotonin is found in many cells that are not neurons, such as platelets, mast cells, and the enterochromaffin cells mentioned above. In fact, only about 1%–2% of the serotonin in the whole body is found in the brain. Nevertheless, because 5-HT cannot cross the blood–brain barrier, it is clear that brain cells must synthesize their own.

For brain cells, the first important step is the uptake of the amino acid tryptophan, which is the primary substrate for the synthesis.

Plasma tryptophan arises primarily from the diet, and elimination of dietary tryptophan can profoundly lower the levels of brain serotonin. In addition, an active uptake process is known to facilitate the entry of tryptophan into the brain, and this carrier process is open to competition from large neutral amino acids, including the aromatic amino acids (tyrosine and phenylalanine), the branch-chain amino acids (leucine, isoleucine, and valine), and others (i.e., methionine and histidine). The competitive nature of the large neutral amino acid carrier means that brain levels of tryptophan will be determined not only by the plasma concentration of tryptophan but also by the plasma concentration of competing neutral amino acids. Thus, dietary protein and carbohydrate content can specifically influence brain tryptophan and serotonin levels by effects on plasma amino acid patterns. Because plasma tryptophan has a daily rhythmic variation in its concentration, it seems likely that this concentration variation could also profoundly influence the rate and synthesis of brain serotonin.

The next step in the synthetic pathway is hydroxylation of tryptophan at the 5 position (Fig. 10-2) to form 5-hydroxytryptophan (5-HTP). The enzyme responsible for this reaction, tryptophan hydroxylase, occurs in low concentrations in most tissues, including the brain, and it was very difficult to isolate for study. After purifying the enzyme from mast cell tumors and determining the characteristic cofactors, however, it became possible to characterize this enzyme in the brain. (Students should investigate the ingenious methods used for the initial assays of this extremely minute enzyme activity.) As isolated from brain, the enzyme appears to have an absolute requirement for molecular oxygen, for reduced pteridine cofactor, and for a sulfhydryl-stabilizing substance, such as mercaptoethanol, to preserve activity in vitro. With this fortified system of assay, there is sufficient activity in the brain to synthesize 1 μg of 5-HTP per gram of brain stem in 1 hour. The pH optimum is approximately 7.2, and the K_m for tryptophan is 3×10^{-4} M. Additional research into the nature of the endogenous cofactor tetrahydrobiopterin yielded a K_m for tryptophan of 5×10^{-5} M, which is still above normal tryptophan concentrations. Thus, the normal

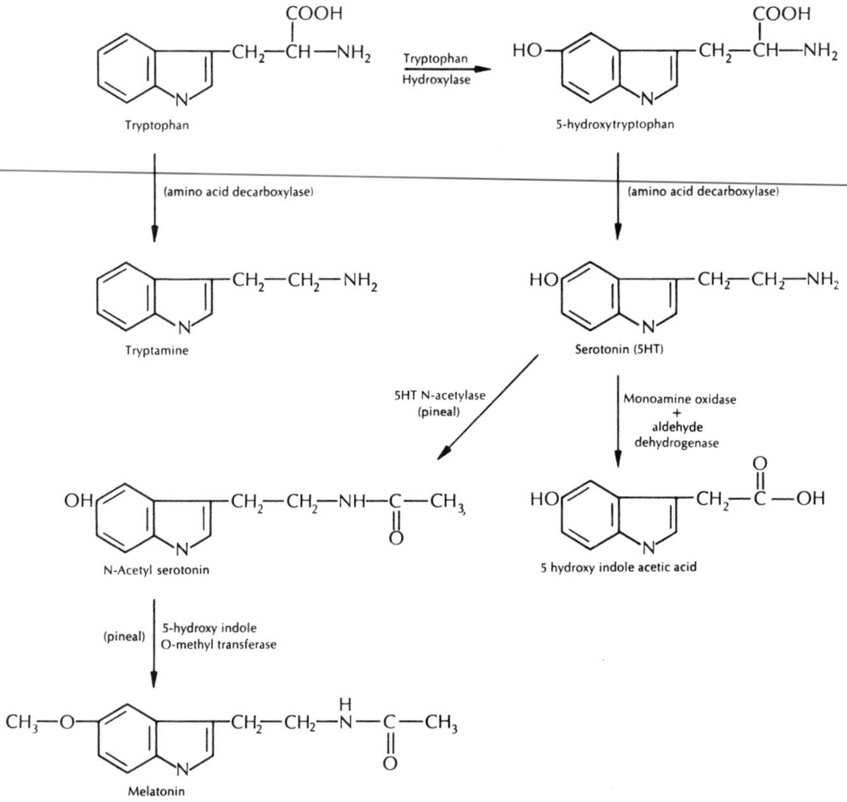

FIGURE 10-2. The metabolic pathways available for the synthesis and metabolism of serotonin.

plasma tryptophan content and the resultant uptake into brain leave the enzyme normally "unsaturated" with available substrate. Tryptophan hydroxylase appears to be a soluble cytoplasmic enzyme, but the procedures used to extract it from the tissues may greatly alter the natural particle-binding capacity. Investigators examining the relative distribution of particulate and soluble tryptophan hydroxylase have reported that the particulate enzyme may be associated with 5-HT-containing synapses, while the soluble form is more likely to be associated with the perikaryal cytoplasm. The particulate form of the enzyme shows the lower K_m and bears an absolute requirement for tetrahydrobiopterin.

Purified tryptophan hydroxylase has a molecular weight of 52,000–60,000. It is a multimer of identical subunits that can be activated by phosphorylation, Ca^{2+} phospholipids, and partial proteolysis.

The cloning and sequencing of cDNAs for tryptophan hydroxylase have recently been reported. Comparison of the rabbit tryptophan hydroxylase sequence with the sequences of phenylalanine hydroxylase and tyrosine hydroxylase demonstrates that these three pterin-dependent aromatic amino acid hydroxylases are highly homologous, reflecting a common evolutionary origin from a single primordial genetic locus. The pattern of sequence homology supports the hypothesis that the C-terminal two-thirds of the molecules constitute the enzymatic activity cores and the N-terminal one-third of the molecules constitute domains for substrate specificity.

The tryptophan hydroxylase step in the synthesis of 5-HT can be specifically blocked by p-chlorophenylalanine, which competes directly with the tryptophan and also binds irreversibly to the enzyme. Therefore, recovery from tryptophan hydroxylase inhibition with p-chlorophenylalanine appears to require the synthesis of new enzyme molecules. In the rat, a single intraperitoneal injection of 300 mg/kg of this inhibitor lowers the brain serotonin content to less than 20% within 3 days, and complete recovery does not occur for almost 2 weeks.

Considerable attention has been directed to the overall regulation of this first step of serotonin synthesis, especially in animals and humans treated with psychoactive drugs alleged to affect the serotonin systems as a primary mode of their action. These studies have made an important general point that seems to apply, in fact, to the brain's response to drug exposure in many other cell systems as well as to serotonin: Because transmitter synthesis, storage, release, and response are all dynamic processes, the acute imbalances produced initially by drug treatments are soon counteracted by the built-in feedback nature of synthesis regulation. Thus, if a drug acts to reduce tryptophan hydroxylase activity, the nerve cells may respond by increasing their synthesis of the enzyme and transporting increased amounts to the nerve terminals.

Mandell and colleagues have provided evidence, for example, that short-term treatment with lithium will initially increase tryptophan uptake, resulting in increased amounts being converted to 5-HT. After 14–21 days of chronic treatment, however, repetition of the measurements shows that while the tryptophan uptake is still increased, the activity of the enzyme is decreased, so that normal amounts of 5-HT are being made. In this new equilibrium state, the neurochemical actions of drugs like amphetamine and cocaine on 5-HT synthesis rates are greatly reduced, as are their behavioral actions (see below). In this way the 5-HT system, by shifting the relationship between uptake and synthesis during Li exposure, can be viewed as more "stable." This factor may be more fully appreciated when one considers that in the treatment of manic-depressive psychosis a minimum of 7–10 days is usually required before the therapeutic action of Li begins, a period during which the reequilibration of the 5-HT synthesizing process could be undergoing restabilization.

Decarboxylation

Once synthesized from tryptophan, 5-HTP is almost immediately decarboxylated to yield serotonin. The enzyme responsible for this conversion was presumed to be identical with the enzyme that decarboxylates dihydroxyphenylalanine (i.e., aromatic amino acid decarboxylase, AADC [EC 4.1.128] or DOPA-decarboxylase). However, this viewpoint has been often debated, since, when the decarboxylating step is examined in brain homogenates, the activity for DOPA differs widely from 5-HTP with respect to pH, temperature, and substrate optima. The cloning of AADC has finally resolved this issue and provided definitive proof that a single enzyme catalyzes the decarboxylation of both L-DOPA in catecholamine neurons and L-5-HTP in serotonin neurons.

Since this decarboxylation reaction occurs so rapidly and since its K_m (5×10^{-6} M) requires less substrate than the preceding steps, tryptophan hydroxylase is the rate-limiting step in the synthesis of serotonin. Because of this kinetic relationship, drug-induced inhibition of serotonin by interference with the decarboxylation step has never proven to be particularly effective.

It is possible to increase serotonin formation by administering 5-HTP and bypassing the rate-limiting tryptophan hydroxylase step. AADC is widespread in distribution; it is found in the peripheral and central nervous systems associated with catecholamine- and serotonin-containing neurons and in the adrenal and pineal glands. This enzyme is also found in the kidney, liver, and various other tissues in which little or no monoamine transmitter is normally produced. Thus, unlike tryptophan administration, which can result in a selective increase in serotonin in serotonin-containing neurons, 5-HTP administration will result in the nonspecific formation of serotonin at all sites containing AADC, including the catecholamine-containing neurons.

Catabolism

The only effective route of continued metabolism for serotonin is deamination by monoamine oxidase. The product of this reaction, 5-hydroxyindoleacetaldehyde, can be further oxidized to 5-hydroxyindoleacetic acid (5-HIAA) or reduced to 5-hydroxytryptophol, depending on the $NAD^+/NADH$ ratio in the tissue. Enzymes have been described in liver and brain by which 5-HT could be catabolized without deamination through formation of a 5-sulfate ester. This could then be transported out of brain, possibly by the acid transport system handling 5-HIAA.

Brain contains an enzyme that catalyzes the N-methylation of 5-HT using S-adenosylmethionine as the methyl donor. An N-methylating enzyme that uniquely preferred 5-methyl tetrahydrofolate (5-MTF) as the methyl donor has been described. Subsequent work by several other groups revealed this reaction to be an artifact of the in vitro assay system (see also Chapter 9) in which the 5-MTF is demethylated, releasing formaldehyde that then condenses across the amino-N and the C-2 of the indole ring to form the cyclic compound tryptoline. Like other derivatives of the 5-HT metabolic product line, the intriguing possibility that one such molecule might be psychotogenic has contributed to as yet unrewarded efforts to find these abnormal metabolites in human psychotics.

Control of Serotonin Synthesis and Catabolism

Although there is a relatively brief sequence of synthetic and degradative steps involved in serotonin turnover, there is still much to be learned regarding the physiological mechanisms for controlling this pathway. At first glance, it seems clear that tryptophan hydroxylase is the rate-limiting enzyme in the synthesis of serotonin, since when this enzyme is inhibited by 80% the serotonin content of the brain rapidly decreases. On the other hand, when the 5-HTP decarboxylase is inhibited by equal or greater amounts, there is no effect on the level of brain 5-HT. These data could only be explained if the important rate-limiting step were the initial hydroxylation. Since this step also depends on molecular oxygen, the rate of 5-HT formation could also be influenced by the tissue level of oxygen. In fact, it can be shown that rats permitted to breathe 100% oxygen greatly increase their synthesis of 5-HT. It is also of interest that 5-hydroxytryptophan does not inhibit the activity of tryptophan hydroxylase.

If the situation for serotonin were similar to that previously described for the catecholamines, we might also expect that the concentration of 5-HT itself could influence the levels of activity at the hydroxylation step. However, when the catabolism of 5-HT is blocked by monoamine oxidase inhibitors, the brain 5-HT concentration accumulates linearly to levels three times greater than controls, thus suggesting that end-product inhibition by serotonin is, at best, trivial. Similarly, if the efflux of 5-HIAA from the brain is blocked by the drug probenecid (which appears to block all forms of acid transport), the 5-HIAA levels also continue to rise linearly for prolonged periods of time, again suggesting that the initial synthesis step is not affected by the levels of any of the subsequent metabolites. Two possibilities, therefore, remain open: The initial synthesis rate may be limited only by the availability of required cofactors or substrate such as oxygen, pteridine, and tryptophan from the bloodstream; or the initial synthesis rate may be limited by the other more subtle control features, more closely related to brain activity. In fact, evidence is beginning to accumulate to suggest that impulse flow may, as in the catecholamine systems, initiate changes in the physical properties of the rate-limiting enzyme tryptophan hydroxylase. Several mechanisms have been postulated for the physiological regulation of tryptophan hydroxylase induced by alterations in neuronal activity within serotonergic neurons. The majority of evidence currently supports the involvement of calcium-dependent phosphorylation in this impulse-coupled regulatory process.

Serotonin Uptake and the Serotonin Transporter

As with the catecholamine-containing neurons, reuptake serves as a major mechanism for the termination of the action of synaptic serotonin. Serotonin nerve terminals possess high-affinity serotonin uptake sites that play an important role in terminating transmitter action and in maintaining transmitter homeostasis. This reuptake of released serotonin is accomplished by a plasma membrane carrier that is capable of transporting serotonin in either direction, depending on the concentration gradient. Although the involvement of transporters in norepinephrine and serotonin clearance has been appreciated for several decades (see Chapter 9), progress in understanding transporter structure and regulation has been slow, mainly because of difficulties associated with transporter protein purification. However, this has recently changed with the successful expression and homology-based cloning of the monoamine transporters (norepinephrine, dopamine, and serotonin) and the realization that they are members of a large gene family comprised of carriers for other transmitters including GABA and glycine (see Chapter 6). The expression of these transporters in nonneuronal cells has established useful model systems for analyzing the structural basis of transporter specificity for transmitters and antagonists. The availability of transporter protein has also enhanced the feasibility of obtaining transporter-specific antibodies and nucleic acid probes to investigate the endogenous mechanism acutely regulating monoamine transporters in vivo and to determine whether chronic alterations in transporter genes underlie neuropsychiatric disorders. To date, however, the cloning of the serotonin transporter has not had as big an impact on the field as the cloning of serotonin receptor subtypes. So far there is little evidence for heterogeneity of the serotonin transporter. Evidence is beginning to accumulate, however, that this transporter might be subject to regulation. Hints of altered serotonin transporter gene regulation following hormonal stimulation suggest that significant information might be acquired from systematic analysis of genomic regulatory elements that control transporter expression. Clearly, a goal for the future is to determine whether hereditary genetic variations in such systems contribute to psychiatric disorders.

PINEAL BODY

The pineal organ is a tiny gland (1 mg or less in the rodent) contained within connective tissue extensions of the dorsal surface of the thalamus. While physically connected to the brain, the pineal is cytologically isolated for all intents, since, as one of the circumventricular organs, it is on the "peripheral" side of the blood permeability barriers (see Chapter 2) and its innervation arises from the superior cervical sympathetic ganglion. The pineal is of interest for two reasons. First, it contains all the enzymes required for the synthesis of serotonin plus two enzymes for further processing serotonin, which are not so pronounced in other organs. The pineal contains more than 50 times as much 5-HT (per gram) as the whole brain. Second, the metabolic activity of the pineal 5-HT enzymes can be controlled by numerous external factors, including the neural activity of the sympathetic nervous system operating through release of norepinephrine. As such, the pineal appears to offer us a potential model for the study of brain 5-HT. The pineal also appears to contain many neuropeptides, however, and thus its secretory role remains as clouded as ever.

Actually, the 5-HT content of the pineal was discovered after the isolation of a pineal factor melatonin, known to induce pigment lightening effects on skin cells. When melatonin was crystallized and its chemical structure determined as 5-methoxy-N-acetyltryptamine, an indolealkylamine, a reasonable extension was to analyze the pineal for 5-HT itself. The production of melatonin from 5-HT requires two additional enzymatic steps. The first is the N-acetylation reaction to form N-acetylserotonin. This intermediate is the preferred substrate for the final step, the 5-hydroxy-indole-O-methyl-transferase reaction, requiring S-adenosyl methionine as the methyl donor.

The melatonin content, and its influence in supressing the female gonads, is reduced by environmental light and enhanced by darkness. The established cyclic daily rhythm of both 5-HT and melatonin in the pineal is driven by environmental lighting patterns through sympathetic innervation. In animals made experimentally blind, the pineal enzymes and melatonin content continue to cycle, but with a rhythm uncoupled from lighting cycles. The adrenergic receptors of the pineal are of the β-type, and—as is characteristic of such receptors—their effect on the pineal is mediated by the postjunctional formation of $3'-5'$ adenosine monophosphate (cAMP). Elevated levels of cAMP occur within minutes of the dark phase and lead to an almost immediate activation of the 5-HT-N-acetyl transferase. The same receptor action also appears to be responsible on a longer time base for tonic enzyme synthesis. Thus, the proposed use of pineal as a model applicable to brain 5-HT loses its luster, since this regulatory step does not seem to be of functional importance in the CNS. Furthermore, the "adrenergic" sympathetic nerves also can accumulate 5-HT (which leaks out of pinealocytes) just as they accumulate and bind norepinephrine (NE). It remains to be shown whether this is a functional mistake (i.e., secretions of an endogenous false transmitter) or is simply a case of mistaken biologic identity. It seems likely, however, that it is the NE whose release is required to pass on the intended communications from the sympathetic nervous system, since only NE activates the pinealocyte cyclase to start the enzyme regulation cascade.

Localizing Brain Serotonin to Nerve Cells

While some of these pineal curiosities may prove to be useful for the study of brain 5-HT, they are no substitute for data derived directly from the brain. Although most early neurochemical pharmacology assumed the brain 5-HT was a neurotransmitter, more than 10 years elapsed before it could be established with certainty that 5-HT in the brain is actually contained within specific nerve circuits. Based on lesion and subcellular fractionation studies, 5-HT content was first related to specific neuronal elements on a very coarse scale. With the introduction of the formaldehyde-induced fluorescence histochemical methodology by Falck and Hillarp, however, it became possible to observe the 5-HT-containing processes directly. Nevertheless, the mapping work has proceeded slowly because the 5-HT fluorophore develops with much less efficiency than that yielded by the condensation with catecholamines and also fades rapidly while viewed in the fluorescence microscope. Although the cell bodies can be seen with relative ease, direct analysis of the nerve terminals of these cell bodies has, until recently, required extreme pharmacological measures such as combined treatment with monoamine oxidase inhibitors and large doses of tryptophan; these methods do not lend themselves to discrete mapping of the projections. The situation has been greatly improved for the 5-HT mapping studies, as it has for most other neurochemically defined systems, through the combined application of immunohistochemistry (directed first toward partially purified tryptophan hydroxylase and most recently to the direct localization of 5-HT itself), of orthograde axoplasmic transport of radiolabeled amino acids microinjected into cells identified as containing 5-HT by fluorescence histochemistry, and of retrograde tracing back to known 5-HT cells from suspected terminal fields (see Chapter 2).

As a result of these extensive efforts by many groups, serotonin-containing neurons are known to be restricted to clusters of cells lying in or near the midline or raphe regions of the pons and upper brain stem (Fig. 10-3). In addition to the nine 5-HT nuclei (B_1–B_9) originally described by Dahlström and Fuxe, recent immunocyto-

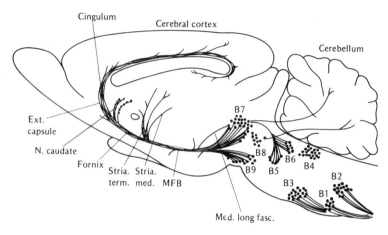

FIGURE 10-3. Schematic diagram illustrating the distribution of the main serotonin-containing pathways in the rat central nervous system. (Modified from G. Breese, *Handbook of Psychopharmacology*, Vol. 1, 1975)

chemical localization of 5-HT has also detected reactive cells in the area postrema and in the caudal locus ceruleus, as well as in and around the interpeduncular nucleus. The more caudal groups, studied by electrolytic or chemically induced lesions, project largely to the medulla and spinal cord. The more rostral 5-HT cell groups (raphe dorsalis, raphe medianus, and centralis superior, or B_7–B_9 [Fig. 10-4]) are thought to provide the extensive 5-HT innervation of the telencephalon and diencephalon. The intermediate groups may project into both ascending and descending groups, but since lesions here also interrupt fibers of passage, discrete mapping has required the analysis of the orthograde and retrograde methods. The immunocytochemical studies also reveal a far more extensive innervation of cerebral cortex, which, unlike the noradrenergic cortical fibers, is quite patternless in general.

In part, these studies could be viewed as disappointing in that most raphe neurons appear to innervate overlapping terminal fields and thus are more NE-like than dopamine-like in their lack of obvious topography. Exceptions to this generalization are that the B_8 group (raphe medianus) appears to furnish a very large component

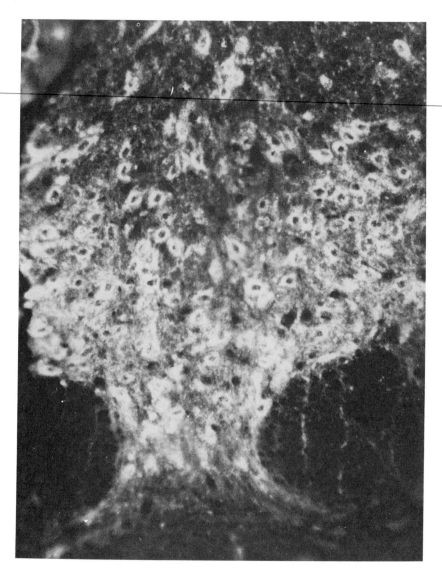

FIGURE 10-4. Fluorescence micrograph of raphe cell bodies in the midbrain of the rat. This rat was pretreated with L-tryptophan (100 mg/kg) 1 hour prior to death. (Courtesy of G. K. Aghajanian).

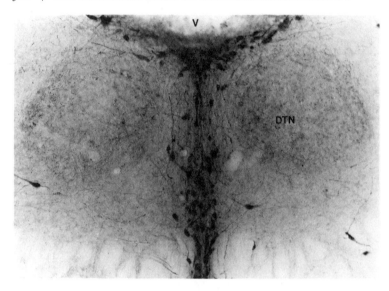

FIGURE 10-5. Photomicrograph of the serotonergic neurons of the caudal portion of the dorsal raphe. Section was stained using an antibody directed against serotonin. The serotonergic innervation of the dorsal tegmental nucleus (DTN) can also be seen. V marks the third ventricle. (Courtesy of A. Y. Deutch)

of the 5-HT innervation of the limbic system, while B₇ (or dorsal raphe) projects with greater density to the neostriatum, cerebral and cerebellar cortices, and thalamus (Fig. 10-5).

In the past, attempts to localize 5-HT-containing terminals relied primarily on the uptake of reactive 5-HT analogues or radiolabeled 5-HT. The use of labeled 5-HT, electron-dense analogues, or 5-HT-selective toxins (like the dihydroxytryptamines) depended for specificity on the selectivity of the uptake process. This situation has been rectified by immunocytochemistry of endogenous 5-HT (Figs. 10-5 and 10-6), employing antibodies directed against serotonin.

With this more sensitive technique, it has become clear that the cerebral cortex in many mammals is innervated by two morphologically distinct types of 5-HT axon terminals. Fine axons with small varicosities originate from the dorsal raphe nuclei, and beaded axons with large spherical varicosity arise from the median raphe nuclei.

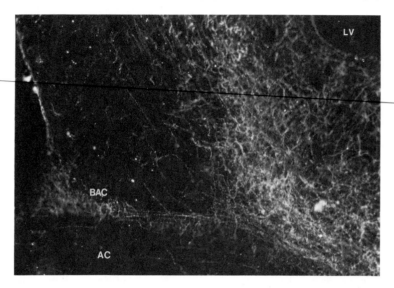

FIGURE 10-6. Darkfield photomicrograph illustrating the serotonergic innervation of the bed nucleus of the stria terminalis, as revealed by immunohistochemical staining with antibody directed against serotonin. The innervation of the bed nucleus of the anterior commissure (BAC) can also be seen above the anterior commissure (AC). LV marks the lateral ventricle. (Courtesy of A. Y. Deutch)

These two types of 5-HT-containing axons have different regional and laminar distributions and appear to be differentially sensitive to the neurotoxic effects of certain amphetamine derivatives, including 3,4-methylenedioxymethamphetamine (MDMA), referred to more commonly as "ecstasy." The fine axons are much more sensitive to the neurotoxic effects than the beaded axons, and the loss of fine axons lasts for months and may be permanent. The beaded axons appear to be resistant and remain unaffected following neurotoxic treatment with MDMA. This finding may be relevant to human studies, which have indicated that individuals using MDMA as a recreational drug may be exposed to dosages approximating those shown to exhibit serotonin neurotoxicity in nonhuman primates. It is noteworthy that a 26% decrease in the serotonin metabolite 5-HIAA was observed in the CSF of MDMA users. This indirect evidence of a decrease in serotonin turnover in brain perhaps reflects destruction or compromised function of this fine serotonin-containing axon system. Further studies of MDMA users seem warranted and could provide important information on the effects of selective loss of this fine axon system in humans. At present, the functional roles played by the fine and beaded axon systems and whether the functions are distinct or similar remain uncertain. In serial section analysis of 5-HT terminals in primate visual cortex, the fine and fat boutons appeared to coexist in the same axon, arguing against distinct 5-HT innervations of this brain region.

CELLULAR EFFECTS OF 5-HT

From the biochemical and morphological data above, we can be relatively certain that the 5-HT of the brain occurs not only within the nerve cells but within specific tracts or projections of nerve cells. We must now inquire into the effects of serotonin when applied at the cellular level. (See Chapter 2.) In those brain areas in which microelectrophoretically administered 5-HT has been tested on cells that exhibit spontaneous electrical activity, the majority of cells decrease their discharge rate. The effects observed typically last much longer than the duration of the microelectrophoretic current. However, in other regions, 5-HT also causes pronounced activation of discharge rate.

Recent electrophysiological analyses of 5-HT have focused on neurons of the facial motor nucleus (cranial nerve VII), where the innervation by 5-HT fibers has been well documented. With intracellular recordings from these cells, 5-HT is found to produce a slow, depolarizing action accompanied by a modest increase in membrane resistance. As might then be anticipated on biophysical grounds, the combination of depolarization and increased membrane resistance facilitates the response of these neurons to other excitatory inputs. In a rigorous sense, such effects are not exactly in keeping with the emerging characteristics of modulatory actions (see Chapter 2), since here iontophoretic 5-HT changes both membrane potential and resistance on its own. It will be of interest to determine if activation of a 5-HT pathway to these cells at levels that do not in themselves directly change membrane properties will nevertheless modify responses of the target cells to other inputs, analogous to effects described for noradrenergic connections.

Several specific 5-HT-containing tracts, investigated electrophysiologically, indicate that 5-HT produces mainly, if not exclusively, inhibitory effects. While the pathways do conduct slowly, as would be expected for such fine-caliber axons, the synaptic mediation process appears to be relatively prompt.

Characterization of 5-HT Receptors

The existence of multiple receptors for serotonin in the CNS has been suggested by physiological studies. Radioligand binding studies demonstrating that H^3-5-HT and H^3-spiperone label separate populations of high-affinity binding sites for 5-HT, termed 5-HT_1 and 5-HT_2, respectively, have also provided evidence for multiple receptors in brain tissue. A high correlation between the potencies of 5-HT antagonists in displacing the binding of H^3-spiperone and inhibiting serotonin-induced behavioral hyperactivity, coupled with a lack of such a correlation of H^3-5-HT binding sites, has led to the proposal that inhibition and excitation induced by serotonin are mediated at 5-HT_1 and 5-HT_2 receptors, respectively. Additional binding studies have further subdivided the 5-HT_1 class of recognition sites into 5-HT_{1A} and 5-HT_{1B} subtypes and compounds selective for these "receptor subtypes" have been identified. Autoreceptors for 5-HT appear to fall into the category of 5-HT_{1A} receptor subtypes. Drugs such as 8-OH DPAT and long-chain substituted piperazines appear to have selective 5-HT_{1A} binding activity, and these 5-HT autoreceptor agonists are effective in inhibiting midbrain raphe neurons. Ketanserin is a new agent that blocks 5-HT receptors that apparently belong to the 5-HT_2 category without having any significant effect on 5-HT_1 receptors. Because of this discrimination, it has been described as a selective 5-HT_2 blocker. However, it is important to note that ketanserin is nonspecific in the sense that it has appreciable affinity for both α_1-adrenergic receptors and histamine

H_1 receptors as well. It does have the advantage that its action on 5-HT receptors is as a pure antagonist.

5-HT Receptors

It is almost 40 years since Gaddum and Picarelli (1957), basing their work on the pharmacological properties of serotonin agonists and antagonists, reported evidence for two separate serotonin receptors in peripheral smooth muscle preparations studied in vitro. One they named the "D" receptor because it was blocked by dibenzyline, and the other they called the "M" receptor because the indirect contractile response to 5-HT mediated by the release of acetylcholine from cholinergic nerves in the myenteric plexus could be antagonized by morphine. The advent of receptor-binding studies in the 1970s revealed the existence of multiple binding sites. But while a correlation existed between the pharmacological profile of the 5-HT_2 binding site and the D receptor originally described by Gaddum, no such correlation existed between the 5-HT_1 receptor subtypes identified by receptor-binding techniques and the M receptor. Since the M receptor originally described in the nerves of the guinea pig ileum is pharmacologically distinct from the 5-HT_1 and 5-HT_2 receptors and their numerous subtypes, it was renamed the 5-HT_3 receptor. With the recent introduction of several extremely potent and highly selective 5-HT_3 antagonists, much attention has been refocused on this receptor and its possible functions in both the periphery and the CNS, as well as the therapeutic potential for these newly developed 5-HT_3 compounds.

At present, some eight subtypes of serotonin receptors in brain tissue have been defined and characterized, based on radioligand binding studies. As noted in Chapter 5, it is premature to characterize as a receptor a binding site defined in this manner. Nevertheless, the majority of the original speculations about serotonergic receptor subtypes generated from radioligand binding experiments appear to have been substantiated by many other types of experiments, including, most recently, the cloning of three of these subtypes. Table 10-1 lists the characteristics of these 5-HT receptor subtypes.

TABLE 10-1. Characteristics of 5-HT Receptor Subtypes

	5-HT₁ Receptors		
	$5\text{-}HT_{1A}$	$5\text{-}HT_{1B}$	$5\text{-}HT_{1D}$
High density regions	Raphe nuclei Hippocampus	SN GP	GP SN BG
Selective agonists	R(1)8-OH DPAT 5-CAT	5-CAT	Sumatriptan 5-CAT
Selective antagonists	Spiperone S(2)-Pindolol WAY 100135	Isamoltane	Isamoltane GR 127935
Membrane effects	Hyperpolar via opening K channels	?	?
Other functional correlates	Basilarartery Thermoreg. Hypotension Sexual behavior 5-HT Syndrome (Somato-dendritic) autoreceptor	Autorec. (Rel. Mod.)	Autorec. (Rel. Mod.)

	5-HT₂ Receptors		
	$5\text{-}HT_{2A}$	$5\text{-}HT_{2B}$	$5\text{-}HT_{2D}$
High density regions	Layer IV Cortex Hippocampus FMH	Stomach Fundus Cortex	Choroid plexus Medulla Hippocampus
Selective agonists	DOI DOB a-Me5-HT	DOI	DOB a-Me-5-HT
Selective antagonists	MDL 100,907 Ketanserin Ritanserin Spiperone	Ketanserin	Ketanserin Spiperone Metergoline
Membrane effects	Depolarization	Depolar.	Depolar. via opening of Ca21 channel

(continued)

TABLE 10-1. Characteristics of 5-HT Receptor Subtypes
(*Continued*)

	$5\text{-}HT_2$ Receptors		
	$5\text{-}HT_{2A}$	$5\text{-}HT_{2B}$	$5\text{-}HT_{2D}$
Other functional correlates	?	Vas. cont. Platelet aggre Paw edema Head twitches	

	Other 5-HT Receptor Subtypes		
$5\text{-}HT_3$	$5\text{-}HT_4$	$5\text{-}HT_6$	$5\text{-}HT_7$
PN EC Area postrema	Hippocampus Colliculi Ileum	Striatum Accumb. Cortex	Thalmus Hypothalmus Amygdala Heart
α-Me-5-HT	SC-53116	5-CAT LSD	LSD
Zacopride Odanserton Granisotron	SDZ 205 557	Lisuride Clozapine	Clozapine Amitryptline
Depolar.			
Transmit. release von Bezold- Jarisch reflex	Smooth musc. relax. Ionotropic		

α-Me-5-HT, α-methyl-5-hydroxytryptamine; BG, basal ganglia; DOB, 2,5-dimethoxy-4-bromoamphetamine; DOI, 2,5-dimethoxy-4-odoamphetamine; EC, entorhinal cortex; 5-CAT, 5-carboxamidotryptamine; FMN, facial motor neurons; GP, globus pallidus; PN, peripheral neurons; and SN, substantia nigra.

Molecular Biology

In the last decade, a vast amount of new information has become available concerning the various 5-HT receptor subtypes and their functional and structural characteristics. This derives from two main research approaches—operational pharmacology employing selective agonists and antagonists and, most recently, molecular biology. With the advent of the latter technique, the field of 5-HT receptors has experienced exceptionally rapid growth over the past several years and the existence of multiple 5-HT receptors has been unequivocally confirmed. Serotonin receptors are highly heterogeneous, and cloning not only has led to the discovery and recognition of previously unknown receptors but has greatly facilitated their classification. As of 1995, there are at least 15 distinct types of 5-HT receptors that have been cloned from mammalian tissue. The majority of the 5-HT receptors belong to the large family of receptors interacting with G proteins except for the 5-HT$_3$ receptors, which are ligand-gated ion channel receptors (see Table 10-2). The 5-HT receptors belonging to the G protein receptor superfamily are characterized by the presence of seven-transmembrane domains and the ability to alter G-protein-dependent processes. The amino acid sequence of the membrane-spanning domains shows the least amount of variability compared with other cloned biogenic amine receptors.

This group of 5-HT receptors can be divided into distinct families based on their coupling to second messengers and their amino acid sequence homology. The 5-HT$_1$ family contains receptors that are negatively coupled to adenylate cyclase. The 5-HT$_2$ family contains three receptors that have striking amino acid homology and the same coupling with second messenger, i.e., activation of phospholipase C. The 5-HT receptors that are positively coupled to adenylate cyclase are a heterogeneous group including the 5-HT$_4$, 5-HT$_6$, and 5-HT$_7$ receptor subtypes. The 5-HT$_5$ group contains two receptors, 5HT$_{5A}$ and 5-HT$_{5B}$, and represent a new family of 5-HT receptors that do not resemble receptors of the 5-HT$_1$ and 5-HT$_2$ families in terms of amino acid sequence, pharmacological profile, and transduction system. They are probably coupled to a

TABLE 10-2. Mammalian 5-HT-Receptor Subtypes and Their
Signal Transduction Pathways

G-Protein-Coupled Receptors	Receptor Subtype	G Protein	Effector Pathway
5-HT$_1$ family	5 HT$_{1A}$	G$_i$	Inhibition of adenylate cyclase
		G$_i$	Opening of K$^+$ channel
		G$_o$	Closing of Ca^{2+} channel
	5HT$_{1B}$	G$_i$	Inhibition of adenylate cyclase
	5HT$_{1D}$	''	''
	5HT$_{1E}$	''	''
5-HT$_2$ family	5HT$_{2A}$	G$_q$	Phosphoinositide hydrolysis
	5HT$_{2B}$		''
	5HT$_{2C}$		''
Others	5 HT$_4$	G$_s$	Activation of adenylate cyclase
	5-HT$_{5A}$, 5-HT$_{5B}$	?	Unknown coupling mech.
	5HT$_6$	G$_s$	Activation of adenylate cyclase
	5HT$_7$	G$_s$	Activation of adenylate cyclase
Ligand-gated ion channels	5-HT$_3$	None	Ligand-gated ion channel

different effector system. In contrast to the G-protein-coupled 5-HT receptors that modulate cell activities via second messenger systems, 5-HT$_3$ receptors directly activate a 5-HT-gated cation channel that rapidly and transiently depolarizes a variety of neurons. Like other ligand-gated ion channels, the 5-HT$_3$ receptor consists of four transmembrane segments and a large extracellular N-terminal region incorporating a cysteine–cysteine loop and potential N-glycosylation sites. Other members of this molecular receptor family include GABA, glutamate, glycine, and nicotinic cholinergic receptors. A third major molecular recognition site for 5-HT is the transporter proteins. These transporter proteins consist of 12 membrane-spanning proteins and represent a large gene family encoding Na^+ and Cl^--dependent transport proteins. The first 5-HT transporter was identified in rat brain, but 5-HT transporters have been well-characterized in other tissues, including platelets, placenta, lung, and basophilic leukemia cells. A number of selective 5-HT uptake blockers have been developed such as fluoxetine, sertraline, citalopram, and paroxetine (see Fig. 10-7), and several of these agents have exhibited clinical utility in the treatment of depression and obsessive-compulsive disorders.

The careful categorization of old and new subtypes of 5-HT receptors should be an important foundation for defining their func-

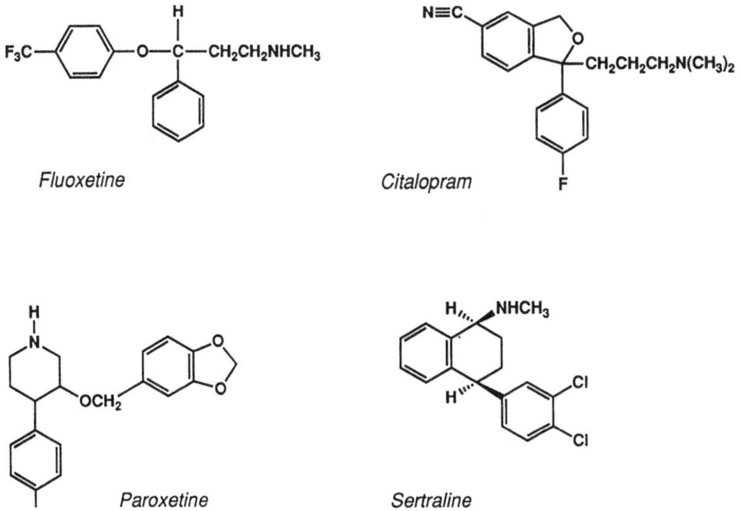

Fluoxetine   Citalopram

Paroxetine   Sertraline

FIGURE 10-7. Selective inhibitors of the serotonin transporter.

tion. The development of selective agents targeting these receptors, coupled with the molecular approaches highlighted in Chapter 3, such as antisense oligonucleotides to decrease steady-state levels of target proteins and transgenic animals in which specific receptors have been "knocked out", will undoubtedly enhance our understanding of the function of central serotonergic systems. The existence of a large number of receptors with distinct signaling properties and expression patterns might enable a single substance like 5-HT to generate simultaneously a large array of effects in many discrete brain structures.

Behavioral Aspects of Serotonin Function

5-HT neurons in the brain stem raphe nuclei exhibit spontaneous monotonic activity, discharging in a clock-like manner with an intrinsic frequency of one to five spikes per second. In the rodent, these monotonic properties are manifested early in development (3 to 4 days before birth). The 5-HT neurons appear to possess a negative feedback mechanism that limits their neuronal activity. As the physiological activity of the 5-HT neuron increases, the local release of 5-HT from dendrites or axonal collaterals acts on somatodendretic 5-HT autoreceptors to inhibit neuronal activity. This autoregulatory mechanism seems to function only under physiological conditions and to be inoperative during periods of low-level activity, but it becomes functional as neuronal activity increases. Dysfunction of this autoregulatory mechanism has been implicated in some forms of human neuropathology so that autoreceptors have become a potentially important site for drug-targeted therapeutic intervention. In fact, the combined use of 5-HT_{1B} autoreceptor antagonists and selective 5-HT uptake blockers holds some promise.

In view of the extraordinarily widespread projections and highly regulated pacemaker pattern of activity that is characteristic of serotonin neurons, a broad homeostatic function has been suggested for serotonergic systems. By exerting simultaneous modulatory effects on neuronal excitability in diverse regions of the brain and spinal cord, the serotonergic system is in a strategic position to coordinate complex sensory and motor patterns during varied behavioral states. Single-unit electrophysiological recordings from serotonergic neurons in unanesthetized animals have shown that serotonergic activity is highest during periods of waking arousal, reduced in quiet waking, reduced further in slow-wave sleep, and absent during rapid-eye-movement sleep. An increase in tonic activity of serotonergic neurons during waking arousal would serve to enhance motor neuron excitability via descending projections to the ventral horn of the spinal cord. A suppression of sensory input, conversely, would serve to screen out distracting sensory cues. A cessation of serotonergic neuronal activity during rapid-eye-movement sleep would tend to impede motor function in this paradoxical state in which internal arousal is associated with diminished motor output. Altered function of serotonergic systems has been reported in several psychopathological conditions, including affective illness, hyper-aggressive states, and schizophrenia. There is mounting evidence for impaired serotonergic function in major depressive illness and suicidal behavior. In this connection, it is interesting that several effective antidepressant drugs appear to act by enhancing serotonergic transmission.

The pathophysiology of major affective illness is poorly understood, but several lines of clinical and preclinical evidence indicate that enhancement of 5-HT-mediated neurotransmission might underlie the therapeutic response to different types of antidepressant treatment (see Chapter 13). Table 10-3 shows the effects of long-term administration of different types of antidepressants on the 5-HT system assessed with electrophysiological techniques. Treatment with all these drugs appears to cause a net increase in 5-HT neurotransmission. Several clinical observations have also provided strong evidence of a pivotal role for 5-HT neurotransmission in depression. For example, a large number of selective 5-HT reuptake inhibitors examined clinically have been found to be effective in major depression. These drugs belong to different chemical families but appear to share a single common property—the ability to inhibit the 5-HT reuptake carrier. In addition, clinical studies show that inhibition of 5-HT synthesis in drug-remitted depressed patients,

TABLE 10-3. Effects of Long-Term Administration of Antidepressant Drugs and ECT on the 5-HT System Assessed Using Electrophysiological Techniques

Antidepressant Treatment	Responsiveness of Somatodendritic 5-HT$_{1A}$ Autoreceptors[a]	Function of Terminal 5-HT Autoreceptors	Function of Terminal α$_2$- Adrenoceptors	Responsiveness of Postsynaptic 5-HT Receptors[a]	Net 5-HT Neurotransmission[b]
Selective 5-HT reuptake inhibitors	Decrease	Decrease	n.c.	n.c.	Increase
Monoamine oxidase inhibitors	Decrease	n.c.	Decrease	n.c. or decrease	Increase
5-HT$_{1A}$ receptor agonists	Decrease	n.c.	n.d.	n.c.	Increase
Tricyclic antidepressants	n.c.	n.c.	n.d.	Increase	Increase
Electroconvulsive shocks	n.c.	n.c.	n.c.	Increase	Increase

n.c., No change; n.d., not determined.

[a]Assessed by microiontophoresis or systemic injection of 5-HT receptor agonists.

[b]Determined from the firing activity of the presynaptic neurons and the effect of stimulating the ascending pathway on post-synaptic neurons.

SOURCE: Modified from Blier and deMontigny, *TIPS 15*, 220, 1994.

using either the tryptophan hydroxylase inhibitor parachlorophenylalanine or the tryptophan depletion paradigm, produces a rapid relapse of depression. It is noteworthy that in the latter paradigm the symptomatology reactivated by the tryptophan depletion was nearly identical to that present before the response to the antidepressant treatment, suggesting a causal relationship.

Neurochemistry

Only recently has it been possible to examine the mechanisms of signal transduction in central 5-HT receptors systems, and evidence is emerging that major transducing systems are linked to different 5-HT receptors in mammalian brain.

Signal Transduction Pathways

In general, two major 5-HT receptor-linked signal transduction pathways exist: a direct regulation of ion channels and a multistep enzyme-mediated pathway. Both pathways require a guanine nucleotide triphosphate (GTP)–binding protein (G protein) to link the receptor to the effector molecule. The 5-HT$_1$ family of receptors is negatively coupled to adenylate cyclase via the G$_1$ family of G proteins. The 5-HT$_{1A}$ receptor is the best characterized of this family. This receptor, in addition to coupling with adenylate cyclase, is linked directly to voltage-sensitive K$^+$ channels via G$_i$-like proteins. This direct coupling with both adenylate cyclase and the K$^+$ channel is a recognized characteristic of G$_i$-linked receptors. Direct coupling to L-type Ca^{2+} channels has also been described as an additional transduction pathway for the G$_i$-linked family of receptors. The other members of the 5-HT$_1$ family of receptors have also been shown to be negatively coupled to adenylate cyclase (Table 10-2). The 5-HT$_2$ family of receptors, in contrast to the 5-HT$_1$ family, is coupled to phospholipase C (PLC). The G protein involved has not been identified but is assumed to be a member of the G$_q$ family. PLC activation induces diverse changes in the cell, leading to regulation of numerous cellular processes. All members of this family appear to be coupled primarily to PLC, leading to phosphoinositide hydrolysis. Although stimulation of adenylate cyclase was the first signal transduction pathway to be linked to 5-HT, the specific receptors mediating activation of adenylate cyclase, the 5-HT$_4$, 5-HT$_6$, and 5-HT$_7$ receptors, were identified only recently. The 5-HT$_4$ receptor is found in rodent brain (hippocampus) and peripheral tissue, including the guinea pig ileum and human atrium. The 5-HT$_7$ receptor is also found in brain and heart (see Table 10-1). Another novel receptor, the 5-HT$_6$ receptor, has also been shown conclusively to couple positively to adenylate cyclase and to have a high affinity for tricyclic antidepressant drugs. The 5-HT$_3$ receptor differs from other 5-HT receptors by forming an ion channel that regulates ion flux in a G-protein-independent manner. This receptor is a member of a large family of ligand-gated ion channels and thus shares more similarities with the nicotinic cholinergic receptor that is the prototype of this superfamily than with the 5-HT$_1$ and 5-HT$_2$ receptor families. The 5-HT$_3$ receptors appear unique among the ligand-gated ion channels in that only a single subunit has at present been found, although alternate splice variants have been described. It is quite likely that more 5-HT$_3$ subunits will be found by molecular cloning in the future.

5-HT$_3$ receptors were first found on peripheral sensory, autonomic, and enteric neurons, where they mediate excitation. The direct demonstration that 5-HT$_3$ receptors in the guinea pig submucosal plexus are ligand-gated ion channels implies a role for 5-HT as a "fast" synaptic transmitter and fits with their function in the periphery. Binding studies in the CNS have indicated that 5-HT$_3$ receptors are primarily localized on nerve terminals, where they regulate the release of neurotransmitters. 5-HT$_3$ receptors are found in entorhinal cortex, frontal cortex, hippocampus, and area postrema. Molecular studies indicated that the cloned 5-HT$_3$ receptor protein forms a homomeric subunit which regulates the gating of cations and thus presumably mediates the rapid and transient depolarization that occurs following 5-HT$_3$ receptor activation. In general, the pharmacological and electrophysiological characteristics of the cloned receptor are largely consistent with the properties of the native receptors. Despite this consistency, however, a number of biochemical, pharmacological, and electrophysiological studies have suggested that CNS 5-HT₃ receptors exhibit heterogeneity with respect to subtypes and intracellular signal transduction mechanisms. Thanks to the pharmacological similarities between the 5-HT-M receptors identified by Gaddum and Picarelli in the late 1950s and the 5-HT₃ receptors, we have numerous pharmacological agents that exert specific effects on 5-HT₃ receptors, and many studies using these agents have addressed the possible functions of 5-HT₃ receptors. Results obtained from these in vivo studies have also suggested the presence of multiple 5-HT₃-like receptors, spurring on the search for other subunit proteins that might explain this pharmacological heterogeneity. The possibility should be entertained that 5-HT₃ receptors may be analogous to the glutamate receptors which were first characterized as inotropic receptors and later shown to exist also as G-protein-linked metabotropic receptors. Clearly, additional studies are required to determine whether the 5-HT₃ receptor heterogeneity can be explained by subtypes of 5-HT₃ receptor from both the inotropic and metabotropic superfamilies. The discovery of multiple subtypes of 5-HT₃ may help to clarify the electrophysiological observation that 5-HT₃-like receptors in the prefrontal cortex elicit a slow depression of cell firing rather than the fast activation expected for a ligand-gated 5-HT₃ receptor.

Recent data suggest that 5-HT₃ receptors are coupled to an ion channel, probably a calcium channel. Thus, the 5-HT₃ receptors share more similarities with the nicotinic cholinergic receptor than with the 5-HT₁ or 5-HT₂ receptor families. The direct demonstration that 5-HT₃ receptors in guinea pig submucosal plexus are ligand-gated ion channels implies a role for 5-HT (and perhaps for other biogenic amines) as a "fast" synaptic transmitter.

Adaptive Regulation

5-HT$_{2A}$ and 5-HT$_{2C}$ receptors adapt to chronic activation by reducing response sensitivity or receptor density as expected, but chronic inactivation does not elicit the opposite adaptive response. Central

5-HT$_2$ and 5-HT$_{2C}$ receptors seem to be relatively resistent to up-regulation. For example, 5-HT$_{2A}$ receptors are not up-regulated after denervation of 5-HT neurons or chronic administration of 5-HT antagonists. In fact, chronic administration of 5-HT antagonists instead results in a paradoxical down-regulation of both 5-HT$_{2A}$ and 5-HT$_{2C}$ receptors.

Physiology

Neurophysiological and behavioral studies have benefited from the development and characterization of selective agents for 5-HT receptors. Electrophysiological studies have clearly demonstrated that 5-HT$_{1A}$ receptors mediate inhibition of the raphe nuclei. The firing of serotonergic neurons is tightly regulated by intrinsic ionic mechanisms (e.g., a calcium-activated potassium conductance), which accounts for the well-known tonic pacemaker pattern of activity of these cells. The intrinsic pacemaker is modulated by at least two neurotransmitters: (1) Norepinephrine acting through adrenergic receptors accelerates the pacemaker; and (2) 5-HT acting through somatodendritic 5-HT$_{1A}$ autoreceptors slows the pacemaker. In the hippocampus, which is another anatomical structure containing a high density of 5-HT$_{1A}$ sites, 5-HT$_{1A}$ agonists hyperpolarize CA$_1$ pyramidal cells by opening potassium channels via a pertussis toxin–sensitive G protein. Electrophysiological studies carried out in *Xenopus* oocytes that express the 5-HT$_{1C}$ receptor reveal that application of 5-HT causes a detectable inward current that is blocked by mianserin. Activation of the 5-HT$_{1C}$ receptor apparently liberates inositol phosphates, raising intracellular Ca^{2+} levels and leading to the opening of Ca^{2+}-dependent chloride channels. In mammalian systems, two specific neurophysiological effects have been attributed to activation of the 5-HT$_2$ receptor. 5-HT facilitates the excitatory effects of glutamate in the facial motor nucleus, an action that is antagonized by 5-HT$_2$ antagonists. Similar data were obtained from intracellular studies that showed that the activation of 5-HT receptors caused a slow depolarization of facial motor neurons and increased input resistance, leading to an increased excitability of the cell, probably through a decrease in resting membrane conductance to potassium. These data suggest that 5-HT$_2$ receptors mediate the 5-HT-induced excitation of facial motor neurons.

It has also been shown that 5-HT causes a slow depolarization of cortical neurons that is associated with decreased conductance. The effect can be desensitized by repeated applications of 5-HT and blocked by the selective 5-HT$_2$ antagonist ritanserin. Thus, the effects of 5-HT on cortical pyramidal neurons share many similarities to the depolarizing effects observed in the facial motor nucleus and appear also to be mediated by 5-HT$_2$ receptors.

5-HT$_3$ receptors in peripheral nervous tissue mediate excitatory responses to 5-HT and are involved in modulating transmitter release. These receptors are also found in the CNS, where they are present in high density in the entorhinal cortex and area postrema. Recent experiments demonstrate that the release of endogenous dopamine by stimulation of 5-HT$_3$ receptors in the striatum occurs in a calcium-dependent fashion. Even though the physiological function of 5-HT$_3$ receptors in the CNS is unclear, the observation that selective 5-HT$_3$ antagonists possess central activity in rats and primates in anxiolytic and antidopaminergic-like behavioral models has generated considerable excitement and speculation about the potential therapeutic use of 5-HT$_3$ antagonists and agonists. From animal experiments it has been speculated that these compounds may be useful in treating schizophrenia, pain, anxiety, drug dependence, and cytotoxic drug-induced emesis; but so far only the analgesic and antiemetic effects have been demonstrated in humans.

Electrophysiology of 5-HT Receptors

Knowledge of the molecular biology of 5-HT receptors has revolutionized electrophysiological approaches to investigating the 5-HT systems in brain since studies can now be directed toward neurons that express specific 5-HT receptor subtypes based on in situ hybridization maps of receptor mRNA expression. This has enabled the diverse electrophysiological actions of 5-HT in the CNS to be categorized according to receptor subtypes and their respective effect or mechanisms of action. Here are a few generalizations that are beginning to emerge from these studies: (1) Inhibitory effects of 5-HT are mediated by 5-HT_1 receptors linked to the opening of K^+ channels or to the closing of Ca^{2+} channels, both via pertussis toxin–sensitive G proteins; (2) facilitory effects of 5-HT that are mediated by 5-HT_2 receptors and involve the closing of K^+ channels can be modulated by the PI second messenger system and PKC acting as a negative feedback loop; (3) other facilitory effects of 5-HT appear to be mediated by 5-HT_4 and 5-HT_7 receptors by a reduction in certain voltage-dependent K^+ currents mediated through the protein kinase A phosphorylation pathway and thus involving a positive coupling of 5-HT response to adenylcyclase; (4) fast excitations are mediated by 5-HT_3 receptors through a ligand-gated cationic ion channel that does not require coupling with a G protein or a second messenger. Thus, it is clear that the electrophysiological actions of 5-HT encompass the two major neurotransmitter gene superfamilies—the G-protein-coupled receptors and the ligand-gated cationic channels. The end effects are determined by the receptor subtype and its anatomical location.

Relevance to Clinical Disorders and Drug Actions

The diversity of receptors and transduction pathways that underlies the actions of 5-HT, together with the differential expression of 5-HT receptor subtypes in different neuronal and effector cell populations, helps to explain how it is possible for a single transmitter to be linked to such a large array of behaviors, clinical conditions, and drug actions. Alterations in 5-HT function have been implicated in a host of clinical states, including affective disorders, obsessive-compulsive disorders, schizophrenia, anxiety states, phobic disorders, eating disorders, migraine, and sleep disorders (see Table 10-4). There is also a wide range of psychotropic drugs that affect 5-HT neurotransmission, including antidepressants (selective 5-HT uptake blockers, e.g., fluoxetine [see Fig. 10-7]); hallucinogens (e.g.,

TABLE 10-4. Clinical Conditions Influenced by Altered Serotonin
Function

Affective disorders	
Aging and	Neuorendocrine regulation
neurodegenerative	Obsessive-compulsive disorder
disorders	Pain sensitivity
Anxiety disorders	Premenstrual syndrome
Carcinoid syndrome	Post-traumatic stress
Circadian rhythm regulation	syndrome
Developmental disorders	Schizophrenia
Eating disorders	Sexual disorders
Emesis	Sleep disorders
Migraine	Stress disorders
Myoclonus	Substance abuse

LSD, psilocin [see Fig. 10-1]); anxiolytics (e.g., buspirone); atypical
antipsychotics (e.g., clozapine); antiemetics (e.g., ondansetron); appetite suppressants (e.g., fenfluramine); and antimigraine drugs (e.g.,
sumatriptan).

Behavioral Effects of 5-HT Systems

One of the first drugs found empirically to be effective as a central
tranquilizing agent, reserpine, was employed mainly for its action in
the treatment of hypertension. In the early 1950s when both brain
serotonin and the central effects of reserpine were first described,
great excitement arose when the dramatic behavioral effects of reserpine were correlated with loss of 5-HT content. However, it was
soon found that NE and dopamine content could also be depleted
by reserpine and that all these depletions lasted longer than sedative
actions. The brain levels of all three amines remained down for
many days while the acute behavioral effects ended in 48–72 hours.
Hence, it became difficult to determine whether it was the loss of
brain catecholamine or serotonin that accounted for the behavioral
depression after reserpine.

More extensive experiments into the nature of this problem were possible when the drugs that specifically block the synthesis of catecholamines or serotonin were discovered. After depleting brain serotonin content with *p*-chlorophenylalanine, which effectively removed 90% of brain serotonin, investigators observed that no behavioral symptoms reminiscent of the reserpine syndrome appeared. Moreover, when *p*-chlorophenylalanine-treated rats, which were already devoid of measurable serotonin, were treated with reserpine, typical reserpine-induced sedation arose. These results again favor the view that loss of catecholamines could be responsible for the reserpine-induced syndrome. Furthermore, when α-methyl-*p*-tyrosine is given and synthesis of NE is blocked for prolonged periods of time, the animals are behaviorally sedated and their condition resembles the depression seen after reserpine. Students will find it profitable to review in detail the original papers describing the results just summarized and the other theories currently in vogue.

Hallucinogenic Drugs

One of the more alluring aspects of the study of brain serotonin is the possibility that it is this system of neurons through which the hallucinogenic drugs cause their effects. In the early 1950s the concept arose that LSD might produce its behavioral effects in the brain by interfering with the action of serotonin there as it did in smooth muscle preparations, such as the rat uterus. However, this theory of LSD action was not supported by the finding that another serotonin blocking agent, 2-bromo LSD, produced minimal behavioral effects in the CNS. And, in fact, it was subsequently shown that very low concentrations of LSD itself—rather than blocking the serotonin action—could potentiate it. However, none of these data could be considered particularly pertinent since all the research was done on the peripheral nervous system and all the philosophy was applied to the CNS.

Shortly thereafter Freedman and Giarman initiated a profitable series of experiments investigating the basic biochemical changes in the rodent brain following injection of LSD. Although their initial studies required them to use bioassay for changes in serotonin, they were able to detect a small (on the order of 100 ng/gm, or less) increase in the serotonin concentration of the rat brain shortly after the injection of very small doses of LSD. Subsequent studies have shown that a decrease in 5-HIAA accompanies the small rise in 5-HT. Although the biochemical effects are similar to those that would be seen from small doses of monoamine oxidase inhibitors, no direct monoamine oxidase inhibitory effect of LSD has been described. This effect was generally interpreted as indicating a temporary decrease in the rate at which serotonin was being broken down, and it could also be seen with higher doses of less effective psychoactive drugs. In related studies by Costa and his coworkers, who estimated the biochemical turnover of brain serotonin, prolonged infusion of somewhat larger doses of LSD clearly promoted a decrease in the turnover rate of brain serotonin.

The next advance in the explanation of the LSD response was made when Aghajanian and Sheard reported that electrical stimulation of the raphe nuclei would selectively increase the metabolism of 5-HT to 5-HIAA. This finding suggested that the electrical activity of the 5-HT cells could be directly reflected in the metabolic turnover of the amine. Subsequently, the same authors recorded single raphe neurons during parenteral administration of LSD and observed that they slowed down with a time course similar to the effect of LSD. Thus, following LSD administration both decreased electrical activity of these cells and decreased transmitter turnover occur.

Of the several explanations originally proposed for this effect, the data now support the view that LSD is able to depress directly 5-HT-containing neurons at receptors that may be the sites where raphe neurons feed back 5-HT messages to each other through recurrent axon collaterals or dendrodendritic interactions; at these receptive sites on raphe neurons, LSD is a 5-HT agonist. In other tests on raphe neurons, iontophoretic LSD will antagonize activity of raphe neurons that show excitant responses to either NE or 5-

HT. However, at sites in 5-HT terminal fields, when LSD is evaluated as an agonist or antagonist of 5-HT, its effects are considerably weaker than on the raphe neurons. These observations have led Aghajanian to suggest that LSD begins its sequence of physiological and neurochemical changes by acting on 5-HT neurons. In this view, the psychedelic actions of LSD must entail a primary or perhaps total reliance upon decreased efferent activity of the raphe neurons.

As Freedman has indicated, however, several points need further clarification before the description of LSD-induced changes in cell firing and cell chemistry can be incorporated in an explanation of the pharmacological and behavioral effects of LSD. If the effects of LSD could be "simply" equated with silencing of the raphe and subsequent disinhibition of raphe synaptic target cells, then many aspects of LSD-induced behavioral changes should be detectable in raphe-lesioned animals, but they are not. Typical LSD effects are also seen when raphe-lesioned animals are administered LSD. When normal animals are pretreated with *p*-Cl-phenylalanine (in dose schedules that antagonize raphe-induced synaptic inhibition) and given LSD, the effect is accentuated. In further contrast to the predictions of the 5-HT silencing effect, the LSD response is further accentuated with concomitant treatment with 5-HTP. The physiological manipulation that most closely simulates behavioral actions of LSD is stimulation of the raphe, leading to decreased habituation to repetitive sensory stimuli. These data are difficult to reconcile with the view that LSD-induced behavior results from inactivation of tonic 5-HT-mediated postsynaptic actions. An important aspect to be evaluated critically in the continued search for the cellular changes that produce the behavioral effects is the issue of tolerance: LSD and other indole psychotominetic drugs show tolerance and cross-tolerance in humans, but these properties have not been seen in animal studies at the cellular level.

The student must realize that it is extremely difficult to track down all of the individual cellular actions of an extremely potent drug like LSD and to fit these effects together in a jigsaw puzzle-like effort to solve the question of how LSD produces hallucinations.

Similar jigsaw puzzles lie just below the surface of every simple attempt to attribute the effects of a drug or the execution of a complex behavioral task like eating, sleeping, mating, and learning (no rank ordering of author's priorities intended) to a single family of neurotransmitters like 5-HT. While it is clearly possible to formulate hypothetical schemes by which divergent inhibitory systems like the 5-HT raphe cells can become an integral part of such diverse behavioral operations as pain suppression, sleeping, thermal regulation, and corticosteroid receptivity, a very wide chasm of unacquired data separates the concept from the documentary evidence needed to support it. The gap is even wider than it seems since we do not at present have the slightest idea of the kinds of methods that can be used to convert correlational data (raphe firing associated with sleeping-stage onset or offset) into proof of cause and effect. While previous editions of this guide to self-instruction in cellular neuropharmacology have attempted a superficial overview of the behavioral implication of 5-HT neuronal actions, we now relinquish that effort until the cellular bases become more readily perceptible.

Actions of Other Psychotropic Drugs

Subsequent studies by Aghajanian and his colleagues have indicated that other drugs that can alter 5-HT metabolism can also disturb the discharge pattern of the raphe neurons. Thus, monoamine oxidase inhibitors and tryptophan, which would be expected to elevate brain 5-HT levels, also slow raphe discharge; while clinically effective tricyclic drugs (antidepressants such as imipramine, chlorimipramine, and amitriptyline) also slow the raphe neurons and could elevate synaptic 5-HT levels locally by inhibition of 5-HT reuptake. At the present time it remains unclear whether these metabolic changes in 5-HT after psychoactive drugs are administered reflect changes in the cells that receive 5-HT synapses or in the raphe neurons themselves. A large question is whether these circuits mediate the therapeutic responses of the drugs or are the primary pathologic site of the diseases the drugs treat. One etiologic view of schizophrenia, for example, might be based on the production of an ab-

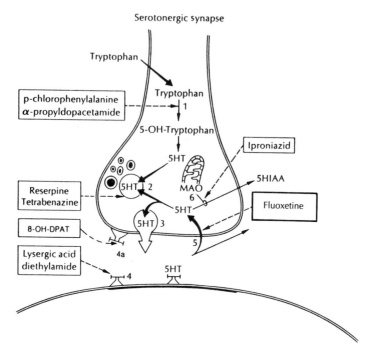

FIGURE 10-8. Schematic model of a central serotonergic neuron indicating possible sites of drug action.

Site 1: *Enzymatic synthesis:* Tryptophan is taken up into the serotonin-containing neuron and converted to 5-OH-tryptophan by the enzyme tryptophan hydroxylase. This enzyme can be effectively inhibited by *p*-chlorophenylalanine and α-propyldopacetamide. The next synthetic step involves the decarboxylation of 5-OH-tryptophan to form serotonin (5-HT).

Site 2: *Storage:* Reserpine and tetrabenazine interfere with the uptake-storage mechanism of the amine granules, causing a marked depletion of serotonin.

Site 3: *Release:* At present there is no drug available which selectively blocks the release of serotonin. However, lysergic acid diethylamide, because of its ability to block or inhibit the firing of serotonin neurons, causes a reduction in the release of serotonin from the nerve terminals.

Site 4: *Receptor interaction:* Lysergic acid diethylamide acts as a partial agonist at serotonergic synapses in the CNS. A number of compounds have also been suggested to act as receptor-blocking agents at serotonergic synapses, but direct proof of these claims at the present time is lacking.

Site 4a: *Autoreceptor interaction:* Autoreceptors on the nerve terminal appear

normal indole, such as the hallucinogenic dimethyltryptamine, a compound that can be formed in human plasma from tryptamine. Some evidence indicates that schizophrenic patients have abnormally low monoamine oxidase activity in their platelets, which could permit the formation of abnormal amounts of plasma tryptamine and consequently abnormal amounts of dimethyltryptamine as well. Figure 10-8 summarizes possible sites of drug interaction in a hypothetical serotonin synapse in the CNS.

In this chapter we have encountered one of the more striking examples of an intensively studied brain biogenic amine, and there is every reason to believe that it is an important synaptic transmitter. Still to be determined are the precise synaptic connections at which this substance accomplishes the transmission of information and the functional role these connections play in the overall operation of the brain with respect to both effective and other multicellular interneuronal operations. The central pharmacology of serotonin, while intensively investigated, remains poorly resolved. Specific inhibition of uptake and synthesis are possible, but truly effective and selective postsynaptic antagonists remain to be developed.

HISTAMINE

Since the beginning of this century, when Henry Dale demonstrated that histamine is an endogenous tissue constituent and a po-

---

to play a role in the modulation of serotonin release. 8-Hydroxy-diproplaminotetraline (8-OH-DPAT) acts as an autoreceptor agonist at serotonergic synapses in the CNS.

Site 5: *Re-uptake:* Considerable evidence now exists to suggest that serotonin may have its action terminated by being taken up into presynaptic terminal. The tricyclic drugs with a tertiary nitrogen such as imipramine and amitryptyline appear to be potent inhibitors of this uptake mechanism.

Site 6: *Monoamine oxidase. (MAO):* Serotonin present in a free state within the presynaptic terminal can be degraded by the enzyme MAO, which appears to be located in the outer membrane of mitochondria. Iproniazid and clorgyline are effective inhibitors of MAO.

tent stimulator of a variety of cells, this substance has been thought to act as a transmitter or neuromodulator. However, direct evidence in support of this idea has accumulated relatively slowly.

The challenge posed by histamine to neuropharmacologists has led to a vigorous chase across meadows of enticing hypotheses surrounded by bogs of confusion and dubious methodology. At last, more than 60 years after its isolation from pituitary by J. J. Abel, the role of histamine in the brain seems to be approaching a resolution that this amine occurs in two types of cells: mast cells and magnocellular neurons in the posterior hypothalamus.

Most of what has been known about the synthesis and degradation of histamine in the brain was based on attempts to simulate in brain tissue data obtained from more or less homogeneous samples of peritoneal mast cells (Fig. 10-9). But since mast cells were not supposed to be found in healthy brains, and since histamine does not cross the blood–brain barrier, it had long been assumed that histamine could also be formed by neurons and hence be considered as a neurotransmitter. In fact, attempts to develop drugs that were histamine antagonists, in the mistaken belief that battle field shock was compounded by histamine release, were a key to the subsequent development of antipsychotic drugs. Furthermore, as every hayfever sufferer knows, present-day antihistamine drugs clearly produce substantial CNS actions such as drowsiness and hunger. Finally, we should realize that in retrospect the increased content of histamine found in transected degenerating sensory nerve trunks was likely to be an artifact of mast cell accumulation rather than a peculiar form of neurochemistry. Viewing the other data on CNS histamine through that same retrospectroscope, as they have been illuminated by the innovative experiments of Schwartz and his associates, we now have a rather compelling case that histamine qualifies as a putative central neurotransmitter in addition to the role the same amine plays in mast cells.

The major obstacles in elucidating the functions of histamine in brain have been the absence of a sensitive and specific method to demonstrate putative histaminergic fibers in situ, the lack of suitable methods to measure histamine and its catabolites, and problems in

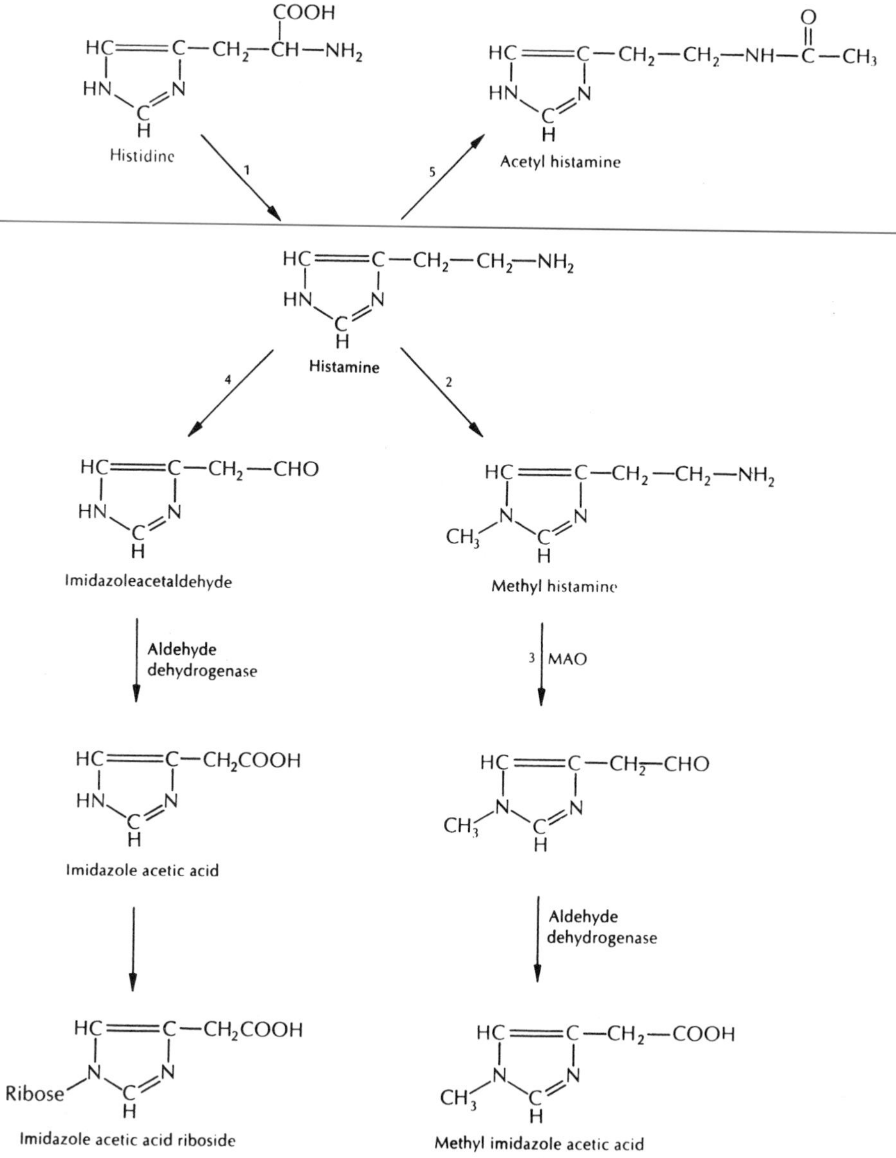

FIGURE 10-9. Metabolism of histamine. (1) Histidine decarboxylase. (2) Histamine methyl transferase. This is the major pathway for inactivation in most mammalian species. (3) Monoamine oxidase. (4) Diamine oxidase (histaminase). (5) Minor pathway of histamine catabolism.

the characterization of histamine receptors in the nervous system. Some progress has been made in recent years to overcome these impediments, and evidence has accumulated to support the hypothesis made more than two decades ago that histamine functions as a neurotransmitter in the brain.

Histamine Synthesis and Catabolism

Because histamine penetrates the brain from blood so poorly, brain histamine arises from histamine synthesis in situ from histidine. Active transport of histidine by brain slice has been demonstrated, and, because histidine loading has been shown to elevate brain histamine, histidine transport could be a controlling factor in brain histamine synthesis.

Two enzymes are capable of decarboxylating histidine in vitro: L-aromatic amino acid decarboxylase (i.e., DOPA-decarboxylase) and the specific histidine decarboxylase. The pH optimum, affinity for histidine, effects of selective inhibitors, and the regional distribution of histamine-synthesizing activity indicate that the specific histidine decarboxylase is responsible for histamine biosynthesis in brain (Fig. 10-9). The instability of histidine decarboxylase and its low activity in adult brain have precluded purification of the enzyme from this source, but fetal liver histidine decarboxylase has been purified to near homogeneity. Studies of the pH optimum, cofactor (pyridoxal phosphate), requirements, inhibitor sensitivity, and antigenic properties have demonstrated that the brain and fetal enzyme have similar properties. Antibodies to histidine decarboxylase have been used to map for the distribution of this enzyme in brain by employing immunohistochemical techniques.

The estimates of the K_m of histidine for brain histidine decarboxylase are much higher than the concentration of histidine in plasma or brain, suggesting that this enzyme is not saturated with substrate in vivo and consistent with the observations that administration of L-histidine increases brain histamine levels.

There are surprisingly few selective inhibitors of histidine decarboxylase. Most of the effective inhibitors act at the cofactor site and, although they reduce histamine formation, they also inhibit other pyridoxal-requiring enzymes.

Recently, a selective inhibitor of histidine decarboxylase, α-fluoromethylhistidine (FMH), has been identified. This agent acts selectively and irreversibly by formation of a covalent linkage with the serine residue at the active site of the enzyme. FMH is more selective and potent than other decarboxylase inhibitors and does not inhibit DOPA and glutamate decarboxylase or other histamine-metabolizing enzymes such as histamine N-methyltransferase and N-acetyl-histamine deacetylase. This agent may prove useful for manipulating in vivo actions of histamine.

Histamine is metabolized by two distinct enzymatic systems in mammals. It is oxidized by diamine oxidase to imidazoleacetaldehyde and then to imidazoleacetic acid and is methylated by histamine methyl transferase to produce methyl histamine. Mammalian brain lacks the ability to oxidize histamine and nearly quantitatively methylates it. Methyl histamine undergoes oxidative deamination by either diamine oxidase or monoamine oxidase. The affinity is higher for diamine oxidase, but, in brains that lack diamine oxidase, methyl histamine is oxidized primarily by monoamine oxidase type B.

The major route of catabolism of orally injested histamine is via N-acetylation of bacteria in the gastrointestinal tract to form N-acetylhistamine.

Histamine-Containing Cells

Although there are fluorogenic condensation reactions that can detect histamine in mast cells by cytochemistry, the method has never been able to demonstrate histamine-containing nerve fibers or cell bodies. Because the histamine content of mast cells is quite substantial, however, it has been possible to use the cytochemical method to detect mast cells in brain and peripheral nerve. From such studies, Schwartz has estimated that the histamine content of brain regions and nerve trunks that show about 50 ng histamine/gm (i.e., every place except the diencephalon) could all be explained on the basis of mast cells. Moreover, the once inexplicable rapid decline of brain histamine in early postnatal development can also be attributed to the relative decline in the number of mast cells in the brain during development.

Mast cell histamine shows some interesting differences from what we shall presume is neuronal histamine: In mast cells, the histamine levels are high, the turnover is relatively slow, and the activity of histidine decarboxylase is relatively low; moreover, mast cell histamine can be depleted by mast cell–degranulating drugs (48/80 and polymyxin B). In brain, only about 50% of the histamine content can be released with these depletors, and for that which remains, the turnover is quite rapid. The activity of histidine decarboxylase is also much greater. Even better separation of the two types of cellular histamine dynamics comes from studies of brain and mast cell homogenates. In the adult brain, a significant portion of the histamine is found in the crude mitochondrial fraction that is enriched in nerve endings and is released from these particles on hyposmotic shock. In cortical brain regions and in mast cells, most histamine is found in large granules that sediment with the crude nuclear fraction, and this histamine, unlike that in the hypothalamic nerve endings, has the slow half-life characteristic of mast cells.

Despite the encouraging result that histamine may be present in fractions of brain homogenates containing—among other broken cellular elements—nerve endings, it is difficult to parlay that information into a direct documentation of intraneuronal storage. More promising steps in that direction were taken in studies where specific brain lesions were made. The student will recall that lesions of the medial forebrain bundle region produced a loss of forebrain NE and 5-HT along a time course parallel to nerve fiber degeneration, in experiments that were done at a time when our understanding of CNS monoaminergic systems was not so far along as it is for histamine today. When such lesions were placed unilaterally and specific histidine decarboxylase activity followed, a 70% decline in forebrain histidine decarboxylase activity was found at the end of 1 week. The decline was not due to the concomitant loss of monoaminergic fibers, since lesions made by hypothalamic injections of 6-OHDA or

5,7-dihydroxytryptamine did not result in histidine decarboxylase loss. The most reasonable explanation would be that the lesion interrupted a histamine-containing diencephalic tract; the histamine and decarboxylating activity on the ipsilateral cortex was reduced only about 40%, suggesting that the pathway may make diffuse contributions to nondiencephalic regions. However, subsequent studies of completely isolated cerebral cortical "islands" show a complete loss of histamine content. As mentioned earlier, antibodies to histidine decarboxylase have also been utilized to map the distribution of this enzyme in rodent brain by immunohistochemical techniques. Steinbusch and coworkers have developed an immunohistochemical method for the visualization of histamine in the CNS using an antibody raised against histamine coupled to a carrier protein. Results obtained with this technique are in general agreement with studies on the immunohistochemical localization of histidine decarboxylase. The highest density of histamine-positive fibers are found in brain regions known to contain high histamine levels, such as the median eminence and the premammillary regions of the hypothalamus.

Lesion and immunohistochemical studies indicate the presence of a major histamine-containing pathway emanating from neurons in the posterior basal hypothalamus and reticular formation and ascending through the medial forebrain bundle to project ipsilaterally to broad areas of the telencephalon (Fig. 10-10). A descending pathway originating from the hypothalamus and projecting to the brain stem and spinal cord has also been suggested.

The majority of histamine-containing perikarya are comprised of a continuous group of magnocellular neurons numbering about 2000 in the rat and confined primarily to the tuberal region of the posterior hypothalamus, collectively named the tuberomammillary nucleus. A similar organization has also been described in humans, although histamine-containing neurons are more abundant (>64,000) and occupy a greater portion of the hypothalamus. Like other monoaminergic neurons, the histaminergic system consists of long, highly divergent, mostly unmyelinated, varicose fibers projecting in a diffuse manner to many telencephalic, mesencephalic, and

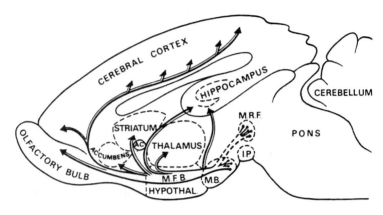

FIGURE 10-10. Schematic illustration of the distribution of histamine-containing neurons in brain. M.R.F., mesencephalic reticular formation; M.B., mammillary bodies; M.F.B., medial forebrain bundle. (Modified from Schwartz et al., 1986)

cerebellar structures (Fig. 10-10). Individual neurons appear to project to widely divergent areas but make few typical synaptic contacts. The histamine neurons are unique in that they are characterized by the presence of an unusually large number of biological markers normally associated with other neurotransmitter systems, including glutaminic acid decarboxylase, adenosine deaminase, Substance P, galanin, met-enkephalin, and brain natriuretic peptide. The tuberomammillary neurons also contain monoamine oxidase B, the major enzyme responsible for deamination of N-methylhistamine, a major metabolite of histamine in the CNS. Unraveling the possible roles of such a large number of putative cotransmitter peptides in a single population of magnocellular neurons in the posterior hypothalamus (the tuberomammillary nucleus) poses an exciting challenge.

Histamine Receptors

Three classes of histamine receptors have been identified in vertebrates and named in the order of their discovery: H_1, H_2, and H_3.

Both H_1 and H_2 receptor DNA have been cloned and found to belong to the superfamily of receptors with seven-transmembrane domains and to be coupled to guanylnucleotide-sensitive G proteins. The molecular structure of the H_3 receptor has not yet been established. However, the observation that the binding of (R)-—methylhistamine is regulated by guanylnucleotides strongly suggests that the H_3 receptor, like the other histamine receptors, belongs to the superfamily of G-protein-coupled receptors.

Histamine H_1 receptors are coupled to inositol phospholipid hydrolysis and are involved in a variety of responses, including contraction of smooth muscle, increased capillary permeability mediated by contraction of terminal venules, hormone release, and brain glycogenolysis induced by a rise in the intracellular concentration of free calcium ions. H_2 receptors are positively coupled to adenylate cyclase via a G_s protein and mediate most of their effects by changes in the intracellular levels of cAMP. The functional responses mediated by H_2 receptors include smooth muscle relaxation, gastric acid secretion, positive chronotropic and inotropic actions on cardiac muscle, and inhibitory effects on the immune system.

A third class of histamine receptors, the H_3 receptors, was initially identified as autoreceptors controlling histamine synthesis and release from histamine-containing neurons in rat cerebral cortex. Subsequently, the H_3 receptor was shown to inhibit the presynaptic release of other monoamines and peptides from brain and peripheral tissue. Functional studies have provided evidence for presynaptic release modulating H_3 receptors on noradrenergic, serotonergic, dopaminergic, cholinergic, and peptidergic neurons. These H_3 receptors are believed to provide a major regulatory mechanism involved in the control of histaminergic activity under physiological conditions. In vivo microdialysis studies have demonstrated that administration of selective H_3 receptor agonists reduces histamine release and turnover. Administration of H_3 receptor antagonists, in contrast, enhances histamine turnover and release, suggesting that autoreceptors are normally under tonic stimulation by endogenous histamine.

H_1 Receptor Agonists and Antagonists

Because of its high affinity and good receptor selectivity, mepyramine remains the agent of choice as a selective H_1 antagonist. A number of H_1 antagonists exist as optical isomers, and in many instances the (+) stereoisomer exhibits higher potency than its (−) stereoisomer counterpart. (+)-Chlorpheniramine is such an example. One of the most potent H_1 antagonists is the geometric isomer transtriprolidine, which has a K_d of 0.1 nM. In humans, a major side effect associated with administration of H_1 antagonists is the high degree of sedation elicited by these agents, which has been attributed to their ability to occupy H_1 receptors in brain. Most of the classic H_1 antihistamines (e.g., diphenhydramine, mepyramine) readily cross the blood–brain barrier and elicit a significant degree of sedation. A number of new compounds, however penetrate poorly into the CNS and thus are relatively nonsedating. Terfenadine (Fig. 10-11) is an example of such an agent. Highly selective agonists do not exist for H_1 receptors. 2-Methylhistamine and 2-thiazoylethylamine show some selectivity, but their relative potency at H_1 receptors differs by no more than an order of magnitude from that exhibited at H_2 receptors.

H_2 Receptor Antagonists

Following the successful use of H_2 receptor antagonists like cimetidine in the treatment of peptic ulcers, a number of these agents have become available. The structures of several H_2 antagonists are illustrated in Fig. 10-12. Although a number of more potent agents have been developed, only ranitidine and tiotidine lack antagonist activity on H_3 receptors. These agents all have limited access to the CNS. The only selective H_2 receptor antagonist that effectively penetrates into the CNS is zolantidine. This compound has been used extensively in experimental animals but has not been marketed for clinical use. Dimaprit is a fairly selective H_2 agonist that discriminates well between H_2 and H_1 or H_3 receptors. Impromidine is one of the most potent H_2 receptor agonists available. Like dimaprit, it has

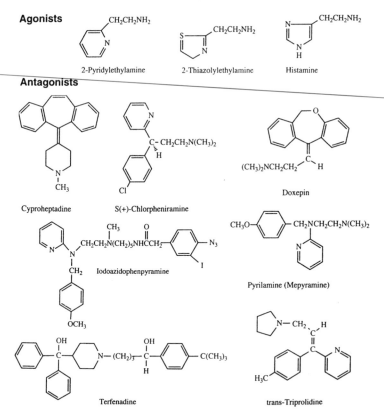

FIGURE 10-11. H₁ agonists and antagonists.

minimal agonist effects on H_1 receptors, but it has more potent H_3 receptor antagonistic properties. Amthamine is also a more potent agonist at H_2 receptors than diamaprit that does not affect H_1 receptors. However, this agent exerts weak agonist activity at the H_3 receptor and thus does not exhibit optimal subtype selectivity.

H_3 receptor antagonists and agonists are illustrated in Fig. 10-13. Thioperamide is a potent H_3 receptor antagonist that exhibits good receptor specificity. This drug, unlike other H_3 antagonists, also crosses the blood–brain barrier and has been useful in evaluating the

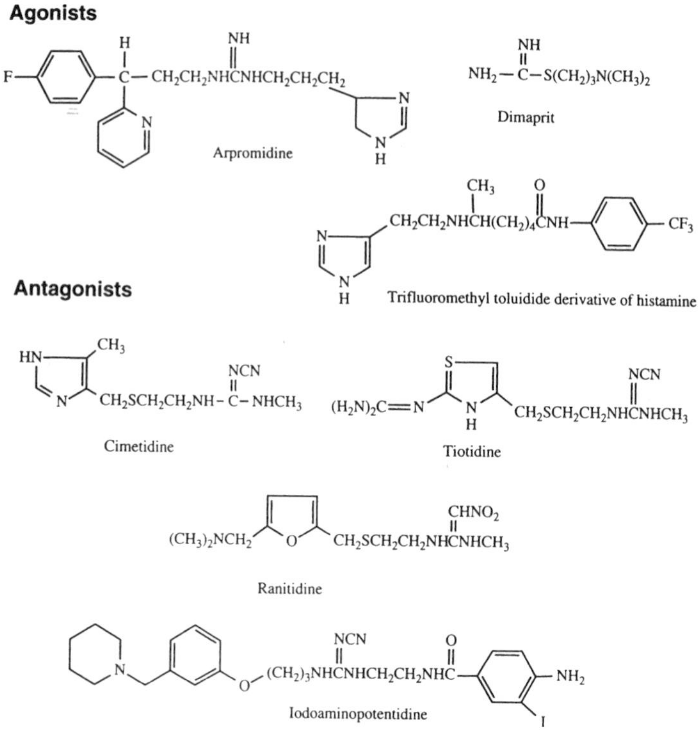

Agonists

FIGURE 10-12. H_2 agonists and antagonists.

behavioral role played by H_3 receptors in the CNS. A new antagonist, clobenpropit, is the most potent H_3 receptor antagonist currently available for investigating these receptors. $N\alpha$-methylhistamine and $N\alpha,N\beta$-dimethylhistamine are both potent H_3 receptor agonists, but neither of these agents shows any marked discrimination between the three histamine receptors. However, substitution of methyl groups in the α or β positions of the side chain of the histamine molecule produces agents that exhibit a high degree of selectivity of the H_3 receptors. For example, $R\alpha$-methylhistamine and $R\alpha,S\beta$-dimethylhistamine both are 15–20 times more potent than histamine as H_3 agonists but possess only about 1% of the activity of histamine on H_1 and H_2 receptors. Imetit is a new potent and selec-

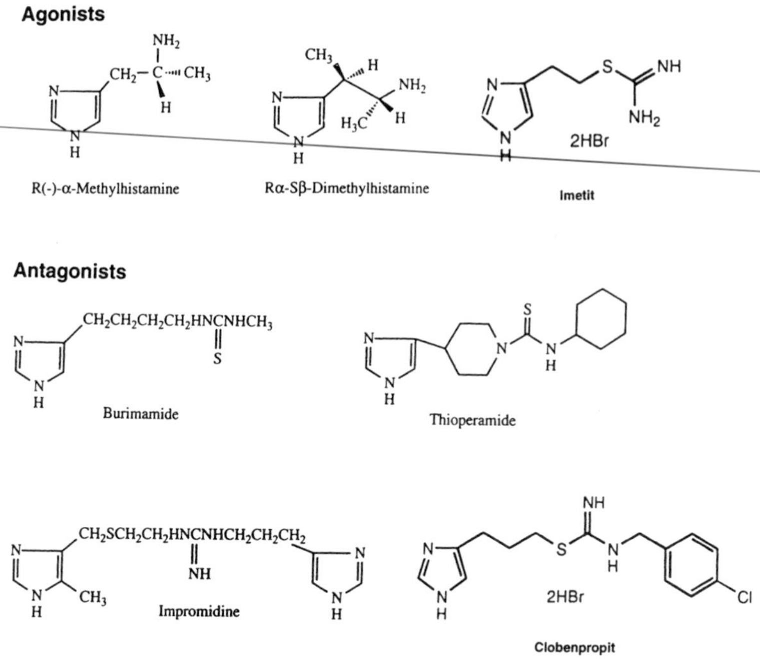

Agonists

R(-)-α-Methylhistamine

Rα-Sβ-Dimethylhistamine

Imetit

Antagonists

Burimamide

Thioperamide

Impromidine

Clobenpropit

FIGURE 10-13. H$_3$ agonists and antagonists.

tive H$_3$ receptor full agonist that is effective both in vivo and in vitro and 60 times more potent than histamine. Table 10-5 summarizes the pharmacology of histamine receptors and their effector pathways.

At present, none of the histamine receptors has been purified or cloned. The second messenger or associated channel of the H$_3$ receptors has not yet been identified. While the second messengers of the H$_1$ and H$_2$ receptors are known, details concerning the intracellular mechanisms they trigger are still poorly understood.

In summary, there is compelling evidence that histamine, in addition to its presence in mast cells, exists in neurons where it probably functions as a neurotransmitter in the CNS. Brain has a similar non-uniform distribution of histamine, a specific histidine decarboxylase, and methyl histamine (the major metabolite of brain histamine).

TABLE 10-5. Pharmacology of Histamine Receptors

Receptor Subtype	H_1	H_2	H_3
Selective agonists	2-(m-Bromophenyl) histamine	Amthamine Impromidine[a] (also an H_3 antagonist)	(R)-α-Methylhistamine[c] (R)-α,(S)β-Dimethylhistamine[c] Imetit[c]
Selective antagonists	Mepyramine[a] Triprolidine[a]	Ranitidine[b] Tiotidine[b]	Thioperamide[c] Clobenpropit[c]
Effector pathways	↑ IP_3/DAG	↑ cAMP	?

[a]See Fig. 10-11.

[b]See Fig. 10-12.

[c]See Fig. 10-13.

The cerebellum has the lowest levels, whereas the hypothalamus is most enriched. Hypothalamic synaptosomes contain histamine and histidine decarboxylase, suggesting a localization in nerve endings. Depolarization of hypothalamic slices causes a calcium-dependent release of histamine, although histamine release from neurons has not been convincingly demonstrated in in vivo. The turnover of brain histamine is quite rapid, as with other biogenic amine transmitters.

The actions of histamine in the CNS appear to be mediated by three classes of receptors. These three types, H_1, H_2, and H_3, can be distinguished in the CNS by their pharmacology, their localization, and the intracellular responses they mediate.

Neuropharmacological studies of brain histaminergic systems have been hampered by the lack of appropriate pharmacological agents. In the last few years, parenterally active drugs, H_2 antagonists, H_3 agonists, and antagonists have become available, and their effects on behavioral, neuroendocrine, and vegetative functions are being studied in animals.

Lesions of the midbrain or caudal hypothalamus cause a progressive reduction in forebrain histamine and histidine decarboxylase, consistent with the existence of an ascending histamine-containing system, which has now been directly visualized by employing immunohistochemical techniques. Continued application of these immunohistochemical techniques for visualization of histidine decarboxylase-like and histamine-like activity in brain should provide more detailed maps of histamine-containing neuron systems in the brain in the near future. However, we are still far from being able to speak with conviction of specific histaminergic circuits and their relation to behavior, and thus the functions of histamine in brain are highly speculative.

The high levels of histamine and the presence of histamine receptors in the hypothalamus, together with the known ability of histamine to alter food and water intake thermoregulation, autonomic activity, and hormone release, indicate a possible role for histamine in these vegetative functions. Also, because of the presence of ascending histamine-containing fibers originating from the reticular formation, coupled with the sedative nature of H_1 antagonists and the activation effects of histamine on some brain cells, a possible role for histamine in arousal has been suggested.

Despite numerous speculations based largely on observations of the response to locally applied histamine, only a few physiological roles for histamine seem to be well-documented. These include arousal, control of appetite, regulation of pituitary hormone secretion, and control of vestibular reactivity. The therapeutic role for histaminergic drugs is fairly limited, and the primary pharmacological actions of these agents are often associated with side effects ascribed to other therapeutic agents. For example, several antipsychotic (e.g., clozapine and thioridazine) and antidepressant (e.g., doxepin and amitriptyline) drugs with sedative and weight gain side effects display potent H_1 antagonist properties. Histamine antagonists have been used therapeutically for only a limited number of conditions. H_1 antagonists like meclizine are the most commonly used anti-motion sickness drugs. However, it is still uncertain whether H_1 blockade is primarily responsible for the effectiveness of these agents. Some H_1 antagonists, notably meclizine and dimenhydrinate, are of benefit in vestibular disturbances such as Meniere's disease. H_1 receptor antagonists are also the ingredients of a number of proprietary remedies sold over the counter as sleeping pills for the treatment of insomnia. These preparations are generally ineffective in the recommended dosages, although some sensitive individuals may derive benefit. H_1 antagonists have a well-established and valued place in the symptomatic treatment of hypersensitivity reactions (allergic rhinitis and conjunctivitis). However, their effect is purely palliative, being confined to the suppression of the symptoms attributed to the antigen–antibody-induced release of histamine. Despite popular belief, H_1 antagonists are without value in treatment of the common cold.

We would be remiss not to mention the therapeutic usefulness of the H_2 antagonists. The main therapeutic use of these agents is in the treatment of gastric and duodenal ulcers. H_2 antagonists are very effective in lowering the basal and nocturnal secretion of gastric acid and that which is stimulated by meals. They both reduce the pain and the consumption of antacids and hasten the healing of the ulcers. The incidence of side effects with these H_2 antagonists (cimetidine and ranitidine) is low and can probably be attributed, in part, to the limited function of H_2 receptors in organs other than the stomach and to the poor penetration of these agents across the blood–brain barrier.

SELECTED REFERENCES

Serotonin

Aghajanian, G. K. (1995). Electrophysiology of serotonin receptor subtypes and signal transduction pathways. In *Psychopharmacology, Fourth Generation of Progress* (F. E. Bloom and D. J. Kupfer, eds.), pp. 451–460, Raven Press, New York.

Barker, E. L. and R. D. Blakely (1995). Norepinephrine and serotonin transporters: Molecular targets of antidepressant drugs. In *Psychopharmacology, Fourth Generation of Progress* (F. E. Bloom and D. J. Kupfer, eds.), pp. 321–333, Raven Press, New York.

Blier, P. and C. deMontigny (1994). Current advances and trends in the treatment of depression. *TIPS 15*, 220–226.

Descarries, L., M. A. Audet, G. Eoucet, et al. (1990). Morphology of central serotonin neurons: Brief review of quantified aspects of their distribution and ultrastructural relationships. In *The Neuropharmacology of Serotonin* (Whitaker-Azmitia and Peroutka, eds.), pp. 81–92, New York Academy of Sciences, New York.

Fozard, J. R. (ed.) (1989). *The Peripheral Actions of 5-Hydroxytryptamine*. Oxford University Press, New York.

Glennon, R. A. and M. Dukat (1995). Serotonin receptor subtypes. In *Psychopharmacology, Fourth Generation of Progress* (F. E. Bloom and D. J. Kupfer, eds.), pp. 415–430, Raven Press, New York.

Heninger, G. R., P. L. Delgado, D. S. Charney, L. H. Price, and G. K. Aghajanian (1992). Tryptophan-deficient diet and amino acid drink deplete plasma tryptophan and induce a relapse of depression in susceptible patients. *J. Chem. Neuroanatomy 5*, 347–348.

Hoyer, D., D. E. Clarke, J. R. Fozard, P. R. Hartig, G. R. Martin, E. J. Mylecharane, P. R. Saxena, and P.P.A. Humphrey (1994). International union of pharmacology classification of receptors for 5-hydroxytryptamine (serotonin). *Pharmacol. Rev. 46(2)*, 157–203.

Lovenberg, T. W., B. M. Baron, L. deLecea, J. D. Miler, R. A. Porsser, M.

A. Rea, P. E. Foye, M. Racke, A. L. Slone, B. W. Siegel, P. E. Danielson, J. G. Sutcliffe, and M. G. Erlander (1993). A novel adenylyl cyclase-activating serotonin receptor (5-HT7) implicated in the regulation of mammalian circadian rhythms. *Neuron* *11*, 449–458.

Mamounas, L. A., C. A. Mullen, E. O'Hearn, and M. E. Molliver (1992). Dual serotonergic projection to forebrain in the rat: Morphologically distinct 5HT axon terminals exhibit differential vulnerability to neurotoxic amphetamine derivatives. *J. Comp. Neurol.* *314*, 558–586.

Maricq, A. V., A. S. Peterson, A. J. Brake, R. M. Myers, and D. Julius (1991). Primary structure and functional expression of the 5HT$_3$ receptor, a serotonin-gated ion channel. *Science* *254*, 432–437.

Monsma, F. J. Jr., Y. Shen, R. P. Ward, M. W. Hamblin, and D. R. Sibley (1993). Cloning and expression of a novel serotonin receptor with high affinity for tricyclic psychotropic drugs. *Mol. Pharmacol.* *43*, 320–327.

Nagatsu, T. (1991). Genes for human catecholaminic synthesizing enzymes. *Neurosci. Res.* *12*, 315–345.

Owens, M. J., and C. B. Nemeroff (1994). The role of serotonin in the pathophysiology of depression: Focus on the serotonin transporter. *Clin. Chem.* *40*, 288–295.

Peroutka, S. J. (1994). Molecular biology of serotonin (5-HT) receptors. *Synapse* *18*, 241–260.

Sanders-Bush, E. and H. Canton (1995). Serotonin receptors: Signal transduction pathways. In *Psychopharmacology, Fourth Generation of Progress* (F. E. Bloom and D. J. Kupfer, eds.), pp. 431–442. Raven Press, New York.

Saudou, F. and R. Hen (1994). 5-Hydroxytryptamine receptor subtypes in vertebrates and invertebrates. *Neurochem. Int.* *25(6)*, 503–532.

Steinbusch, H.W.M. and A. H. Mulder (1984). Serotonin-immunoreactive neurons and their projections in the CNS. In *Handbook of Chemical Neuroanatomy, Classical Transmitters and Transmitter Receptors in the CNS*, Vol. 3, Part II (A. Björklund, T. Hökfelt, and M. Kuhar, eds.), Elsevier, New York.

Histamine

Hill, S. J. (1990). Distribution, properties, and functional characteristics of three classes of histamine receptor. *Pharmacol. Rev.* *42*, 45–83.

Panula, P. and M. S. Airaksinen (1991). The histaminergic neuronal system as revealed with antisera against histamine. In *Histaminergic Neurons: Morphology and Function* (T. Watanabe and H. Wada, eds.), pp. 127–144, Boca Raton, FL, CRC Press.

Schwartz, J. C., J. M. Arrang, M. Garbarg, H. Pollard, and M. Ruat (1991). Histaminergic transmission in the mammalian brain. *Physiol. Rev. 71*, 1–51.

Schwartz, J. C., J. M. Arrang, M. Garbarg, and E. Traiffort (1995). Histamine. In *Psychopharmacology, Fourth Generation of Progress* (F. E. Bloom and D. J. Kupfer, eds.), pp. 397–406, Raven Press, New York.

Schwartz, J. C. and H. L. Haas (eds.), (1992). *The Histamine Receptor.* New York, Wiley-Liss.

Staines, W. A. and J. I. Nagy (1991). Neurotransmitter coexistence in the tuberomammillary nucleus. In *Histaminergic Neurons: Morphology and Function* (T. Watanabe and H. Wada, eds.), pp. 163–176, Boca Raton, FL, CRC Press.

Steinbusch, H.W.M. and A. H. Mulder (1984). Immunohistochemical localization of histamine in neurons and mast cells in brain. In *Handbook of Chemical Neuroanatomy*, Vol. 3, Part II, *Classical Transmitters and Transmitter Receptors in the CNS.* (A. Björklund, T. Hökfelt, and M. J. Kuhar, eds.), pp. 126–140, Elsevier, New York.

Tohyama, M., R. Tamiya, N. Inagaki, and H. Takagi (1991). Morphology of histaminergic neurons with histidine decarboxylase as a marker. In *Histaminergic Neurons: Morphology and Function* (T. Watanabe and H. Wada, eds.), pp. 107–126, Boca Raton, FL, CRC Press.

Wouterlood, F. G. and H.W.M. Steinbusch (1991). Afferent and efferent fiber connections of histaminergic neurons in the rat brain: Comparison with dopaminergic, noradrenergic and serotonergic systems. In *Histaminergic Neurons: Morphology and Function* (T. Watanabe and H. Wada, eds.), pp. 145–162, Boca Raton, FL, CRC Press.

11 | Neuroactive Peptides

It is with no little amazement that we again note the continuing allure of neuropeptides for neuropharmacologists and for those hoping to devise new medications to enhance or antagonize their actions. The rapid advances in molecular biological understanding of gene structure and regulation of gene expression, the determination of peptide sequences, the characterization of their receptors by cloning and transfection studies (see Chapter 3), and the successful synthesis of both peptidic and nonpeptidic antagonists for these receptors fuel this attraction. The new student entering this field might assume that the peptide story must be nearly all told. Alas, this appearance is misleading. As in most of the fields discussed in this book, better data beget better and more complex questions. Nevertheless, readers can test here their comprehension of how to identify a neurotransmitter and characterize the role of the circuits that contain it.

Almost all peptides made by neurons will affect specific target cells of the central and peripheral nervous systems: neuronal, glial, smooth muscle, glandular, and vascular. The census of peptides already exceeds several dozen, with no obvious end in sight. However, the actual molar concentrations of any given peptide in the brain is maximally two to three orders of magnitude lower than that of the monoamines, acetylcholine, and amino acids, and their receptors respond at correspondingly lower concentrations. The specific signaling advantages of peptides are not yet obvious. Furthermore, with very few exceptions, no peptides have yet been linked selectively to a given functional system in the brain or correlated in any exclusive manner with any pathological states. Thus, the challenge of devising new drugs for neurological or psychiatric disorders based on these peptides is considerably more difficult than *just* designing potent congeners that can tweak their receptors on or off: It requires that we have some reason to do that. Nevertheless, there are strong implications that the biological properties evoked by peptide-mediated signals will influence future pharmacological development.

SOME BASIC QUESTIONS

What's Different about Peptides?

Peptide-secreting neurons differ in their biology from neurons, using amino acids and monoamines in ways other than the molecular structure of their transmitter. The amino acid or monoamine transmitters are formed from dietary sources by one or two intracellular enzymatic steps; the end product of these enzymatic actions is the active transmitter molecule, which is then stored in the nerve terminal until release. After release, the transmitter (or choline in the case of ACh) can be reaccumulated back into the nerve terminal by the energy-dependent active reuptake property, thus conserving the requirement for de novo synthesis. Peptide-secreting cells employ a somewhat more formidable approach: Synthesis directed by mRNA can take place only on ribosomes and thus only in the perikaryon or dendrites of a neuron. Furthermore, all known peptides arise from larger prohormone forms (inactive precursors). The biologically active, secreted form is cleaved from these prohormones by the actions of selective proteolytic enzymes.

Thus, the process for peptides starts with ribosomal synthesis of the prohormone, which is then packaged into vesicles in the smooth endoplasmic reticulum, and transported from the perikaryon to the nerve terminals for eventual release. These steps in the molecular processing of secretory products are described in greater detail in Chapter 3.

Insofar as present data reveal, peptide-releasing neurons share certain basic properties with all other chemically characterized neurons: Transmitter release is Ca dependent, and some of the postsynaptic effects may be mediated by directly altering ion channel conductances or indirectly regulating ion channels through second messengers such as Ca^{2+}, cyclic nucleotides, or inositol triphosphates (see Chapter 6). Furthermore, like the monoamines, peptides identical to those extracted from the nervous system are often made by nonneural secretory cells in endocrine glands and the mucosa of the hollow viscera. Some observers of the peptide scene would like to emphasize characteristics of neuropeptide action that are extrapolations from the actions of endocrine peptide hormones: targets that are distant from release sites, release into portal or other circulatory systems, higher potency, and a longer time course of effects. Although endocrine hormones (peptides or amines) can produce effects in the nanomolar range or below, we do not yet know the time course or thresholds for the same neuropeptides to act on their target cells in the brain or ganglia. In amphibian autonomic ganglia, at least one peptide does act in such a local endocrine, or *paracrine*, fashion, producing synaptic potentials that last for hundreds of milliseconds. These effects of an endogenous peptide that is very similar to hypothalamic luteinizing hormone-releasing hormone are produced on a nonsynaptic neuronal target by diffusing at least 10–100 μm from preganglionic synaptic release sites, without apparently acting at all on the immediate postsynaptic neuron. Whether these particular paracrine properties are ways around the simple amphibian autonomic anatomy or a prototype for all central and peripheral peptide actions remains debatable.

How Are New Peptides Found?

Before considering the interesting relationships among the many known peptides, we can anticipate future events by examining the recurrent pattern of events that seems to underlie every wave of peptide discovery, and characterization. From this historical overview, the student can see that new peptides do not just happen into existence. The primary necessity for the discovery of a new peptide has classically been the recognition of a biological function controlled by an unknown factor; this biological action is then employed as the bioassay for the subsequent steps of the discovery process. Although this approach was used for the characterization of gastrointestinal peptide hormones in the 1940s and 1950s, the process was perfected and greatly increased in power by Guillemin and Schally and their teams during the 1960s and 1970s, as they pursued the hypophysiotrophic hormones by which the hypothalamus regulates the secretion of adenohypophysial (anterior pituitary) hormones for control of the major peripheral endocrine tissues. The starting bioassay is used as a screen for chemical manipulations of tissue extracts, and through chemical manipulations of the extract (i.e., differential extractions, molecular sieves, etc.) the active factor is enriched and ultimately purified. For some factors, the purification of nanograms of active peptide required hundreds of thousands of brains. The peptidic nature of the purified material is inferred by the loss of biological activity after treatment with peptidases (carboxypeptidases or endopeptidases). The cleaved peptide fragments are then sequenced, and the entire factor is eventually reconstructed.

When a sequence has been achieved and matched against the amino acid composition of the pure factor, it is then necessary to synthesize the peptide and determine if the synthetic replicate matches actions and potencies with the purified natural factor. Only when the replicate's properties match those of the factor can the natural peptide be considered to have been "identified." At this point the factor is given a name that reflects its actions in the original bioassay.

However, this is far from the end of the discovery process. When the chemical structure of a peptide has been identified and synthetically replicated, two new opportunities arise: Large amounts of the synthetic material can be prepared for examination of physiological and pharmacological properties; at the same time, antibodies can be prepared against the synthetic material. The antibodies are used for radioimmunoassay and for immunocytochemistry. The quantitative measurements made possible by the radioimmunoassay and the qualitative distribution of the peptide assessed by microscopic cytochemical localizations then frequently reveal that the peptide has a much broader distribution, and, therefore, ostensibly has a much more general series of actions than was originally presumed. Autoradiographic mapping of potential response sites (first by ligand binding for peptides and more recently by in situ hybridization for the mRNAs of their receptors) is generally the last step in the molecular and cellular characterization. Curiously, these maps often flagrantly disregard the cellular distribution of the recognized forms of the endogenous peptides; this mismatch strengthens in some minds the concept of paracrine diffusion for peptide actions. Thereafter, the endogenous peptide or an analog of it is ordinarily proposed to be the cause or cure of a major mental illness. In many of the more recent reports, it is frequently observed that the peptide isolated and found to be active is, in fact, not the only active form contained in the tissue: In many cases better extraction methods confirmed by radioimmunoassay reveal larger molecules with full or even greater potency. Although the general rule is that active peptides are synthesized from larger precursor hormones, processing can sometimes lead to active larger peptides. These cannot really be viewed as "precursors," since they have potent activity in their own right.

As more and more peptides have been identified within the nervous system and from other sources, their internal structural similarities allow us to consider them as members of one or another family grouping of peptides. The family groupings are useful because they indicate that certain sequences of amino acids have had considerable evolutionary conservation, presumably because they provide unambiguous signals between secreting and responding cells.

Two aspects of peptide family groupings deserve recognition: (1) families in which separate propeptide genes or mRNAs sharing one or more short peptide sequences are expressed by different cell groups in the same species; (2) families in which the peptides expressed by homologous cell groups share generally identical peptide sequences with minor to moderate variation. The first category, exemplified by the opioid peptide and the tachykinin peptide families, implies that the natural ligand's receptors may diverge into subtypes that can discriminate the fine differences in peptide sequences. Although some individual peptides have already earned their own receptor subtypes (vasopressin, somatostatin), in other cases, such as the tachykinins and the opioid peptides, the multiple receptor sub-types predicted the later definition of multiple natural ligands. In those cases in which the propeptide form may give rise to multiple different agonists, it is possible—but by no means yet broadly supported—that more than one peptide messenger is released from the possible array. Some dreamy-eyed pundits have theorized that different branches of a neuron might have the option of changing the mix of cleavage products available for release, perhaps under steroid-driven conditions.

The latest additions to the list of neuroactive peptides have been discovered by strategies that depart from the classic bioassay strategy purification schemes and instead rely either on specific common general features, such as a C-terminally amidated sequence, or on the detection of potential cleavage fragments when the propeptide sequence has been deduced by molecular cloning of the mRNA or gene. Although these historical discovery cycles are individually entertaining stories, the punch lines—determining what the peptides do and how to make useful drugs based on these actions—still remain elusive.

Transmitters or Not?

In Chapter 3 we considered the kinds of evidence needed to demonstrate that a substance found in nerve cells is the transmitter that those cells secrete at their synapses. To recapitulate, it must be shown that the substance is present within the nerve cell and specifically in its presynaptic terminals, that the nerve cell can make or accumulate the substance, and that it can release that substance when activated; when the substance is released, it must be shown to mimic in every aspect the functional activity following stimulation of the nerves that released it, including the magnitude and quality of changes in postsynaptic membrane conductances. A criterion especially pertinent to the objectives of this book is that drugs that can simulate or antagonize the effects of nerve fiber activation must have an identical influence on the effects of the substance applied exogenously to the target cell.

In the case of peptides, the opportunities for supplying these data would at first glance seem better than they are with the many categories of small molecules we have discussed in previous chapters. Once the peptide has been isolated and its structure determined, immunocytochemical localization of the peptide usually follows promptly, as does the demonstration that, with the right sensitive assay, the peptide can be shown to be released from brain slices by a Ca^{2+}-dependent, depolarization process. The development of a suitable synthetic ligand or receptor mRNA in situ probe helps define receptor distribution, possible actions on presumptive synaptic targets, their mechanisms of regulation, and the discrimination of receptor subtypes. In several peptide families, synthetic agonist and antagonist peptide analogs have been employed to document behavior-altering actions of the exogenous forms, and, by inference, a role of the endogenous form in behavior and possibly in disease states. Recent advances here include short-term inactivation of peptide effects by intracerebral injection of antisense oligonucleotides and longer-term regulation of peptide synthesis in transgenic animals by overexpression or gene knock-outs. Although these molecular tools have accelerated the early partial fulfillment of transmitter criteria for peptides generally, there is only one case within the CNS in which the complete array of neuronal actions transmitted by a peptide-containing circuit have been shown to be completely attributable to the peptide. There are important reasons for this unsatisfying state of peptide transmitter documentation, and lack of effort is not one of them.

They Are Not Alone

Previous editions of this book were organized according to a once-prevalent concept in neuropharmacology: A given neuron operated by one and only one transmitter. According to this concept, one aspect of a neuron's phenotype, as classic as its size, shape, location, circuitry, and synaptic function, was its neurochemical designation as "GABAergic" or "cholinergic" or "adrenergic" or whatever. All card-carrying pharmacologists knew that autonomic neurons came in only two color-coded categories: adrenergic and cholinergic. And then a curious finding began to repeat itself: Neuroactive peptides started showing up in autonomic neurons where there was no need for additional transmitters. Initially viewed by many as an oddity, the idea of coexisting peptides in central as well as peripheral neurons that are already occupied by an amino acid or monoamine or even another neuropeptide has now gained wide acceptance (see Table 11-1).

The notion of coexistence of other transmitters with peptides leads one to ask whether there is any such thing as a truly "peptidergic" neuron, or rather only neurons in which a peptide or two is there to expand the armamentarium of messenger molecules available.

A neuron might then be able to transmit an *enriched* message by employing more than one type of transmitter for its fully refined signal content. Although the latter concept is far from a neuroscientific consensus, the data in hand are certainly compatible, recognizing the enormous void in our knowledge in which the majority of neurons have not yet had either their peptidic or nonpeptidic transmitters identified.

Such a revolutionary concept then forces us to ask: If the peptide is not the sole signaling molecule, but rather a minor, fractional percent of messenger-signaling capacity of the nerve terminal, how might peptides affect signals transmitted by the coexisting amino acid or monoamine? Except for some recent work with opioid peptide circuits in hippocampus (see below), "We don't know" remains the easy answer. However, in the autonomic nervous system, where Tomas Hökfelt and his peptide liberation team have managed to find at least one peptide in every sympathetic, parasympathetic, or enteric neuron they have studied, some solid peptide–monoamine interactions are emerging. Vasoactive intestinal peptide (VIP), released when cholinergic nerves discharge at high frequency, augments parasympathetic control of salivation by increasing glandular blood flow. Neuropeptide Y (NPY), found in many sympathetic neurons, sensitizes smooth muscle target cells to adrenergic signals.

TABLE 11-1. Examples of Neuroactive Peptides Coexisting with Other Transmitters

Transmitter	Peptide	Location
GABA	Somatostatin	Cortical and hippocampal neurons
	Cholecystokinin	Cortical neurons
Acetylcholine	VIP	Parasympathetic and cortical neurons
	Substance P	Pontine neurons
Norepinephrine	Somatostatin	Sympathetic neurons
	Enkephalin	Sympathetic neurons
	NPY	Medullary and pontine neurons
	Neurotensin	Locus ceruleus
Dopamine	CCK	Ventrotegmental neurons
	Neurotensin	Ventrotegmental neurons
Epinephrine	NPY	Reticular neurons
	Neurotensin	Reticular neurons
Serotonin	Substance P	Medullary raphe
	TRH	Medullary raphe
	Enkephalin	Medullary raphe
Vasopressin	CCK, dynorphin	Magnocellular hypothalamic neurons
Oxytocin	Enkephalin	Magnocellular hypothalamic neurons

Thus, peptide agonist or antagonist actions might well alter neuronal information flow by disturbing the normal symbiotic relationships between the signals.

Other peptide actions, although not necessarily on coexisting amino acid or monoamine transmitters, have been reported with irregularity: increasing or decreasing the amount of transmitter released by a given stimulus, or enhancing or diminishing the response of a common target cell (in either amplitude or duration or both) to combined transmitter signals. For example, VIP greatly accentuates the cAMP response of cortical neurons to low doses of norepinephrine, and galanin changes the release of acetylcholine in ventral hippocampus. Several peptides have been reported to alter the affinity of postsynaptic receptors for their nonpeptidic ligands.

One way to recast these players in the proper perspective might be to consider that a peptide usually embellishes what the primary transmitter for a neuronal connection seeks to accomplish. Such an effect may be to strengthen or prolong the primary transmitter's actions, especially when the nerve is called on to fire at higher than normal frequencies. The peptide may provide a part of the intraneuronal signal to alter the rate of production of the primary transmitter, and there may still be places in which the peptide itself can fulfill all of the effects of a primary transmitter. Of course, accepting coexistences leads to new logistical problems for peptide-using neurons: how to coordinate synthesis of the peptide with the small-molecule transmitters (recall that the latter can be made within the synapse and are often re-accumulated after release but that peptides can only be made with ribosomes and not in nerve terminals) and how to decide when the secreted transmission should include the peptide?

Several other important concepts for pharmacology also emerge: the effects of peptide-directed drugs will depend not only on the location of the receptors and their actions, but on the context of its coexisting signals. Thus, the pharmacology of NPY or neurotensin may be most appropriately understood as a part of the total picture of central noradrenergic pharmacology, given the degree to which these two peptides may participate in that monoaminergic neuron's transmission. Given the relatively modest number of response mechanisms on which peptide messengers may operate (e.g., direct regulation of Na, K, or Ca channels, or their second messenger-mediated regulation), a great enrichment of signaling possibilities becomes attainable through the interplay between frequency-dependent and diffusion-dependent release and response sites. (In terms of the development of drugs active on the nervous system, research on the peptides has given us a humbling view of the riches that may await mining in the inner depths of the brain.) The rest of this chapter reviews pharmacologically relevant data for the grand peptide families (oxytocin/vasopressin, the tachykinins, glucagon-related peptides, pancreatic polypeptide-related peptides, and opioid peptides) a few individual peptides (CCK, somatostatin, angiotensin, calcitonin-gene-related peptide, and CRF), and a new large family of neuroactive and "glia-active" peptides, the *cytokines*, which may figure prominently in pharmacological attacks on neurodegenerative disorders.

EMERGING TRENDS IN NEUROPEPTIDE RESEARCH

Observers of the active peptide parade may find it useful to monitor certain recurring themes in several of the brain and autonomic peptide systems. They illustrate potential principles that could apply to peptide signaling systems generally.

Peptides are dynamically regulated. In the past few years, the dynamic synthesis and catabolism of neuropeptides has been estimated by measuring simultaneously mRNA, the precursor propeptide, and the signaling form. These studies have helped clarify the effects on peptide synthesis of drug treatments or other experimental perturbations that alter the levels of the coexisting amines in a given class of neurons. Furthermore, examination of the genes for some neuropeptides has revealed regulatory genetic elements that make gene transcription sensitive to gonadal hormones (oxytocin, vasopressin, and pro-enkephalin) or to intracellular second messengers (vasoactive intestinal polypeptide, pro-enkephalin, neurotensin, corticotropin releasing hormone, and prosomatostatin), providing direct proof of independence from their cotransmitters.

Post-release cleavage products may provide different signals. Based on the absence of any specific data, it had been assumed that despite the enormous logistical prelude to constructing a propeptide and then cleaving and storing the active form, its secretion was the end of the line. In several peptide systems, however, pharmacological evidence suggests that catabolic forms of the secreted peptide may have real signal value, both on the neuron that secreted the full peptide and on the postsynaptic neuron. The data suggest that such shorter forms may participate in fine-tuning the levels of interaction between the peptides and other coexisting transmitters.

Nonpeptidic agonist and antagonist drugs. The third emerging trend is less an example of peptide physiology than an expression of appreciation for the achievements of medicinal chemists in synthesizing rigid nonpeptidic molecules that can act as agonists and (experimentally more important) as antagonists of the peptide's receptors. While it was clear that mother nature could do this in the form of the basic morphine molecule by making an agonist for opioid peptide receptors, extension of this principle to other peptides was slow in coming. However, new molecules are coming forth, and, as nonpeptidic antagonists are created for peptide systems, the ability to dissect out their functional roles will almost certainly be improved.

THE GRAND PEPTIDE FAMILIES

Vasopressin and Oxytocin

These two highly similar nonapeptides (see Fig. 11-1) with internal 1,6 disulfide bridges are the original mammalian peptide "neurohormones": They are synthesized in the neuronal perikarya found in the large neurons of the supraoptic and paraventricular nuclei, and they are stored in the axons of these neurons in the neurohypophysis, from which they are released into the bloodstream. Each peptide is synthesized as part of a larger propeptide (see Chapter 3) with

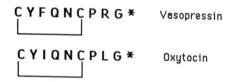

FIGURE 11-1. Molecular sequence of oxytocin and vasopressin, in which amino acid names are symbolized by the standard single letter symbols, and * indicates an amidated C terminus. (Key to amino acid single letter symbols: A = Ala; R = Arg; D = Asp; N = Asparagine; C = Cys; Q = glutamine; E = Glu; G = Gly; H = His; I = Ile; L = Leu; K = Lys; M = Met; F = Phe; P = Pro; S = Ser; T = Thr; W = Trp; Y = Tyr; V = Val.)

which it is stored and released and from which it is cleaved as part of the release process. In the kidney, vasopressin (also known as antidiuretic hormone) facilitates distal tubular water reabsorption, while oxytocin stimulates epididymal and uterine muscle contraction.

The oxytocin and vasopressin peptides are expressed within the classic magnocellular neurons of the paraventricular and supraoptic nuclei whose axons form the neurohypophysis, where the peptides are secreted into the bloodstream for their peripheral targets. The same peptides are also expressed within the parvocellular neurons of these two nuclei for secretion into the pituitary portal circulation, where vasopressin can act synergistically with corticotropin releasing hormone (see below) to release ACTH. In addition, vasopressin is also expressed within a subset of neurons of the suprachiasmatic nucleus; outside the hypothalamus, vasopressin is also expressed within some neurons of the bed nucleus of the stria terminalis (which project to lateral septum, medial amygdala, and periaqueductal gray, among others), the medial amygdala (which project to the ventral hippocampus), the septum, and reputedly within neurons of the locus ceruleus. While neurons in these target regions are definitely responsive to the peptides infused locally, documentation of synaptic actions is still lacking. An interesting but mysterious aspect of the biology of these neurons is their capacity to transport the mRNAs for their peptides into their axons, especially during periods of functional load (postpartum or salt load) and then back again to the perikaryon where the messages can be translated.

The behavioral actions of vasopressin are quite impressive, although the mechanisms accounting for them remain unclear. These effects are described in greater detail in Chapter 13.

Analogs of vasopressin and oxytocin have now been developed with selective agonist or antagonist properties; these molecules, often shorter and more stable than the native peptide, should make useful pharmacological probes. Based on the selective actions of vasopressin analogs that act as agonists or antagonists, two subtypes of peripheral receptors have been characterized: The V_1 receptors mediate responses of some analogs on arteriolar walls, and the V_2 receptors mediate the effects of other analogues on the renal tubules. A V_1 receptor agonist form of AVP has been reported to maintain the long-term potentiation that can be induced in neurons of the lateral septal nucleus by high-frequency stimulation of the fimbria, a two-way pathway between the septum and the hippocampal formation (see Chapter 13). The dynamic regulation of oxytocin and vasopressin expression is sensitive to gonadal steroids, thus providing a gender-specific neuronal expression pattern that is turned on at puberty and that can be further regulated through reproductive function. Novice neuropeptide researchers have an opportunity left behind by the marauding peptide pioneers: What is the function of oxytocin in the male brain?

The effects of vasopressin analogs on either memory or learning behaviors or on secretory responses mediated through the median eminence do not adhere to either receptor class and may indicate that more subtypes will be characterized, and at least four AVP receptors have been cloned.

The Tachykinin Peptides

In 1931, von Euler and Gaddum discovered an unexpected pharmacologically active substance in extracts of brain and intestine, which they later named Substance P, because it was present in the dried acetone powder of the extract. Although studied intermittently through bioassays, Substance P remained somewhat shaded in obscurity until almost 40 years later, when Leeman and her coworkers purified a sialogogic peptide from hypothalamic extracts, while looking for the still-elusive corticotropin releasing factor. This sialogogic peptide turned out to have an amino acid content and pharmacological profile identical to Substance P, and, when finally purified, sequenced, and synthesized, it was identified as Substance P, an undecapeptide (Fig. 11-2). The knowledge of the structure and the availability of large amounts of the synthetic material permitted the development of radioimmunoassays and immunocytochemical tests that were then used to map the brain and assay its Substance P content. Substance P is present in small neuronal systems in many parts of the CNS, and on subcellular fractionation it is found in the vesicular layer. It is especially rich in neurons projecting into the substantia gelatinosa of the spinal cord from the dorsal root ganglia and has even been proposed as the transmitter for primary afferent sensory fibers. While it is very potent as a depolarizing substance in direct tests of spinal cord excitability (about 200 times stronger on a molar basis than GLU), its long duration of action prompted caution in accepting Substance P as the primary sensory transmitter. Other peptides have also been identified in dorsal root ganglion cells, and GLU released from cultures of these cells can also produce a prompt, powerful excitatory action on spinal dorsal horn

The TACHYKININ Peptides

R P K P Q Q F F G L M *   Substance P

D V P K S D Q F V G L M *   Kassinin

H K T D S F V G L M *   Neurokinin A

pE P S K D A F I G L M *   Eledoisin

D M H D F F V G L M *   Neurokinin B

FIGURE 11-2. Structural homologies between the peptides of the tachykinin family, presented according to the schema of Fig. 11-1. pE indicates pyroglutamate.

neurons. If GLU and Substance P were cotransmitters for some sensory fibers, Substance P could be viewed as prolonging and intensifying the transmission into the cord.

Radioimmunoassays and immunocytochemistry also show brain regions other than the spinal cord to be rich in Substance P, especially the substantia nigra, caudate-putamen, amygdala, hypothalamus, and cerebral cortex. Tests with iontophoretic application in these regions generally indicate excitatory actions, again with a long duration. The dynamic regulation of Substance P within the rat striatum has been informative. Substance P–containing neurons of the striatum project to the dopamine neurons of the substantia nigra in reciprocal circuitry. Dopamine antagonists or 6-OHDA treatment of the substantia nigra lead to a substantial drop in Substance P content; analysis of the peptide, propeptide, and mRNA indicates that this drop is due to decreased gene expression and decreased release. After treatment with methamphetamine, Substance P synthesis is accelerated. After axotomy, the levels of Substance P and several of its coexisting neuropeptides (see Fig. 11-4, below) plunge in dorsal root ganglion neurons, while alternative neuropeptides are now expressed. In the human neurologic disease Huntington's chorea, characterized by profound movement disorders and psychological changes, Substance P levels in the substantia nigra are considerably reduced.

Based on smooth muscle bioassays of N-terminally shortened Substance P analogs and early antagonist analogs, an initial subtyping of receptors suggested a P type, where Substance P was as potent or more potent than other nonmammalian analogues (see Fig. 11-2), and an E-type receptor, where one such amphibian peptide, *eledoisin*, was potent, and another, *kassinin*, was an even more potent agonist. The search for kassinin-like peptides, using such smooth muscle receptor subtypes as a bioassay, revealed two forms that were clearly not Substance P and have been termed Substance K and neuromedin K. Using these three peptides, receptor subtypes were defined in which substance P was either the most or the least potent; however, all the receptors showed some effects of all the peptides.

Tachykinin research has been one of the windfall areas benefitting from the application of the molecular biological methods. The cloning of the precursor forms of the three main tachykinin peptides in the mammalian CNS and their first receptor subtypes has helped to clarify their relationships to each other and to other neurotransmitters. There are two mammalian tachykinin genes: The neurokinin A gene (on human chromosome 7) can produce three forms of mRNA through alternative splicing of its many exons; the least prevalent form encodes only substance P, and the other two forms encode both Substance P and neurokinin A (the term the tachykinin nomenclature committee prefers for the peptide previously called Substance K). In general, tissue and cellular distributions of Substance P and neurokinin A have been similar. A second tachykinin gene (located on human chromosome 12) encodes the neurokinin B precursor, and the peptide previously called neuromedin K has been designated neurokinin B. The receptors for all three peptides have been cloned, and their sequence suggests they are all members of the seven-transmembrane domain, G protein, regulatory type (see Chapter 6). The receptors are called NK_1, NK_2, and NK_3. Substance P is more potent than NKA or NKB at the NK_1 site, but much less potent than NKA or NKB at the NK_2 site. At the NK_3 site, NKB is slightly more potent than NKA or Substance P. Highly selective nonpeptidic agonists and antagonists have helped to define potential behavioral actions of substance P and have indicated that inositol phosphate breakdown is a major transducing pathway for at least the NK_1 and NK_2 receptors. However, drugs acting at the human NK_3 receptor do not work well on the rat NK_3, eliminating the animal model frequently relied upon.

VIP-Related Peptides

Vasoactive intestinal polypeptide (VIP), a peptide composed of 29 amino acids, was originally isolated from porcine intestine and named for its ability to alter enteric blood flow. Establishment of its sequence revealed marked structural similarities to glucagon, secretin, and another peptide of gastric origin that inhibits gastric muscular contraction (GIP). These structural similarities constitute the VIP-related peptide family (see Fig. 11-3), named for the member most enriched in brain. The development of synthetic VIP for preparation of immunoassays and cellular localization revealed that VIP exists independently of its cousins in gut and pancreas and showed that VIP was prominent in many regions of the autonomic and central nervous systems. In parasympathetic nerves to the cat salivary gland, VIP coexists with acetylcholine and is apparently released with ACh as part of an integrated command to activate secretion and to increase blood flow through the gland. In the CNS, VIP-reactive neurons are among the most numerous of the chemically defined cells of the neocortex, and they exhibit there a very narrow radial orientation, suggesting that they innervate cellular targets located wholly within a single cortical column assembly (see Fig. 11-4). In peripheral structures, VIP-reactive nerves innervate gut, especially sphincter regions, as well as lung and possibly even the thyroid gland. Preliminary tests on neuronal activity suggest that VIP excites; biochemical tests show that it is able to activate synthesis of cAMP, an action that in the gut is blocked by somato-

The Vasoactive Intestinal Polypeptide Family

HSDAVFTDNYTRLRKOMAYKKYLNSI LN*	VIP
HSDG I **FTD**SYSRYRKQMA V**KKYL**AA VL*	PACAP
HADGVFTSDFSRLLGQL S**AKKYLES**LI*	PHI 27
HADGVFTSDFSRLLGQL S**AKKYLES**MI*	PHM 27
YADI**FT**NSYRKYLGQL S**ARKKLL**QD...	GHRH 1-24
HSQG**TFTSD**YSKYLDSR RAQDFVQWLMNT	Glucagon
HSDGTFYSELSR L**RD**SARL QRLLQGLY*	Secretin

FIGURE 11-3. The VIP-related peptide family represented by their single-letter amino acid symbols. The sequences of PHI-27, PHM-27, growth hormone releasing hormone 1–24, and glucagon and secretin that match those of VIP are indicated in bold letters. For an interesting exercise, the reader may wish to construct a complementary table in which matches to glucagon are highlighted.

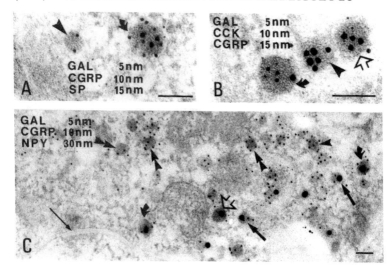

FIGURE 11-4. Electron micrographs of large dense-core vesicles (LDCV) in calcitonin gene-related peptide (CGRP) immunoreactive primary afferent terminals in lamina II of the dorsal horn of rat lumbar spinal cord after triple-immunogold-staining for galanin (GAL)-, CGRP-, and Substance P (SP)-LIIs (A), or for GAL-, cholecystokinin (CCK)-, and CGRP-LIs (B), or for GAL-, CGRP-, and neuropeptide Y (NPY)-LIs (C). The size of gold particles for each peptide is indicated in the micrographs. The different examples of colocalization of neuropeptides in LDCVs are shown. (A) Curved arrow indicates 3-labeled LDCV. Arrowhead, CGRP alone. (B) Curved arrow indicates 3-labeled LDCV. Arrowhead, CGRP alone; open arrow, CGRP plus GH. (C) Curved arrow indicates 3-labeled. Arrowhead, GAL and CGRP; double arrowheads, CGRP alone; open arrow, GAL + NPY; thick arrows, CGRP + NPY; thin arrow points to a synapse. Bars indicate 100 nM. (From X. Zhang and T. Hökfelt)

statin. A very pernicious gut cancer that actively secretes VIP has been described, but no other known disease states involving this peptide have yet been documented.

Additional members were added to this important family of peptides when Guillemin and colleagues isolated the long-sought growth hormone releasing hormone (GHRH) peptide and recognized the homologous shared amino acid sequences. Two later members of the family are termed PHI-27 and PHM-27 to reflect their size, 27 amino acids long, and the one-letter initials of their N-terminal (H = histidine) and C-terminal (isoleucine or methionine) peptides (where P stands for "peptide," not proline). PHI-27 was initially isolated by Mutt and associates from gut extracts while searching for other C-terminally amidated peptides, a property all active members of this family share. PHM-27 was identified from the deduced sequence of the pro-VIP mRNA in cloning experiments. GHRH has a relatively limited neuronal distribution concentrated in the hypothalamus and median eminence. No cellular maps for PHI or PHM in brain have yet been reported.

The most recent addition to the VIP-related peptide family is "PACAP" or "pituitary adenylate cyclase activating peptide," which has a 1000-fold greater capacity to activate adenylate cyclase in pituitary cultures relative to VIP. PACAP is a C-terminally amidated, 27-residue peptide with a very high conservation of the VIP sequence (see Figure 11-3) that is also expressed in brain; a 38-residue, C-terminally extended form has also been isolated from brain and in fact may be the predominant form. Two forms of PACAP receptor have been cloned, and all receptors for peptides in this family show strong structural similarities.

To reverse a trend, molecular cloning of the mRNA for the precursor of secretin, another member of this peptide superfamily, indicates that this specific peptide is not actually found in the brain, as had been proposed based on immunocytochemistry with polyclonal antibodies some years ago.

Pancreatic Polypeptide-Related Peptides

The pancreatic polypeptides were recognized in extracts of pancreatic islets in the mid-1970s, and those from pigeon, pig, and human pancreas were all found to be highly homologous 36-residue peptides with amidated C termini (see Fig. 11-5). Antisera against these peptides recognized cells and fibers in the autonomic and central nervous systems, but these immunocytochemical observations were rightly qualified as "pancreatic polypeptide-like immunoreactivity," because, when the same sera were used in their highly dilute form for radioimmunoassay, only scant extractable material was detected.

PANCREATIC POLYPEPTIDES

```
YPSKPDNPGEDAPAEDLARYYSALRHYINL ITRQRY*    NPY
YPAKPEAPGQNASPQQLSRYYASLRHYLNLVTRQRY*     PYY
GPSQPTYPGDDAPVEDLI RFYDNLQQYLNVVTRHRY*    APP
APLEPVYPGDNATPEQMAQYAADLRRYINMLTRPRY*     HPP
```

FIGURE 11-5. The pancreatic polypeptide family represented by their single-letter amino acid symbols. The sequences of peptide YY (PYY), avian pancreatic polypeptide (APP), and human pancreatic polypeptide (HPP) that match those of NPY are indicated in bold letters.

When Mutt and Tatemoto implemented their chemical isolation strategy centered on the detection of novel C-terminally amidated gut and brain peptides, they soon were able to announce the amino-acid sequences of two more members of this family. The first one, found in gut, was given the name PYY (a neuropeptide with tyrosine (single-letter amino acid symbol "Y") residues at both the N and C termini). However, the one found in brain also had the same length and the same N- and C-terminal tyrosines, with slight internal amino acid substitutions, and has been termed NPY. In rapid order, this peptide's distribution was described in rodent and human brains, and its regional content was quantified, showing NPY to form one of the most extensive of the peptide systems, with very high amounts in the hypothalamus, limbic system, and neocortex— not to mention some loosely correlated losses in some schizophrenics. In many places, including the autonomic nervous system, NPY coexists with either norepinephrine or epinephrine. NPY increases the sensitivity of sympathetically innervated smooth muscle to nor-epinephrine and is one of the most potent natural vasoconstrictors known. The so-called pancreatic polypeptide family should really be renamed the NPY family, since it is the most prevalent form expressed in any tissue. Upon sequencing of the gene in many species, from humans to fish, the structure of the NPY message is one of the most highly conserved genes ever reported, surpassing even the conservation seen in insulin. Within the C terminus of the pre-NPY peptide, a second potential cleavage product known as the C-terminal peptide of NPY (or CPON) is almost as highly conserved. Since CPON will be formed to liberate NPY, it is frustrating that there are no reports of its actions.

NPY, especially in the autonomic nervous system, also exhibits the phenomenon of post-release signals arising from shortened versions of the secreted peptide. Three types of receptors have been detected pharmacologically. The Y1 requires the full NPY 1–36 for action and is largely postsynaptic to enhance responses to co-released norepinephrine; it has been cloned. The Y2 reacts as well to N-terminally truncated forms of NP and even fragments as short as 22–36 have significant effects. The potentially anxiolytic effects of NPY on central administration have been partially confirmed by combinations of in vitro and in vivo antisense strand pharmacology in which repeated intraventricular injections of an antisense RNA oligonucleotide for the Y1 receptor, but not the one for Y2, produced animals that were excessively anxious in an elevated maze task.

Opioid Peptides

An "endorphin" is any "endogenous substance" (i.e., one naturally formed in the living animal) that exhibits pharmacological properties of morphine. When this term was first coined in mid-1975, it was a useful abstraction to cover "morphine-like factors" from brain extracts and spinal fluids that were active in opiate ligand-displacement assays or opiate-sensitive smooth muscle assays. Within a year, a highly competitive effort resulted in the purification, isolation, sequencing, and synthetic replication not only of one but of nearly a half-dozen peptides that deserved the term endorphin. Subsequent research has greatly clarified the molecular and genomic relationships between the three major branches of the opioid peptide greater family (see Fig. 11-6): the proopiomelanocortin (POMC)-derived peptides, the proenkephalin-derived peptides, and the prodynorphin-derived peptides. Other structurally related natural peptides lack opioid receptor activity (such as the invertebrate cardioacceleratory peptide FMRF-amide); furthermore, some opi-

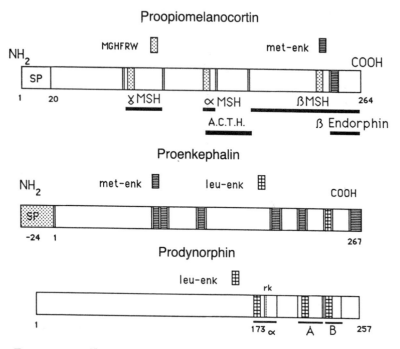

FIGURE 11-6. Structural relationships among the prohormone, precursor forms of the three major branches of the opioid peptides, depicted as a bar diagram, whose length in amino acid residues is indicated by the small number at the corresponding C termini. The location of the repeating peptide sequences are indicated. Basic amino acid sequences that constitute consensus cleavage sites for processing are indicated as single or double vertical lines within the bar. Pro-opiomelanocortin and pro-enkephalin have consensus signal peptides indicated at their N termini.

oid-acting peptides have been found "exogenously" in milk and in plant proteins and have been called "exorphins." These developments have lead activists in the field to give the entire group of structures a more general name, such as endogenous opioid peptides. This also spares those who prefer to study enkephalin and dynorphin peptides from the burden of seeming to carry the endorphin banner.

The incredible explosion of work on this class of peptides began with attempts to isolate and characterize the receptor to opiates as part of a molecular biological approach to the question of narcotic addiction. When specific binding assays were developed, substantial evidence was accumulated to indicate that a high-affinity binding site in synaptic membranes showed stereo-selective opioid recognition properties. The receptor showed Na-sensitive allosteric-like affinity shifts (i.e., in the presence of Na, agonists show reduced affinity) but was not altered in number or affinity during narcotic dependence or withdrawal.

A whole new approach to neurotransmitter identification took shape when Hughes, Kosterlitz, and their colleagues in Aberdeen demonstrated that extracts of brain contain a substance that can compete in the opiate receptor assays and show opioid activity in in vitro smooth muscle bioassays. In late 1975, this endogenous opioid activity was attributed to two pentapeptides, named enkephalins, which shared a common tetrapeptide sequence, YGGF, varying only in the C-terminal position: hence, they were called Met^5-enkephalin and Leu^5-enkephalin. Perhaps even more dramatic than the announcement that the brain contained not one but two opioid peptides was the realization that the entire structure of Met^5-enkephalin is contained within a 91-amino-acid pituitary hormone, β-lipotropin (β-LPH), whose isolation and sequence was reported several years earlier by C. H. Li, but whose function was never really very clear.

Several groups then reported isolation, purification, chemical structures, and synthetic confirmation of three additional endorphin peptides: α-endorphin (β-LPH$_{61-76}$), γ-endorphin (β-LPH$_{61-77}$) and β-endorphin (β-LPH$_{61-91}$; also called "C fragment"). Numerous claims and counterclaims were parried across the symposium stages for several months as to one person's peptide being another person's artifact or precursor. Eventually, this was clarified.

Chemical and Cellular Relationships Among the Opioid Peptides

1. *The proopiomelanocortin (POMC) peptides* are expressed independently in the anterior pituitary, the intermediate lobe of

the pituitary, and one main cluster of neurons in the area of the arcuate nucleus of the hypothalamus. The major endorphin agonist produced from POMC is the 31-amino-acid C-terminal fragment β-endorphin, the most potent of the natural opioids. N-terminal fragments of β-endorphin are much less potent, and analogs with no N-terminal tyrosine (so-called des-Tyr versions) lose all opioid activity, although in some hands such des-Tyr peptides are reputed to be active behaviorally. In the corticotropin-secreting cells of the anterior pituitary, POMC is processed largely to corticotropin and to an inactive form of β-endorphin; whereas in intermediate lobe cells and arcuate neurons the same precursor is processed to α-MSH and active β-endorphin. A third MSH-containing heptapeptide component, discovered during the cloning and sequencing of the POMC mRNA, suggests that yet another end product may be possible.

2. *The enkephalin pentapeptides* Met^5-enkephalin and Leu^5-enkephalin are expressed in wholly separate neuronal systems from the POMC neurons and are much more pervasively distributed throughout the central and peripheral nervous systems, including the adrenal medulla and enteric nervous system. The cloning and sequencing of the mRNA for the pro-enkephalin, starting with mRNA from adrenal medullary tissue, produced an unexpected dividend in that the precursor exhibits multiple copies of the two peptides in almost exactly the 6:1 ratio of Met^5- to Leu^5-enkephalins that had been described in regional brain and gut assays: This solved the mystery of the two similar peptides.

3. *The prodynorphin peptides* consist of C-terminally extended forms of Leu^5-enkephalin arising from a different gene and from a different mRNA that encodes for production of four major peptides, termed dynorphin A, dynorphin B, and two neoendorphins, α and β. These C-terminally extended peptides all act as potent opioid agonists without cleavage down to the enkephalin pentapeptide form, and

on mapping they were found to represent a third separated series of rather generally distributed central and peripheral neurons.

Each of the separate classes of neurons containing β-endorphin, enkephalin, or dynorphin peptides have distinct morphological features (see Fig. 11-7). The β-endorphin-containing neurons are long projection systems that fall within the general endocrine-oriented

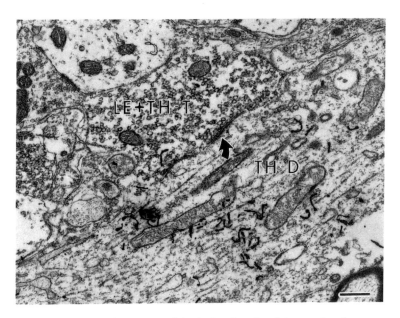

FIGURE 11-7. An ultrastructural basis for functional interaction between central opioid-containing neurons and central catecholamine neurons is provided by this dual label, autoradiography and immunocytochemistry. Nerve terminals exhibit Leu-enkephalin immunoreactivity (immunoperoxidase reaction around many small lucent vesicles in terminals labeled LE-T), and one also exhibits tyrosine hydroxylase immunoreactivity (small autoradiographic grains; terminal labeled LE + TH-T). The latter is making a symmetric synaptic contact on a dendrite (TH-D) bearing many autoradiographic grains, indicating extensive tyrosine hydroxylase content. Scale bar lower right = 0.5 m. (Modified from Milner, Pickel, and Reis, *J. Neurosci. 9*, 2114–2130, 1989)

systems of the medial hypothalamus, diencephalon, and pons. The pro-enkephalin-derived peptides and the prodynorphin-derived peptides are generally found in neurons with modest to short projections, groups of which are widespread. In some regions the enkephalin-derived and dynorphin-derived peptides show intriguing relationships. For example, enkephalin-containing neurons project from the entorhinal cortex to the molecular layer of the dentate gyrus of the hippocampus, while dynorphin-containing neurons project from the dentate gyrus to the CA_3 pyramidal cells. In the spinal cord, intrinsic interneurons contain dynorphin peptides, while descending long axons from the pons and medulla contain enkephalin peptides. In addition, all these peptide systems presumably contain other nonpeptidic transmitters that have not yet been rigorously identified as well as some peptides that have. The most clear-cut evidence suggests functional relationships between opioid peptide systems and catecholamines in the medullary systems regulating cardiovascular function (see Fig. 11-7).

CELLULAR EFFECTS

Traditional (i.e., pre-cloning) pharmacology produced at least three schemes of endorphin–receptor classifications based on (1) the comparative actions of morphine or of morphine-like drugs with mixed agonist–antagonist actions on the dog spinal cord (the μ, κ, and σ scheme of Martin and Gilbert) or (2) the relative effects of endogenous and synthetic opioids on various smooth muscle in vitro assays (the μ and δ scheme of Kosterlitz or the μ, δ, and ε scheme of Herz), or (3) GTP-modifiable ligand binding. One of the most remarkable features of the partially characterized endorphin "receptor" schemes is how poorly they equate with the endorphin or enkephalin-like peptide circuits and how little any of them are affected by chronic exposure to morphine. The best concordance between receptor subtypes and opioid peptide systems stems from the studies of dynorphin-derived peptides by Goldstein and Chavkin, which emphasized the ability of these peptides to act on the κ type of opioid receptor site in both gut and brain.

The membrane mechanisms transducing the various classes of opioid receptors are not well understood, but several categories of opioid action have been well characterized at the molecular level: inhibition of adenylate cyclase, inhibition of the N-type Ca^{2+} conductance that underlies transmitter release, enhancement of the M current K^+ conductance, as well as the activation of the inward rectifier, and delayed rectifier K^+ conductances. Intracellular recordings of gut neurons and brain neurons in vitro indicate that these cells respond by hyperpolarization to acute administration of opioids mediated by an increased K^+ conductance that secondarily depresses Ca-dependent spike activity. In the dorsal horn, neither membrane potential nor conductance is impressively altered, but responsiveness to depolarizing synaptic potentials is reduced.

Iontophoretic tests with enkephalins and β-endorphin suggest that neurons throughout the CNS can be influenced by these peptides at naloxone-sensitive receptors; in general, responses tend to be depressant, except in the hippocampus, where excitatory actions are so profound that hippocampal seizures can be induced. However, the mechanism of this excitation is the exception that proves the rule, as it is based on inhibition of *inhibitory* interneurons and produces excitation by disinhibition. In such tests, all endorphins show qualitatively equivalent effects and similar onset of actions; effects of β-endorphin appear to be more potent and longer lasting, which may be due to its slower hydrolysis. Analogs in which cleavage of the N-terminal dipeptide bond is delayed by substitution of a D-Ala are more long-lasting, while peptides with no N-terminal Tyr are much weaker.

This picture of opioid receptor pharmacology changed irrevocably within a few short months beginning in the fall of 1992 when first one (the δ receptor) and then all three of the opioid peptide receptor subtypes were cloned (see Uhl et al., 1994, for details). As predicted, all were similar G-protein-coupled receptors with some very surprising similarities to other peptide receptors outside the opioid peptide family. Considering all the potent and receptor subtype-specific agonists and antagonists, and their assessment in the hippocampus, it was only a matter of time before the well-defined opioid peptide-containing circuits there were subjected to analysis of their roles in synaptic transmission.

The results of these studies (see review by Wagner and Chavkin, 1994) have paid off in what is arguably the best evidence yet for central synaptic actions of any neuropeptide. High-frequency stimulation of the perforant path to activate dynorphin-containing dentate granule cells leads to a naloxone-sensitive suppression of IPSPs recorded from CA3 pyramidal cells, but this initial claim of synaptic action was soon attributed to presynaptic opioid suppression of norepinephrine release. A second synaptic effect was then detected as a blunted excitatory response to perforant pathway stimulation in the dentate granule cells themselves (the ones containing the dynorphin), but this effect turned out to be a highly novel mechanism in which dendritic dynorphin acted as a retrograde transmitter to reduce presynaptic release of glutamate. However, it was in the extended analysis of the oft-reported ability of naloxone to block hippocampal long-term potentiation (LTP; see Chapter 12) that the best evidence for more conventional synaptic actions emerged: Here μ-antagonists will block the LTP induced in CA3 neurons by mossy fiber stimulation, and both μ- and δ-receptor antagonists will block the LTP produced by stimulation of the lateral perforant pathway (a pro-enkephalin-containing pathway). Furthermore, LTP produced in dentate granule cells by high-frequency stimulation in the hilus of the hippocampus is enhanced with κ agonists, and this effect is blocked with anti-dynorphin antibodies or κ-selective antagonists.

Lastly, the true endorphin fanatic will want to follow the emergence of another series of peptides purified from brain and adrenal medulla on the basis of cross-reactivity with the invertebrate tetrapeptide FMRF-amide. Although FMRF-amide is not expressed as such in mammals, it is potent in many whole animal and in vitro smooth muscle assays. Searches for the endogenous ligands whose receptors could generate these effects led to neuropeptide FF (an octapeptide with phenylalanines at the N and C termini) and a second peptide termed A18Famide (an octadecapeptide with N-terminal alanine and C-terminal phenylalanine). The reason for interest in what might seem to be peptide trivia is that these peptides are expressed with some density in the spinal cord, periaqueductal grey matter, and pons and medulla of all mammals, and they are potent antiopiates when administered with morphine or opioid peptides.

A minor disappointment in the field of opioid peptide research has been its failure to deliver on the promise of providing insight into the nature of drug dependence. Despite many attempts, little significant change in the concentrations of any of the opioid peptides or their receptors can be detected with drug dependence. Recent studies of morphine-dosed animals show initial increases in β-endorphin mRNAs, leading to a C-terminally shortened form that may be either a very weak agonist or, more interestingly, an opioid antagonist. Surprisingly, NMDA antagonists can block opiate tolerance formation. Chronic naltrexone treatments, which up-regulate ligand binding sites for the shorter opioid peptides, also increase their mRNAs by severalfold in the rat striatum and to lesser degrees in other opioid-sensitive brain regions; however, the concentration of peptides encoded by the mRNAs is not changed.

BEHAVIORAL EFFECTS

Although the behavioral effects of these peptides once attracted much attention, the area has not maintained that momentum. The enkephalins produce only transient analgesia after direct intracerebroventricular injection; in such tests β-endorphin is 50–100 times more potent than morphine on a molar basis. The fact that blood-borne peptides do not penetrate into the brain well or survive the gauntlet of peptidases to which they are exposed in the process has spurred the search for more effective analogs, although surpassing the morphine analogs seem a tough prize to claim. The sheer number of whole animal effects that have been interpreted as endorphinergic physiology on the basis of naloxone effects has grown considerably but remains open to more comprehensive analysis. Among the proposed physiological properties that may be regulated by one or another of the endorphin substances are blood pressure, temperature, feeding, sexual activity, and lymphocyte mitosis, along with pain perception and memory. All this seems like a lot to ask of one peptide family. But then, who are we to ask what a peptide may—or can—do for us? As we said before, don't blink or you'll miss the next blazing development.

INDIVIDUAL PEPTIDES WORTH TRACKING

We conclude this chapter with brief comments on a few of the other neuroactive peptides that are under active investigation. Around those we have selected, bodies of research are consolidating, and the neuropharmacological implications seem strong.

Somatostatin (Somatotropin Release-Inhibiting Factor)

In 1971, workers began to look for other goodies in their hard-won hypothalamic extracts. Vale, Brazeau, and Guillemin tried their extracts for potency in releasing growth hormone from long-term cultured anterior pituitary cells and found to their amazement a factor that inhibited even basal growth hormone release in minute amounts. They named this factor somatostatin. Isolation and purification studies eventually culminated in the characterization of this molecule as a tetradecapeptide with disulfide bridge between Cys^3 and Cys^{14} (Fig. 11-8). Radioimmunoassays, immunocytochemistry, and whole-animal tests of the synthetic peptide made it clear that somatostatin was doing more than "just" inhibiting growth hormone release. Somatostatin was found to be widely distributed in the gastrointestinal tract and pancreatic islets, localized in the latter tissue to the δ cells by immunocytochemistry. Somatostatin in islets can apparently suppress the release of both glucagon and insulin, and in diabetic humans, who have no insulin, suppression of glucagon can be an important element in determining insulin requirements. The mechanisms of this suppression are not known but may be similar to the effect of somatostatin on growth hormone release from pituitary, an action accompanied by suppression of TSH release.

When somatostatin is injected intracerebroventricularly, animals show decreased spontaneous motor activity, reduced sensitivity to

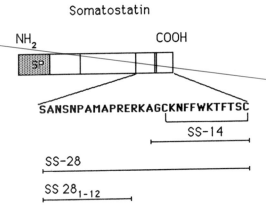

FIGURE 11-8. Structure of the prohormone form of somatostatin, with the location and sequence of the 28-amino-acid residue form at the C terminus and the relative sequences and locations of the different forms of somatostatin indicated.

barbiturates, loss of slow wave and REM sleep, and increased appetite. Radioimmunoassays and immunocytochemistry of rat brain regions show that somatostatin is largely concentrated in the mediobasal hypothalamus, with much smaller amounts present in a few other brain regions (see Fig. 11-9). Immunocytochemical studies led to the surprising finding that somatostatin-reactive cells and fibers are also present in dorsal root ganglia, in the autonomic plexi of the intestine, and in the amygdala and neocortex. Avian and amphibian brains contain considerably more somatostatin than mammalian brains: Physiological tests on the isolated frog spinal cord suggest that it may facilitate the transmission of dorsal root reflexes after a long latent period. Iontophoretic tests on rat neurons indicate a relatively common depressant action that is brisk in onset and termination. At the level of cellular electrophysiology, somatostatin hyperpolarizes neurons in a potent manner and can dynamically open the so-called "M" current channel that cholinergic muscarinic receptors close in order to excite hippocampal and other neurons. The net result of somatostatin's actions in the intact brain is to enhance responsiveness to acetylcholine. The important missing

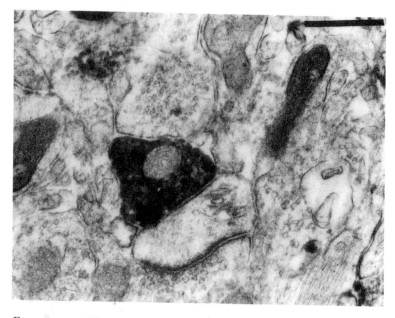

FIGURE 11-9. Ultrastructural immunoperoxidase localization of somato-statin in a nerve terminal making synaptic contact with an unreactive small dendritic spine in the rat neocortex. Calibration bar = 0.5 μm. (Unpublished micrographs from Battenberg and Bloom, Scripps Clinic)

datum is what somatostatin does to the responses to GABA, its frequent but not universal coexistence partner.

An endogenous large somatostatin, extended at the N terminus by an additional 14 amino acids, also exists in both brain and gut and shares equipotency for many somatostatin actions. In rodent and primate neocortex, the N-terminally extended forms reveal a far more extensive neuronal density, which at least overlaps with the intrinsic GABAergic cortical neurons. In early-onset Alzheimer's disease, cortical somatostatin content is depleted, and somatostatin-28 (SS-28) immunoreactive neuritic processes are involved in the formation of the hallmark Alzheimer's lesion, the plaque. Cysteamine, a drug developed to treat the metabolic disorder known as cystinosis, will selectively deplete somatostatin-14 (SS-14) without altering brain levels of SS-28 or SS-28 [1–12], suggesting that these peptides may be stored in separable subcellular compartments.

As we have seen with the other peptides, receptor subtypes are the norm, and here cloning studies have raised the ante well above what was expected from the physiology: Five different cloned receptors have now been reported. Two of them align fairly well on distributional maps: the SIRF1 was expressed in dentate granule cells, striatum, locus ceruleus, and deeper layers of neocortex and showed homologous desensitization; while the SIRF2 was seen more diffusely in cortex, was more abundant in the CA1 field of the hippocampus, and did not desensitize. The third cloned form matches neither, while the fourth form binds SS-28 with greater affinity than SS-14.

Although somatostatin was an unexpected by-product of the race to identify the positive regulators of anterior pituitary release, its pharmacological and diagnostic potentials promise an important future. Somatostatin deficits in CSF have been linked to affective disorder and to Alzheimer's disease. Long-lasting analogs of somatostatin have already been clinically applied in the control of pituitary hypersecretion of growth hormone. The widespread distribution of somatostatin in the brain strongly suggests that it may be difficult to influence receptors in a selective manner for disorders based on regional alterations; but, then, whoever said it would be easy to make a good new drug?

Cholecystokinin (CCK)

Among the gut hormones with the longest histories are gastrin and cholecystokinin (CCK). Subsequently, when their molecular structures were determined, a striking homology was revealed between the C termini (Fig. 11-10). After some confusing interludes, it became recognized that the gastrin-like material extractable from nervous tissue was caused by antisera cross-reacting with the C-terminal octapeptide of CCK, whose sole tyrosine is sulfated. As it turns out, there is more CCK-8 in the brain than there is in the gut, but the N-terminally extended forms found in plasma, especially after eating, all arise from the periphery. CCK is among the most

SQEPPISLDLTFHLLREVLEMTKADQLAQQAHSNRKLDIA   C R F (OVINE)
SEEPPISLDLTFHLLREVLEMARAEQLAQQAHSNRKMEII   C R F (RAT)
pEGPPISIDLSLELLRKMIEIEKQEKEKQQAANNRLLDTI   SAUVAGINE

SCNTATCVTHRLAGLLSRSGGVVKDNFVPTNVGSEAF*   "CGRP"

DYMGWMDF*   CCK$_8$

FIGURE 11-10. The homologies between ovine and rodent corticotropin-releasing factor and sauvagine are indicated by bold single-letter amino acid symbols. Below that are indicated the amino acid sequences of calcitonin gene-related peptide (CGRP) and of the C-terminal octapeptide of cholecystokinin, with a sulfated tyrosine at position 2. PE indicates pyroglutamate.

promiscuous of the coexisting peptides, living with dopamine and neurotensin in the substantia nigra and ventrotegmental area, with VIP, NPY, and GABA in thalamocortical and thalamostriatal connections and with Substance P and 5-HT in medullary neurons. It is also seen in the dorsal root ganglia with several other peptides. In both cortex and thalamus, CCK coexists with pro-enkephalin-derived peptides; interestingly, when animals are given morphine, although enkephalins do not change, *brain* CCK is depleted while *CSF* CCK rises. Exogenous CCK has opioid antagonist properties.

Two (and only two—a rarity in this business) forms of CCK receptors were characterized pharmacologically for their responsivity to the sulfated (CCK$_A$, known to be rich in pancreas) or unsulfated (CCK$_B$, known to be rich in stomach) forms of gastrin and CCK. Both forms have now been cloned and are also expressed in brain. In fact, CCK$_B$ is the predominant brain form, while CCK$_A$ is expressed by neurons in the nucleus of the solitary tract, the area postrema, the posterior hypothalamus, and the nucleus accumbens. CCK pharmacology is by far the leader in de novo development of nonpeptidic antagonists, with an extreme degree of selectivity between the two receptor subtypes. Antagonists of the CCK$_A$ form have been linked functionally to enhancement of opiate analgesia and to centrally elicited postprandial satiety. Antagonists of the CCK$_B$ form have been linked functionally to anxiety disorder, which is induced by intravenous CCK tetrapeptide or CCK-8. Low doses of systemic or iontophoretic benzodiazepines block CCK-induced cell firing.

One interesting, but as yet unexplained, pharmacological interaction deserves continued monitoring. In iontophoretic tests on spontaneously active hippocampal pyramidal cells, CCK-8 shows a potent ability to activate firing, an action that is compatible with the CCK contained within the dentate granule cell mossy fiber synapses to these neurons.

Neurotensin

Neurotensin, a tridecapeptide (pELYENKPRRPYIL) and one of the few that are not C-terminally amidated, became quite interesting to neuropsychopharmacologists after its inauspicious origins as a by-product discovery during the search for corticotropin-releasing factor. In screening for what this structurally independent peptide might do, researchers found that its potent neuropharmacological properties after intracisternal administration included potent analgesia, hypothermia, accentuation of barbiturate and ethanol sleeping time, and increased release of growth hormone and prolactin. Mapping studies showed that the largest amounts of neurotensin are present in the anterior and basal hypothalamus, the nucleus accumbens and septum the midbrain dopamine neurons (along with cholecystokinin), and selected scattered neurons in the spinal cord and brain stem. This led to the speculation that with its profile of sedation, hypothermia, and analgesia, neurotensin might be an "endogenous antipsychotic peptide." However, biochemical neuropharmacologists really began to pay attention when it was found that treatment of rats with conventional antipsychotic drugs increased levels of neurotensin and its mRNA in both striatum and nucleus accumbens, while comparable treatment with the so-called atypical antipsychotic drugs (i.e., those that do not cause the extrapyramidal symptom complex known as "tardive dyskinesia"—see Chapter 13) only increased neurotensin peptide and mRNA in nucleus accumbens.

Furthermore, this change in gene expression would occur whether D_2 dopamine receptors were blocked or not; D_1 agonists produced similar effects. While it has been reported that CSF levels of neurotensin are reduced in unmedicated schizophrenics, and recover somewhat with treatment, the story remains incomplete as to its role in either pathogenesis or treatment. The mRNA for neurotensin contains a partial repeat of the C-terminal tetrapeptide PYIL within a hexapeptide (named neuromedin N) released when proneurotensin is cleaved. The human and rat neurotensin receptors have been cloned and show similar pharmacology, with inositol triphosphate hydrolysis as the transducing mechanism. Cellular tests of neurotensin action suggest it may produce excitation, but the mechanisms are not yet defined. It is quite likely that the development of potent bio-available antagonists of neurotensin, such as SR 48692, may reveal the functions regulated by neurotensin. Studies in humans and dogs show marked increases in neurotensin-like material circulating in plasma after fatty meals.

Calcitonin Gene-Related Peptide

Calcitonin is a 32-amino-acid peptide hormone secreted by special cells in the thyroid gland whose function is to oppose parathormone and increase the urinary excretion of Ca^{2+} and PO_4. The cells that make this hormone, the C cells, are embryologically regarded as arising from neural crest. However, this ancestral lineage would not have merited inclusion of the hormone here were it not for the results of studies originally carried out on a very undifferentiated thyroid carcinoma as part of a plan to characterize the genomic relationships between the segments of DNA encoding the calcitonin prohormone. Under certain conditions of growth in culture, the thyroid carcinoma cells were observed to produce additional mRNA forms that contained part of a series of amino acids encoded by the procalcitonin mRNA but that differed at their C terminus, where the calcitonin occurs. Analysis of the genomic relationship between the exons encoding the segments of the mRNA revealed that the new form of the calcitonin gene-related prohormone contained a potentially novel peptide, termed calcitonin gene-related peptide (CGRP), which bore no significant structural similarity to calcitonin. Antisera were then raised to a synthetic version of the predicted 37-amino acid-long peptide (see Fig. 11-10). The antisera as well as the cDNA forms of the CGRP mRNA were then used to determine whether this previously unexpected peptide was actually expressed in normal tissues. Tests on the hypothalamus and on the trigeminal sensory ganglion were positive, just as they were in other neuronal systems when the rat nervous system was mapped more completely. The concept of "alternative" RNA processing may underlie one part of the peptide coexistence puzzle and clearly represents a mechanism by which a limited number of peptide-encoding gene sequences can be variably arranged into different end products in different cell types.

CGRP has been functionally linked with at least three types of activity, and two CGRP receptors have been cloned. The peptide is expressed in spinal motorneurons and may assist with trophic actions in the expression of acetylcholine receptors by muscle. It is also richly expressed in sensory neurons of the trigeminal and dorsal root ganglia and in the central nucleus of the amygdala, where it is proposed to regulate some aspects of social function.

Corticotropin-Releasing Factor

The search for the corticotropin-releasing factor ended in 1981 when Wylie Vale and colleagues completed the purification and sequence analysis of the peptide taken from extracts of sheep hypothalamus. Thus ended a three-decade-long search for a mystery factor that put the secretion of corticotropin and other anterior pituitary hormones under neuronal control and allowed the brain to replace the pituitary as the master gland of the body. The remaining member of the original four horsemen of the hypophysiotrophic hypothalamic control system was lassoed a year later when Vale and Roger Guillemin independently reported the isolation, sequencing, and synthetic replication of growth hormone-releasing hormone. Nonparticipants in this highly competitive struggle probably had mixed emotions when these quests ended, since along the way the world of neuropeptide discovery was at least triply blessed first by Substance P, then somatostatin, and eventually neurotensin, all of which were discovered by teams intent on CRF.

When CRF was in hand, it turned out to be an interesting 41-amino-acid peptide (see Fig. 11-10) structurally related to a rather obscure frog skin peptide, sauvagine, and to another peptide, termed *urotensin*, obtained from an even more obscure organ, the urohypophysis of certain fish. The cellular origin of the mammalian CRF was traced to that portion of the paraventricular nucleus that projects to the median eminence, but not into the posterior pituitary. Later, these and other CRF-containing circuits were identified as innervating a very extensive group of neurons in the pons and medulla, cortex, and amygdala, perhaps representing a stress-related circuitry. CRF is an extremely potent ACTH secretogogue, but its effects are significantly augmented by vasopressin (produced by the adjacent magnocellular paraventricular neurons) and by norepinephrine and angiotensin.

In addition to its premier action in regulation of corticotropin secretion, synthetic CRF has equally potent effects on neurons in vitro and in vivo, increasing the frequency of action potentials in cells that fire in bursts, such as the hippocampal pyramidal cells. Even modest doses of the peptide can induce seizure-like activity within the limbic system. At still smaller doses given intracerebroventricularly, CRF is a potent activator of spontaneous locomotion and can also produce an "anxiogenic" response (opposite to the effects of antianxiety drugs like benzodiazepines and ethanol) in different behavior tests. Clearly, the complete physiological effects of CRF at the pituitary and at other neuronal sites represent a much more comprehensive basis for the mobilization of bodily systems during stress. The Vale group has since developed a very potent CRF antagonist peptide (an N-terminally shortened CRF 9-41) that will reverse many of the locomotor, thermoregulatory, and autonomic effects of CRF. CRF may well be one of the most overworked of the brain's potent peptides and a fair target for still more selective drug development.

Initial efforts to define CRF receptors focused on ligand binding autoradiography, which more or less replicated the tissue patterns of CRF immunoreactivity with highest binding in cortex, amygdala, hippocampus, pons, and in a yet unexploited circuit crying for functional analysis—a projection from inferior olive to cerebellar cortex containing CRF and glutamate. Then a remarkable discovery was made when scientists recognized that in the final days of pregnancy, maternal blood is very high in apparently high molecular weight forms of CRF believed to arise in placenta and yet without signs of excessive ACTH secretion generally. Efforts to solve this puzzle led to the definition of a CRF-binding protein, also well expressed in brain, although with a different distribution than the CRF fibers. More recently the picture has been extended by cloning and mapping of the (first?) CRF receptor and its mapping pattern in brain, which is distinct from that of the CRF-binding protein. To help determine the functional importance of these various forms of binding, transgenic mice overexpressing CRF or with that gene knocked out have recently been developed. Drugs directed toward CRF receptors may well play a role in depression, stress, and even some forms of inflammation within the CNS.

A READER'S GUIDE TO PEPTIDE POACHING

Until more is learned about the functions, sites, and mechanisms of action of any one peptide, we will have trouble formulating even tentative roles for these substances in neurotransmitter–neurohormonal regulation of central drug actions.

Only the stern hand of our stingy editor prevents us having the space to discuss in any depth galanin (a free ice cream cone to the first reader who deduces how it got its name; clue: it has alanine at its C terminus) or the cytokines. However, the cytokines are too important just to pass by altogether. The reader should know this much: Cytokines are a large and diverse family of polypeptide regulators, produced widely throughout the body by cells of many embryological lines. In general, the cytokines interact as a network with synergistic, additive, or opposing actions. Within the immune system, macrophages and activated T-lymphocytes are the major producers of the macrophage-derived cytokines interleukin (IL) 1_A, $IL1_B$, and IL6 and tumor necrosis factor (TNF)α. These are the cytokines that have received the most attention for potential regulatory roles in nervous system inflammation (as with the early dementia of CNS human immunodeficiency virus infection) and in recovery from traumatic injury. Had they been discovered in brain by neuropharmacologists they might have "growth factor" names instead of cytokine numbers.

As the story of neuroactive peptides unfolds (the sheer number of peptides available for pursuit makes the going slower owing to the division of the available work force), interested students should be alert for answers to the following questions: Are there general patterns of circuitry? Are specific peptidases amenable to selective pharmacological intervention, or do neuronal and glial peptidases read only the dipeptides, whose bonds they are about to cleave? Do the peptides act presynaptically, postsynaptically, or at both sites? How are peptidases specifically activated either to release neuroactive peptides from precursors or to terminate the activity of the peptide? Can the peptides modulate the release or response to transmitters of other neurons or those with which they coexist in a given neuron? Can the promise of peptide chemistry deliver useful antagonists to prove identity of receptors with nerve pathway stimulation? Does the presence of receptors to other centrally active drugs indicate that still more endogenous peptides should be sought?

Regulatory peptides first discovered in neurons also work within some of the complex regulatory cascades in the immune system. Furthermore, some peptide signals of the immune system may also be expressed by neurons. For the moment, this renders arguments about cross-talk between immune cells and neurons less appealing than the idea of a universal system of peptide-mediated molecular events that serve to fine-tune function within the cells that compose many of the body's complex regulatory networks. Stay tuned, the data flow fast.

SELECTED REFERENCES

General Sources

Bloom, F. E. (1990). Peptides: Regulators of cell function in brain and beyond. In: *FIDIA Research Foundation Neuroscience Award Lectures*, Vol. 4, pp. 229–268, Raven Press, New York.

Campbell, I. L., C. R. Abraham, E. Masliah, P. Kemper, J. D. Inglis, M.B.A. Oldstone, and L. Mucke. (1993). Neurologic disease induced in transgenic mice by cerebral overexpression of interleukin 6. *Proc. Natl. Acad. Sci. U.S.A. 90*, 10061–10065.

Hökfelt, T., M.-N. Castel, P. Morino, X. Zhang, and Å. Dagerlind (1994). General overview of neuropeptides. In *Psychopharmacology: The Fourth Generation of Progress* (F. E. Bloom and D. J. Kupfer, eds.), pp. 483–492, Raven Press, New York.

Sherman, T. G., H. Akil, and S. J. Watson (1989). The molecular biology of neuropeptides. *Disc. Neurosci. 6*, 1–58.

Toggas, S. M., E. Masliah, E. M. Rockenstein, G. F. Rall, C. R. Abraham, and L. Mucke. (1994). Central nervous system damage produced by expression of the HIV-1 coat protein gp 120 in transgenic mice. *Nature 367*, 188–193.

Oxytocin and Vasopressin

Barberis, C., S. Audigier, T. Durroux, J. Elands, A. Schmidt, and S. Jard. (1992). Pharmacology of oxytocin and vasopressin receptors in the central and peripheral nervous system. *Ann. N.Y. Acad. Sci. 652*, 39–45.

De Vries, G. J., H. A. al-Shamma, and L. Zhou. (1994). The sexually dimorphic vasopressin innervation of the brain as a model for steroid modulation of neuropeptide transmission. *Ann. N.Y. Acad. Sci. 743*, 95–120.

de Wied, D. (1983). The importance of vasopressin memory. *Trends Neurosci. 7*, 62–64.

Gash, D. M. and G. J. Thomas (1983). What is the importance of vasopressin in memory processes? *Trends Neurosci. 6*, 197–198.

Kasting, N. W. (1989). Criteria for establishing a physiological role for brain peptides. A case in point: The role of vasopressin in thermoregulation during fever and antipyresis. *Brain Res. Rev. 14*, 143–153.

Jirikowski, G. F., P. P. Sanna, D. Macjiewski-Lenoir, and F. E. Bloom. (1992). Reversal of diabetes insipidus in Brattelboro rats: Intrahypothalamic injection of vasopressin mRNA. *Science 255*, 996–998.

Koob, G. F., R. Dantzer, F. Rodriguez, F. E. Bloom, and M. Le Moal (1985). Osmotic stress mimics the effects of vasopressin on learned behaviour. *Nature 315*, 750–752.

Pittman, Q. J. and B. Bagdan. (1992) Vasopressin involvement in central control of blood pressure. *Prog. Brain Res. 91*, 69–74.

Swanson, L. W. and P. E. Sawchenko (1983). Hypothalamic integration: Organization of the paraventricular and supraoptic nuclei. *Annu. Rev. Neurosci. 6*, 269–324.

Neurotensin

Balasubramaniam, A., S. Sheriff, M. E. Johnson, M. Prabhakaran, Y. Huang, J. E. Fischer, and W. T. Chance. (1994) [D-TRP32]neuropeptide Y: A competitive antagonist of NPY in rat hypothalamus. *J. Med. Chem. 37*, 811–815.

Bean, A. J., M. J. During, A. Y. Deutsch, and R. H. Roth (1989). Effects of dopamine depletion on striatal neurotensin: Biochemical and immunohistochemical effects. *J. Neurosci. 9*, 4430–4438.

Carraway, R. and S. E. Leeman (1975). The amino acid sequence of a hypothalamic peptide, neurotensin. *J. Biol. Chem. 250*, 1907–1912.

Deutch, A. Y., and D. S. Zahm (1992). The current status of neurotensin–dopamine interactions. Issues and speculations. *Ann. N.Y. Acad. Sci. 668*, 232–252.

Erwin, V. G. and B. C. Jones (1993). Genetic correlations among ethanol-related behaviors and neurotensin receptors in long sleep (LS) × short sleep (SS) recombinant inbred strains of mice. *Behav. Genet. 23*, 191–196.

Kobayashi, R. M., M. Brown, and W. Vale (1977). Regional distribution of neurotensin and somatostatin in rat brain. *Brain Res. 126*, 584–590.

Merchant, K. M., D. J. Dobie, and D. M. Dorsa (1992). Expression of the proneurotensin gene in the rat brain and its regulation by antipsychotic drugs. *Ann. N.Y. Acad. Sci. 668*, 54–69.

Nemeroff, C. B., B. Levant, B. Myers, and G. Bissette (1992). Neurotensin, antipsychotic drugs, and schizophrenia. Basic and clinical studies. *Ann. N.Y. Acad. Sci. 668*, 146–156.

Tachykinin Peptides

Kotani, H., M. Hoshimura, H. Nawa, and S. Nakanishi (1986). Structure and gene organization of bovine neuromedin K precursor. *Proc. Natl. Acad. Sci. U.S.A. 83*, 7074–7078.

Lundberg, J. M., A. Saria, E. Brodin, S. Rosell, and K. Folkers (1983). A substance P antagonist inhibits vagally induced increase in vascular permeability and bronchial smooth muscle contraction in the guinea pig. *Proc. Natl. Acad. Sci. U.S.A. 80*, 1120–1124.

Maggio, J. E. (1988). Tachykinins. *Annu. Rev. Neurosci. II*, 13–28.

Nawa, H., H. Kotani, and S. Nakanishi (1984). Tissue specific generation of two preprotachykinin mRNAs from one gene by alternative RNA splicing. *Nature 312*, 729–734.

Nicoll, R. A., C. Schenker, and S. E. Leeman (1980). Substance P as a transmitter candidate. *Annu. Rev. Neurosci. 3*, 227–268.

Nilsson, G., T. Hökfelt, and B. Pernow (1974). Distribution of Substance P-like immunoreactivity in the rat central nervous system as revealed by immunohistochemistry. *Med. Biol. 52*, 424–448.

Substance P in the nervous system (1982). *Ciba Foundation Symposium 91*, Pitman Publishers, London.

Vaught, J. L. (1988). Substance P antagonists and analgesia: A review of the hypothesis. *Life Sci. 43*, 1419–1431.

Watling, K. J., S. Guard, S. J. Boyle, A. T. McKnight, and G. N. Woodruff (1994). Species variants of tachykinin receptor types. *Biochem. Soc. Trans. 22*, 118–122.

VIP-Related Peptides

Hosoya, M., H. Onda, K. Ogi, Y. Masuda, Y. Miyamoto, T. Ohtaki, H. Okazaki, A. Arimura, and M. Fujino (1993). Molecular cloning and functional expression of rat cDNAs encoding the receptor for pituitary adenylate cyclase activating polypeptide (PACAP). *Biochem. Biophys. Res. Commun. 194*, 133–143.

Gottschall, P. E., I. Tatsuno, and A. Arimura (1994). Regulation of interleukin-6 (IL-6) secretion in primary cultured rat astrocytes: Synergism of interleukin-1 (IL-1) and pituitary adenylate cyclase activating polypeptide (PACAP). *Brain Res. 637*, 197–203.

Harmar, T., and E. Lutz. (1994) Multiple receptors for PACAP and VIP. *Trends Pharmacol. Sci. 15*, 97–99.

Fahrenkrug, J. and P. C. Emson (1982). Vasoactive intestinal polypeptide: Functional aspects. *Br. Med. Bull. 38*, 26S–70.

Guillemin, R., P. Brazeau, P. Bohlen, F. Eseh, N. Ling, and W. B. Wehrenberg (1982). Growth hormone releasing factor from a human panereatic tumor that eaused acromegaly. *Science 218*, 585–587.

Lundberg, J. M., B. Hedlund, and T. Bartfai (1982). Vasoactive intestinal polypeptide enhances muscarinic ligand binding in cat submandibular gland. *Nature 295*, 147–149.

Magistretti, P. J. and M. Schorderet (1985). Norepinephrine and histamine potentiate the increases in cyclic adenosine $3':5'$-monophosphate elicited by vasoactive intestinal polypeptide in mouse cerebral cortical slices: Mediation by αI-adrenergic and H-histaminergic receptors. *J. Neurosci.* 5, 362–368.

Matsumoto, Y., M. Tsuda, and M. Fujino (1993). Regional distribution of pituitary adenylate cyclase activating polypeptide (PACAP) in the rat central nervous system as determined by sandwich-enzyme immunoassay. *Brain Res.* 602, 57–63.

Morrison, J. H., P. J. Magistretti, R. Benoit, and F. E. Bloom (1984). The distribution and morphologieal characteristics of the intracortical VIP-positive cell: An immunohistochemical analysis. *Brain Res.* 292, 269–282.

Rivier, J., J. Spiess, M. Thorner, and W. Vale (1982). Characterization of a growth hormone releasing factor from a human pancreatic islet tumor. *Nature 300*, 276–278.

Said, S. I. (1984). Vasoactive intestinal polypeptide (VIP): Current status. *Peptides 5*, 143–150.

Tatemoto K. and V. Mutt (1981). Isolation and characterization of the intestinal peptide porcine PHI (PHI-27), a new member of the glucagon-secretin family. *Proc. Natl. Acad. Sci. U.S.A. 78*, 6603–6607.

Pancreatic Polypeptide-Related Peptides

Allen, Y. S., T. E. Adrian, J. M. Allen, K. Tatemoto, T. J. Crow, S. R. Bloom, and J. M. Polak (1983). Neuropeptide Y distribution in the rat brain. *Science 221*, 877–879.

Mutt, V. (1983). New approaches to the identification and isolation of hormonal polypeptides. *Trends Neurosci.* 6, 357–360.

Sheikh, S. P., R. Hakanson, and T. W. Schwartz (1989). Y1 and Y2 receptors for neuropeptide Y. *FEBS Lett.* 245, 209–214.

Wahlestadt, C., and M. Heilig (1994). Neuropeptide Y and related peptides. In *Neuropsychopharmacology: The Fourth Generation of Progress* (F. E. Bloom and D. J. Kupfer, eds.), pp. 543–551, Raven Press, New York.

Wahlestedt, C., E. M. Pich, G. F. Koob, F. Yee, and M. Heilig (1993). Modulation of anxiety and neuropeptide Y-Y1 receptors by antisense oligodeoxynucleotides. *Science 259*, 528–531.

Opioid Peptides

Bloom, F. E. (1983). The endorphins: A growing family of pharmacologically pertinent peptides. *Annu. Rev. Pharmacol. Toxicol. 23*, 151–170.

Goldstein, A., S. Tachibana, L. I. Lowney, and L. Hold (1979). Dynorphin—(1–13), an extraordinarily potent opioid peptide. *Proc. Natl. Acad. Sci. U.S.A. 76*, 6666–6670.

Herkenham, M. and S. McLean (1988). The anatomical relationship of opioid peptides and opiate receptors in the hippocampi of four rodent species. *NIDA Res. (Monogr) 82*, 33–47.

Hughes, J., T. W. Smith, H. W. Kosterlitz, L. A. Fothergill, G. A. Morgan, and H. R. Morris (1975). Identification of two related pentapeptides from the brain with potent opiate agonist activity. *Nature 258*, 577.

Kivipelto, L., A. Aarnisalo, and P. Panula (1992). Neuropeptide FF is colocalized with catecholamine-synthesizing enzymes in neurons of the nucleus of the solitary tract. *Neurosci. Lett. 143*, 190–194.

Kivipelto, L., H. A. Majane, H.-Y.T. Yand, and P. Panula (1989). Immunohistochemical distribution and partial characterization of FLFQPQRFamide-like peptides in the central nervous system of rats. *J. Comp. Neurol. 286*, 269–287.

Nakanishi, S., A. Inoue, T. Kita, A.C.Y. Chang, S. Cohen, and S. Numa (1979). Nucleotide sequence of cloned cDNA for bovine corticotropin—lipotropin precursor. *Nature 257*, 238–240.

Nicoll, R. A., G. R. Siggins, N. Ling, F. E. Bloom, and R. Guillemin (1977). Neuronal actions of endorphins and enkephalins among brain regions: A comparative microiontophoretic study. *Proc. Natl. Acad. Sci. U.S.A. 74*, 2584.

Tempel, A., J. A. Kessler, and R. S. Zukin (1990). Chronic naltrexone treatment increases expression of preproenkephalin and preprotachykinin mRNA in discrete brain regions. *J. Neurosci. 10*, 741–747.

Terman, G. W., J. J. Wagner, and C. Chavkin (1994). Kappa opioids inhibit induction of long-term potentiation in the dentate gyrus of the guinea pig hippocampus. *J. Neurosci. 14*, 4740–4747.

Trujillo, K. A., and H. Akil. (1994) Inhibition of opiate tolerance by noncompetitive *N*-methyl-D-aspartate receptor antagonists. *Brain Res. 633*, 178–188.

Uhl, G. R., S. Childers, and G. Pasternak (1994). An opiate receptor gene family reunion. *TINS 17*, 89–93.

Wagner, J. J., and C. Chavkin (1994). Neuropharmacology of endogenous opioid peptides. In *Psychopharmacology: The Fourth Generation of Progress*, (F. E. Bloom and D. J. Kupfer, eds.), pp. 519–529, Raven Press, New York.

Wagner, J. J., G. W. Terman, and C. Chavkin (1993). Endogenous dynorphins inhibit excitatory neurotransmission and block LTP induction in the hippocampus. *Nature 363*, 451–454.

Somatostatin

Bell, G. I., and T. Reisine (1993). Molecular biology of somatostatin receptors. *Trends Neurosci. 16*, 34–38.

Bissette, G., and B. Myers (1992). Somatostatin in Alzheimer's disease and depression. *Life Sci. 51*, 1389–1410.

de Lima, A. D. and J. H. Morrison (1989). An ultrastructural analysis of somatostatin-immunoreactive neurons and synapses in the temporal and occipital cortex of the macaque monkey. *J. Comp. Neurol. 283*, 212–227.

Effendic, S. and R. Luft (1980). Somatostatin: A classical hormone, a locally active polypeptide, and a neurotransmitter. *Ann. Clin. Med. 12*, 8794.

Moore, S. D., S. G. Madamba, M. Joels, and G. R. Siggins (1988). Somatostatin augments the M-current in hippocampal neurons. *Science 239*, 278–280.

Raynor, K., and T. Reisine (1992). Somatostatin receptors. *Crit. Rev. Neurobiol. 6*, 273–289.

Cholecystokinin

Albus, M. (1988). Cholecystokinin. *Prog. Neuropsychopharmacol. Biol. Psychiatry 21*.

Baile, C. A., C. L. McLaughlin, and F.M.A. Della (1986). Role of cholecystokinin and opioid peptides in control of food intake. *Physiol. Rev. 66*, 172–234.

Beinfeld, M. C. (1985). Cholecystokinin (CCK) gene-related peptides: Distribution, and characterization of immunoreactive pro-CCK, and an amino terminal pro-CCK fragment in rat brain. *Brain Res. 334*, 351–355.

Dourish, C. T., D. Hawley, and S. D. Iversen (1988). The novel CCK antagonist L364,718 abolished caerulein but potentiates morphine-induced antinociception. *Eur. J. Pharmacol. 152*, 163–166.

Lotti, Y. J., R. G. Pendleton, R. J. Gould, H. M. Hanson, R. S. Chang, and B. V. Clineschmidt (1987). In vivo pharmacology of L-364,718, a new potent nonpeptide peripheral cholecystokinin antagonist. *J. Pharmacol. Exp. Ther. 241*, 103–109.

Rehfeld, J. F., N. R. Goltermann, L.-I. Larsson, P. M. Emson, and C. M. Lee (1979). Gastrin and cholecystokinin in central and peripheral neurones. *Fed. Proc. 38*, 2325–2329.

White, F. J. and R. Wang (1984). Interactions of cholecystokinin octapeptide and dopamine on nucleus accumbens neurons. *Brain Res. 300*, 161–166.

Zetler, G. (1985). Neuropharmacological profile of cholecystokinin-like peptides. *Ann. N.Y. Acad. Sci. 448*, 448–469.

Calcitonin Gene-Related Peptide

Aiyar, N., E. Baker, P. Nambi, G. Feuerstein, and R. Willette (1994). Characterization of CGRP receptors in various regions of gerbil brain. *Neuropeptides 26*, 313–317.

Quirion, R., D. Van Rossum, Y. Dumont, S. St.-Pierre, and A. Fournier. (1992). Characterization of CGRP1 and CGRP2 receptor subtypes. *Ann. N.Y. Acad. Sci. 657*, 88–105.

Rosenfeld, M. G., J.-J. Mermod, S. G. Amara, L. W. Swanson, P. E. Sawchenko, J. Rivier, W. W. Vale, and R. M. Evans (1983). Production of a novel neuropeptide encoded by the calcitonin gene via tissue-specific RNA processing. *Nature 304*, 129–135.

Rosenfeld, M. G., S. G. Amara, and R. M. Evans (1984). Alternative RNA processing: Determining neuronal phenotype. *Science 225*, 1315–1320.

Saria, A., G. Bernatzky, C. Humpel, C. Haring, G. Skofitsch, and J. Panksepp (1992). Calcitonin gene-related peptide in the brain. Neurochemical and behavioral investigations. *Ann. N.Y. Acad. Sci. 657*, 164–169.

Corticotropin-Releasing Hormone

Behan, D. P., E. Potter, S. Sutton, W. Fischer, P. J. Lowry, and W. W. Vale (1993). Corticotropin-releasing factor-binding protein. A putative peripheral and central modulator of the CRF family of neuropeptides. *Ann. N.Y. Acad. Sci. 697*, 1–8.

Heinrichs, S. C., E. M. Pich, K. A. Miczek, K. T. Britton, and G. F. Koob. (1992). Corticotropin-releasing factor antagonist reduces emotionality in socially defeated rats via direct neurotropic action. *Brain Res. 581*, 190–197.

Hernandez, J. F., W. Kornreich, C. Rivier, A. Miranda, G. Yamamoto, J. (1993). Andrews, Y. Tache, W. Vale, and J. Rivier. Synthesis and relative potencies of new constrained CRF antagonists. *J. Med. Chem. 36*, 2860–2867.

Koob, G. F., S. C. Heinrichs, E. M. Pich, F. Menzaghi, H. Baldwin, K. Miczek, and K. T. Britton (1993). The role of corticotropin-releasing factor in behavioral responses to stress. *Ciba Found Symp 172*, 277–289, 290–295.

Lovenberg, T. W., C. W. Liaw, D. E. Grigoriadis, W. Clevenger, D. T.

Chalmers, E.B.D. Souza, and T. Oltersdorf (1995). Cloning and characterization of a functionally distinct corticotropin-releasing factor receptor subtype from rat brain. *Proc. Natl. Acad. Sci. U.S.A.* 92, 836–840.

Menzaghi, F., R. L. Howard, S. C. Heinrichs, W. Vale, J. Rivier, and G. F. Koob (1994). Characterization of a novel and potent corticotropin-releasing factor antagonist in rats. *J. Pharmacol Exp Ther.* 269, 564–572.

Potter, E., D. P. Behan, E. A. Linton, P. J. Lowry, P. E. Sawchenko, and W. W. Vale (1992). The central distribution of a corticotropin-releasing factor (CRF)-binding protein predicts multiple sites and modes of interaction with CRF. *Proc. Natl. Acad. Sci. U.S.A.* 89, 4192–4196.

Sawchenko, P. E., T. Imaki, E. Potter, K. Kovacs, J. Imaki, and W. Vale. (1993). The functional neuroanatomy of corticotropin-releasing factor. *Ciba Found Symp* 172, 5–21, 21–29.

Stenzel-Poore, M. P., S. C. Heinrichs, S. Rivest, G. F. Koob, and W. W. Vale (1994). Overproduction of corticotropin-releasing factor in transgenic mice: A genetic model of anxiogenic behavior. *J. Neurosci.* 14:2579–2584.

Vale, W., J. Spiess, J. Rivier, and C. Rivier (1981). Characterization of a 41-residue ovine hypothalamic peptide that stimulates secretion of corticotropin and beta-endorphin. *Science 213*, 1394–1397.

12 | Cellular Mechanisms in Learning and Memory

In other chapters we have noted when specific behavioral changes correlated with either biochemical differences in specific transmitters or functional changes in the activity of biochemically defined neuronal systems. This level of correlation avoids the more difficult question of just what role the specific systems play in initiating, maintaining, or terminating the behavior. In this chapter we consider an issue that represents one of the most complex behavioral achievements of nervous systems—the ability to learn and remember.

Four editions ago, we observed that "unequivocal causal relations between a change in cell structure or function and learning still remain to be demonstrated" and concluded that this subject probably did not yet merit intellectual investment by neuropharmacologists. Subsequently, however, the courageous pioneers of neuroscience moved into this void with a vengence. Although we cannot yet provide inquiring minds with certified maps to today's Holy Grail in drug development—the cognitive enhancer designed to delay or reverse the ravages of Alzheimer's disease—there are now definite leads.

Part of our dilemma in presenting the neuropharmacology of learning and memory has been that there were no data establishing that a change in a specific neurotransmitter action at a specific synaptic location was necessary and sufficient for a behavioral change in a living organism. The experimental and theoretical reports that make such claims fall into four large classes:

1. *Cellular and molecular* research, which seeks to attribute the changes in cell–cell interactions to specific molecular events in their trans-synaptic operations;

2. *Neuronal process* research, which seeks to define either the functional changes in a brain network or the set of neural pathways that are necessary and sufficient to account for the behavioral changes observed in an experimental learning paradigm;

3. *Behavioral* research in which intracerebral or parenteral drug treatments or other perturbations of brain structure and function are used to disrupt the ability of an animal to modify its behavior in a predictable way in specific environmental settings;

4. *Model systems* research in which the object is to determine, from simulations of neurobiological events, the minimum number of hypothetical operations required to explain an equally abstractly defined memory or learning phenomenon; this model network of hypothetical "neurons" is then used to predict either how brains work or how better computers ought to.

These lines of research more or less independently address the issues of learning and memory by setting their questions *within* one of the main hierarchical levels of the neuroscience enterprise: molecular, cellular, multicellular, or behavioral (see Chapter 1). The complex interactions between these levels make it difficult and often misleading to connect small changes at one end of this hierarchy with changes at the other. Over the last decade, however, many new data have appeared that relate the abstract and essentially behaviorally defined operations of memory and learning in whole animals to specific synaptic events. It is these lines of experimental evidence that emphasize a vertically connected series of data between molecules, cells, and behavior that we now take as our starting point.

FORMULATING EXPERIMENTS IN MEMORY AND LEARNING

Biological analysis of memory and learning typically requires two parallel lines of work: a *behavioral* component in which the subject is trained, given a training recess of variable intervals, and then tested for retention of the trained response and a *functional, structural,* or *biochemical* component intended to define the essential changes, sites, or molecular sequelae underlying these events. Such two-track experiments thus offer—in our view—a better approach than earlier studies in which the biological substrates for the learning were inferred from the effects of systemic drug treatments (e.g., protein synthesis inhibitors or hormone antagonists) administered between the learning and testing phases.

Scientists working in this field commonly distinguish two categories of learning: "non-associative" and "associative." Non-associative learning describes the functional changes that ensue when an organism interacts with a single form of stimulus. For example, the decreased responses that occur when the same sensory stimulus is repeated sequentially, termed "habituation," is a non-associative form of learning, as is the reversal of the habituated response, termed "sensitization," by a different intensity or quality of sensory stimulus. Associative learning, as the name implies, is the form of learning in which the subject associates a previously neutral stimulus with a response normally generated by another cue, the sort made famous by Pavlov's classic experiments that triggered dog salivation with a bell instead of raw meat.

An exciting prospect for such work has been opened by the development of experimental test systems in invertebrates—and more recently in vertebrates as well—to show that the behavioral performance of the organism depends on synaptic events in specific locations and then to define the changes that occur at these locations as the organism learns and performs. Such possibilities for molecular and cellular explanations exist, it should be recognized, only because of the extensive data that have been obtained over this same interval.

The great neurocytologist Ramón y Cajal held strongly to the view that learning was a synaptic event. He regarded the synapse as the primary site of interneuronal communication. The subsequent advances in biochemistry, anatomy, and physiology that are emphasized in this text led to the notion that drugs should be able to regulate the critical sites of synaptic transmission. Given this premise, the manipulations of memory processes by specific drugs used as tools for dissecting the roles played by specific transmitters and defined sites should not only illuminate the underlying process but also provide a means toward eventual therapeutic interventions.

CELLULAR MODELS OF LEARNING IN INVERTEBRATE NERVOUS SYSTEMS

Studies carried out in the visceral ganglia of invertebrates, particularly the marine mollusc *Aplysia*, have already provided information on molecular mechanisms of the cellular and synaptic changes that accompany simple forms of behavior modification, such as habituation and sensitization. Sensitization is the ability of nonspecific but strongly arousing sensory stimulation to enhance the effectiveness of synaptic transmission in other neuronal pathways. When the gill-withdrawal reflex is studied for this effect, strong electrical stimulation of the connections between the visceral ganglion and the head ganglion facilitates the withdrawal. The strong stimulation also diminishes habituation of the reflex if tested with identical repetitive sensory stimuli. As judged by the analysis of postsynaptic potentials, the facilitating stimulation increases the amount of transmitter released by the afferent limb of the withdrawal reflex (i.e., the sensory nerves). Once initiated in the *Aplysia* ganglion, the effects of the sensitization can last from several minutes to hours.

Further pursuit of the presynaptic mechanism of the enhanced release became possible when it was found that the sensitizing stimulation also increased the ganglionic content of cAMP and that exposure of the ganglion both to serotonin (known to activate cAMP production in the ganglion) and to cAMP (applied to the whole ganglion or injected intracellularly into the sensory nerve cell) replicated the effects of the sensitizing stimulation. In this case, the molecular mediation sequence of the second messenger hypothesis (see Chapter 6) suggests that a serotonin-secreting interneuron is activated by the sensitizing stimulation, leading to increased presynaptic levels of cAMP and a consequent enhancement of the sensory transmitter release.

Subsequent studies have shown that the short-term memory for sensitization of the *Aplysia* gill- and siphon-withdrawal reflex is in fact distributed across at least four sites of circuit modification. Each of these invokes a slightly different circuitry and a different type of synaptic modification (presynaptic facilitation, presynaptic inhibition, post-tetanic potentiation, and increased tonic firing rate), all of which result in facilitation. Nevertheless, all four of these mechanisms seem to be mediated by a common modulatory transmitter (serotonin), to utilize the same cAMP-mediated second messenger system, and to require a phosphorylation event (affecting either intravesicular proteins or ion channels) to enhance transmitter release. These short-term changes, lasting from minutes to hours, do not require protein synthesis.

More recently, Kandel and his associates have extended their definition of the molecular mechanisms involved in the long-term changes in the plasticity underlying the gill withdrawal reflex. Having established that such long-term functional changes are indeed dependent on gene transcription and protein synthesis, (while the short-term facilitatory changes are not) several of the early and late genes involved specifically in the synaptic events have been defined. As with the short-term changes, cAMP generation would appear to be a key step in long-term facilitation, leading to both protein and structural synaptic changes. In the most recent phases of this work, attention has been focused on the cAMP response element binding protein (CREB), in which it can be shown by gene transfer experiments on cultured neurons that cumulative responses to 5-HT lead to cAMP synthesis, activation of the cyclic AMP-dependent protein kinase (PKA), a translocation of the activated enzyme to the nucleus, and phosphorylation of Serine 119 on CREB. The phosphorylated CREB presumably activates the transcription of immediate early genes, one of which has in fact been specifically incriminated as a key player. The *Aplysia* CCAAT enhancer-binding protein (ApC/EBP) was found to be induced rapidly by both 5-HT and cAMP, even in the presence of protein synthesis inhibitors. Furthermore, immunologically blocking the function of ApC/EBP blocks long-term facilitation selectively without affecting short-term facilitation. These two themes, namely, that the short-term and long-term changes rely on different consequences of the 5-HT activation of cAMP synthesis and that the functional consequences of specific immediate early gene systems are important also epitomize recent work in mammalian models of learning and plasticity.

The scientific search for the cellular and molecular bases of learning was launched by Kandel, Tauc, Gershenfeld, Strumwasser, Alkon, Levitan, Kaczmarek, and other explorers of invertebrate psychobiology. They thought these smaller nervous systems could be fathomed, and they set themselves the task of working out the rules of connectivity by which these systems adapt to environmental conditions. Their meticulous descriptive work has given us a logical, consistent, and compelling account of the cellular changes accompanying behavioral modification and the molecular basis of these changes.

The monoamine hypothesis of memory and learning in mammals advanced by Kety in the late 1960s could now be viewed as anticipating, in a more primitive mechanistic way, the serotonin and cAMP-mediated presynaptic facilitation model derived from the studies of invertebrate sensitization. Kety's model called for a brain system that could mediate the ability of arousal to consolidate experiences into adaptive behavioral mechanisms. To reconcile the two situations, we must first substitute the mammalian transmitter norepinephrine (NE) for the role played by 5-HT in the invertebrate. The central NE neurons and their synapses have now been well mapped, and electrophysiological recordings in behaviorally responding rats and monkeys have established that the NE cells fire when novel sensory events occur in the external environment. These data are logical and internally consistent with a role of NE neurons in mediating arousal. However, there are clearly many more neurotransmitter systems at work in mammalian mechanisms of learning and memory, especially when one begins to focus on discrete synaptic changes within precisely constrained regions of the brain in vitro or in vivo. We will examine some of these specific synaptic sites below. Learning and memory events in mammalian brain function—as in the invertebrate—are probably too critical to rely on only one transmitter system.

THE RABBIT NICTITATING MEMBRANE REFLEX AND ASSOCIATIVE LEARNING

The rabbit's nictitating membrane has provided another molecular and cellular model for memory and associative learning. This mammalian neuronal system permits the direct observation of an animal's performance during tests of learning and memory and also the direct determination of exactly where in the circuitry the learning events have occurred.

The basic studies, largely done independently by John Harvey's and Richard Thompson's groups, showed that the rabbit blink reflex could be associatively conditioned. Rabbits will normally blink (i.e., cover their cornea with the nictitating membrane) every time a puff of air is directed at the cornea. Repeated presentation of a loud tone just before the air puff conditioned the rabbits to blink in response to the tone. Eventually the rabbits will blink every time they hear the tone, with no air puff being required.

At this point, chemical and electrolytic lesions as well as microstimulations were used to define the pathways that relay the unconditional (air puff-induced) blink and the conditional blink triggered by the tone (see Fig. 12-1). The air puff stimulus travels through the sensory fibers of the trigeminal nerve to activate motor neurons of the abducens nucleus, retracting the eye into the orbit, and then causing the nictitating membrane to extend over the eye. The acoustic stimulus travels through the cochlear nucleus to the pontine nucleus and seems quite separate at its early stages from the corneal sensory pathway. However, the researchers also noted that even before the rabbits learned to associate the sound with the blinking, the acoustic stimulus that was rather loud could often cause an increased response amplitude to the normal air puff stimulus. This suggested that the acoustic stimulus could perhaps also sensitize the unassociated air puff stimulus.

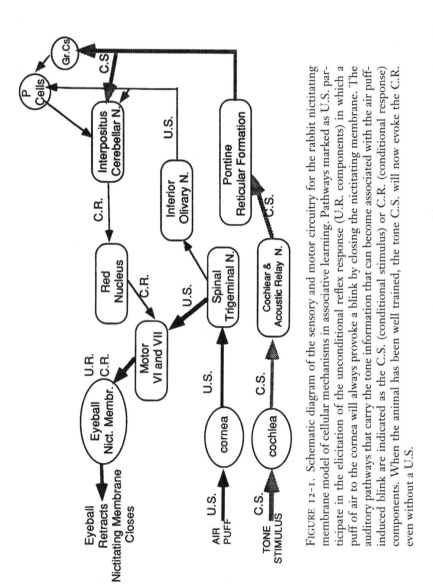

FIGURE 12-1. Schematic diagram of the sensory and motor circuitry for the rabbit nictitating membrane model of cellular mechanisms in associative learning. Pathways marked as U.S. participate in the elicitation of the unconditional reflex response (U.R. components) in which a puff of air to the cornea will always provoke a blink by closing the nictitating membrane. The auditory pathways that carry the tone information that can become associated with the air puff-induced blink are indicated as the C.S. (conditional stimulus) or C.R. (conditional response) components. When the animal has been well trained, the tone C.S. will now evoke the C.R. even without a U.S.

Based upon comparisons of rabbits in which discrete lesions were made along the known anatomical paths within the two systems of sensory processing, the critical site for producing the association seems to be located in the interpositus nucleus, one of the deep cerebellar nuclei. The efferent limb of the motor reflex then courses through the red nucleus directly to the abducens nerve, where it activates the eyeball retraction reflex (see Fig. 12-1).

Since most of the transmitters operating in these hierarchical controlling chains have not yet been defined, it would seem that the model was not quite ready for transmitter-specific pharmacology to be a useful analytical tool. The only sites that have been probed so far are the synapses within the interpositus and red nuclei; here the GABA antagonists bicuculline and picrotoxin will block the conditional reflex. Chemical lesions of the inferior olivary nucleus (the source of the climbing fiber projection to cerebellar Purkinje neurons) will completely disrupt the acquisition and retention of the conditional reflex. Since glutamate is presumed to be the transmitter of this pathway, this implies that GLU is the transmitter for this final step. Harvey and colleagues have used the associative eye blink model to show that low doses of amphetamine as well as LSD will enhance this form of learning, while a wide range of drugs from pimozide to atropine and scopolomine, as well as agonists for κ and σ opioid receptors, will decrease acquisition at doses that do not affect the basic reflex. Because these latter drugs were all given systemically, however, it was not clear where (within the cranial nerve–cerebellar circuitry or elsewhere) they were acting to produce these effects.

In more recent experiments, forebrain regions traditionally viewed as participating in memory and learning functions have also been implicated in the nictitating membrane associational reflex. For these experiments, the rabbits received "classic conditioning" of the nictitating membrane response; the tone stimulus and the air puff were separated by a pause (or "trace") of 500 msec. The short interruption made the simple learning task just a bit more complicated. The extra time delay allowed the investigators to distinguish nonspecific arousal—possibly sensitizing the reflex—from more specific interactions.

Thompson and colleagues found that lesions of the hippocampus or cingulate/retrosplenial cortex would disrupt acquisition of the interrupted conditioned response but that neocortical lesions did not. Neither lesion affected the acquisition of the noninterrupted pairing of tone and air puff. When animals with hippocampal or cingulate/retrosplenial cortex lesions were switched to a standard delay paradigm in which the conditional stimulus and the unconditional stimulus were given at the same time, rabbits acquired the association in about the same number of trials as naïve animals.

In the interrupted protocol, however, macroelectrode recordings showed substantial increases of multi-unit activity in the CA1 region of the hippocampus that began during the tone and persisted through the trace interval, even before the rabbit showed consistent association. As the rabbit's performance in responding to the tone stimulus improved, the hippocampal activity shifted to later and later in the trace interval. Although the transmitter has not been identified for this activation, it is a safe presumption that GLU is involved; in fact, changes in AMPA subtype receptor binding, but not in NMDA receptor kinetics, have been reported to accompany the learning.

These data thus directly relate behavioral demonstration of associational learning with discrete neuronal circuitry. Such an experimental system should eventually lend itself to pharmacological and electrophysiological analyses of the changes within the critical neurons that alter synaptic function. Student volunteers are welcome.

LONG-TERM POTENTIATION: A VERTEBRATE MODEL OF SYNAPTIC PLASTICITY

Next we turn to an increasingly popular form of mammalian forebrain synaptic plasticity in which electrophysiological changes at the synaptic level have been probed extensively by drugs to identify the transmitter systems responsible for discrete transductive mechanisms. The term "long-term potentiation" (LTP) refers to a long-lasting enhancement of synaptic transmission (10 minutes to days, depending critically on the conditions used to evoke the response and where it is tested). The effect is measured as increased amplitudes of excitatory postsynaptic potentials (or the currents generated by these potentials) in specific circuits after high-frequency, high-intensity activation of the same circuits or other discrete paths. Because the phenomenon was originally described with macroelectrode recordings in the hippocampus of intact animals, it has been a candidate to link learning and cellular changes in vivo. However, the most intense recent studies have more often involved analysis of the pertinent connections in vitro in the now classic slice preparations of the hippocampus or neocortex in normal animals or in transgenic animals with specific genes mutated.

This special form of enhanced synaptic transmission was first demonstrated by brief, high-frequency stimulation of the entorhinal cortex, through the perforant path (see Fig. 12-2), to enhance activation of granule cells of the hippocampal dentate gyrus by subsequent single stimuli of the perforant path. In awake cats, guinea pigs, and rats this enhanced transmission could be seen for periods of days to weeks. The transmitter for this pathway is now thought to be glutamate (GLU) or possibly aspartate (ASP), although a role of at least one family of coexisting peptides, such as the pro-enkephalin-derived opioid peptides, has been reported (see Chapter 11).

Long-term potentiation has also been observed at two other sites within the hippocampal formation: (1) at the synapses between the dentate granule cells and the targets of their mossy fiber synapses, the CA3 pyramidal neurons, and (2) between the Schaffer collaterals of CA3 pyramidal neurons, as well as associational fibers from the contralateral hippocampus, and the CA1 pyramidal neurons (see Fig. 11-3). The transmitter for these two pathways is thought to be ASP, GLU, or possibly homocysteate; coexisting peptides (pro-dynorphin derived, as well as CCK) remain candidates for some part of the action here, at least for the mossy fiber connection to CA3 pyramidal neurons. Although the basic phenomenon of long-term

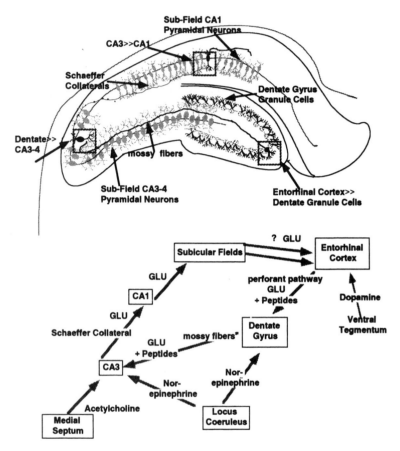

FIGURE 12-2. The synaptic circuits of the rodent hippocampal slice preparation in which long-term potentiation (LTP) can be elicited are indicated in the upper panel, while the neurotransmitters associated with these circuits or with the target cells capable of expressing the LTP response are shown in the lower panel. Each of the three major hippocampal circuits (entorhinal cortex to dentate, dentate to CA3, and CA3 to CA1) shows a generally similar LTP, but the pharmacology of the modification varies with the specific location and pathways. Glutamate-mediated synaptic events figure prominently, as indicated in the text.

potentiation after a high-frequency, high-intensity period of afferent pathway activation is similar in all three synaptic sites, the relative importance of specific transmitters and their responsible transductive mechanisms differ on a site-by-site basis.

The essential pharmacological advance came with the recognition that the N-methyl-D-aspartic acid (NMDA) subtype of excitatory amino acid (EAA) receptor (see Chapter 7) was responsible for the induction phase of the long-lasting synaptic enhancement. Antagonists of this form of EAA receptor (such as aminophosphovaleric acid) will prevent LTP induction without interfering with previously potentiated transmission or with normal low-frequency transmission in the pathway under study. The critical link established by Nicoll and colleagues and several other groups is that the required high-frequency, high-intensity stimulation acts via the NMDA receptor to couple the depolarization with the increased Ca^{2+} entry. This combined response is then thought to depolarize the small dendritic domains that are the site of convergent afferents in the pathway under study and to increase their subsequent transmission. In the perforant path-to-dentate circuit, drugs that block μ-opioid receptors or β-adrenergic receptors will also block LTP. The latter antagonists are also very effective in blocking LTP of CA3 neurons, although not that of CA1 neurons. These drug effects coincide nicely with the regional patterns of monoamine innervation of the hippocampal formation (see Fig. 11-2).

The NMDA receptor has been accepted as a critical step in initiating LTP, as well as in a related intrahippocampal process called long-term depression that has so far received much less scientific examination. Recent evidence suggests that the NMDA receptor must be in the appropriate state for LTP to be induced and that a molecular switch triggered by the metabotropic GLU receptor is required for the proper setting. A minor glitch in validating this evolving consensus emerged when disruption of the NMDA receptor NMDAR1 led to early postnatal death of the transgenic mice, with grossly maldeveloped brains. However, subsequent efforts to knock out only one of the NMDA subunits (the $\varepsilon 1$ of the NR2 subunit family) were more successful, leading to normal brain cellular structure, normal growth and development, but with significantly reduced LTP in the hippocampal CA1 subfield and reduced (albeit moderately so) spatial learning. Disrupted LTP induction in the transgenic mouse mutant *fyn*, in which one tyrosine kinase isoform is disrupted (but not three others), and in mice lacking the PKCγ isoform, indicates that very specific protein phsophorylations can regulate the thresholds for LTP induction. Despite the lack of LTP in the latter mice, however, spatial learning deficits can only be seen under rather extreme test conditions.

The NMDA-regulated step alone does not seem to be adequate for the long-term maintenance of the enhanced transmission. Given the critical role established in the *Aplysia* and other invertebrate models for a cAMP-regulated process in longer-term plasticity, that process has also been explored in the hippocampal model. Again, transgenic mice were employed, in this case expressing mutated knock-out constructs of the CREB α and δ isoforms. These mice responded normally to short-term conditional fear associative learning, but were unable to retain the learned fear for more than two hours (which the wild-type mice could readily do). Interestingly, LTP induction was also normal in the CREB knock-out mice, but returned to baseline in two hours as well. Lest the reader think it is now safe to memorize these schemata, if not the underlying facts, we must deliver a cautionary note. The CREB had been previously knocked out in a couple of alternate ways—one leading to dwarf mice when the CREB could not be phosphorylated and another with full CREB knock-out leading to no deficits at all. Attempts to reconcile these observations lead to the recognition that CREB belongs to a family of transcription factor regulatory proteins, which can often compensate for each other. Thus, there is an alternative interpretation of the LTP and memory problems of the CREB knock-out mice (pointed out to me by a clever young molecular neurobiologist who prefers anonymity until this idea is tested): namely, the knock-out CREB should have been compensated for by another of those transcription factors, and it is perhaps that failure to compensate which causes the problems observed. Life gets confusing, doesn't it!

In allowing the world to spin a few times between editions, we may have spared the reader at least one apparently abortive chapter in the LTP saga: Nitric oxide found one of its many overnight stardom roles as a potential retrograde transmitter leading to the longer-term increases in presynaptic transmitter release, based in part on the ability of nitric oxide synthase (NOS) inhibitors to block LTP, at least at certain nonphysiological temperatures. However, when LTP was found to be unchanged in mice in which the neuronal NOS had been knocked out and in which enzymatic activity was virtually immeasurable, and yet that retained LTP could still be blocked by the NOS inhibitors, covert side effects of those blockers pulled NO from center stage. You can bet this isn't over by a long shot. Thus, much remains to be accomplished with this system, including the demonstration that such events within single pathways of the hippocampal formation are necessary and sufficient for the animal or the hippocampus to "learn" something.

Readers intrigued by these memory model synaptic systems should consult the selected bibliography for other models in which the circuitry has been spelled out and awaits the application of creative pharmacological analysis: the olfactory memory pathways, long-term depression of hippocampal and cerebellar units, and the amygdaloid complex.

ARE THERE ANY NATURAL MEMORY DRUGS?

Back in the third edition, in our last outing onto the sea of memory modulators, we mentioned the growing literature on the ability of natural hormones such as vasopressin and adrenocorticotropin (ACTH), as well as "endocrinologically inert" fragments derived from them, either to repair learning deficiencies in hypophysectomized rats or to delay or accelerate the extinction of a previously learned performance. Unfortunately, as the pathways containing these peptides were more clearly defined in their projections to targets other than the posterior pituitary, and as the known barriers to diffusion of these peptides from the bloodstream into the brain were shown to apply to all of them, this once-promising area became a source of contention. However, this body of research remains an important case study for scholars of the neuropharmacology of behavior. Extended analyses of the behavioral responses regulated by systemic peptide hormones have demonstrated that visceral afferent nerves can influence the brain even if the peptides are excluded by the blood–brain barrier. Furthermore, there is a strong possibility that secrets leading to the development of memory-enhancing drugs may be discovered among the intracerebral pharmacologic effects of what were once considered to be exclusively endocrine peptides.

Therefore, despite the murky nature of this particular research strategy, students should persist. For example, several studies have shown that "low doses" of subcutaneously administered vasopressin can enhance acquisition of a behavioral response, such as maze running, or extinction of active avoidance in a shuttle box. Since vasopressin was named for its potent ability to increase blood pressure, it is not surprising that while "low" in dose weight, the behaviorally active systemic doses of vasopressin also cause systemic hypertension. Moreover, vasopressin analogues that block the peripheral vascular receptors will also block the "memory performance" effects of the hormone. Koob has concluded that all behavioral effects of vasopressin have been blocked by antagonists of the vascular receptor and that the apparent central effect is indirectly mediated by an arousal secondary to the inappropriate hypertension. On the other hand, deWied and his colleagues report that substituted and truncated versions of vasopressin lacking any vascular activity can still alter memory performance. Since the latter demonstrations of effectiveness have been obtained with memory tasks driven by aversive stimuli, their interpretation is still open.

Regardless of the arguments as to how or where the peripherally injected vasopressin acts, still smaller doses of vasopressin (1/1000th the systemic dose) have similar effects on memory performance when given intracerebroventricularly or intracerebrally. These low-dose (nanogram or picomolar) central effects are also reduced or prevented by central administration of the peripheral vascular vasopressin antagonist. Furthermore, opposite effects (i.e., memory performance impairment) have been reported in some tasks after central administration of vasopressin antagonists that are thought to act at "vascular" receptors. This result would suggest that somewhere within the central vasopressin-containing pathways is a circuit whose release of vasopressin is critical in eliciting proper performance and which has a receptor highly similar to those on blood vessels. Whether this means that the vascular receptor is also present on the central targets, or that the central targets are actually on the blood side of the blood–brain barrier, is an unanswered question. In fact, without directly testing a role for vasopressin, Thompson and colleagues have recently reported that water deprivation increased the magnitude of hippocampal LTP (perforant path to dentate) and, in parallel experiments, facilitated contextual fear conditioning.

The concept of vasopressin as simple natural memory hormone would seem to be seriously challenged by recent studies of the Brattleboro rat, whose genetic defect produces a defective messenger RNA that cannot be translated into a secretable vasopressin. Although initially reported as being memory deficient, the Brattleboro rat has supporters who regard its memory ability as normal, especially when salt and water balance are maintained by oral or systemic loading. Furthermore, humans with diabetes insipidus (the human disease of vasopressin deficiency) are not reported to have significant memory impairment.

Without some rigorous identification of the precise cellular sites and specific receptor transductive mechanisms on which either blood-borne or cerebrospinal fluid-borne vasopressin acts, this line of work faces difficulties. The possibility that vasopressin-derived subfragments (whose informational value has been suggested but never defined) can influence neuronal operations critical to the acquisition or retention of the memory tasks that have been studied remains ripe for further study. Future research will probably establish the superficiality of such interpretations as the following: (1) vasopressin acts directly on "memory processes"; (2) vasopressin can be an aversive hormone that when given at nonphysiological doses arouses the animal, who then learns better. We await eagerly the answers to this mind–drug–behavior puzzle, but they may not be found in the next edition, either.

Approaching the Neuropharmacology of Human Memory

In the selected animal models reviewed here, several key ingredients were required for the partial resolution of how selective drugs could modify the process of memory and learning: well-defined and relatively simple memory tasks, well-defined neuronal circuits whose transmitters and transmitter receptors were identified, and drugs that could be administered in ways that restricted their access to the sites being observed. An ultimate goal of such studies would be to develop a specific neuropharmacology to improve human memory function, either in normal subjects or in those in the early stages of dementing illnesses such as Alzheimer's disease or AIDS. Until recently, almost none of the essential facts were in place. Furthermore, the prospects of inferring from noninvasive techniques which sites in a human's brain were participating in memory-related functions or how systemically administered drugs with multiple sites of interaction affected these events seemed dim not all that long ago.

However, dealing with memory processes in humans or even nonhuman primates has recently become far more attractive, partly because of the ability of methods such as positron emission tomography to reveal very brief shifts in blood flow within the brain while humans perform complex cognitive processing, thereby indicating which locations were active with specific kinds of higher mental activities (see Posner et al., 1988). These methods, combined with experimental neurosurgical approaches to the rapidly unfolding description of cortical circuits in the monkey brain, have helped clarify a long-studied area of human memory research, namely, the amnesias that follow discrete toxic or traumatic lesions of the human brain. These combined strategies have allowed investigators to return to a point made earlier in the chapter (namely, that memory functions in more complex brains are likely to involve multiple systems), to ask how the primate brain is organized for memory functions, and to determine which neuronal systems are involved.

As exciting as this work is, an interesting paradox has been recognized: while very discrete lesions of the human hippocampus can lead to profound anterograde amnesia, the hippocampus is not among the many cortical sites in which "functional activation" (i.e., blood flow or glucose utilization) occurs when normal humans are performing complex memory-forming or memory-testing tasks. However, selected regions of the frontal cortex are activated in these tasks.

From this body of work has emerged the hypothesis that information processing is partly tied to the specific processing areas of the brain that are engaged during learning and that these "memory" events are stored within the same neural systems that also participate in the processes of perceiving, analyzing, and processing the sensory information (see Squire et al., 1990). Thus, in the visual system, the temporal cortical region involved in analyzing complex visual patterns (like faces) seems also to be involved in processing the information and in storing it. These and other examples of complex sensory systems suggest that some aspects of learning can be localized to specific subtasks, but that many parts of the brain are involved with the overall process through intricate parallel-distributed neuronal circuits.

An indication of future prospects are recent efforts to bolster short-term memory abilities in aged rhesus monkeys (see Arnsten et al., 1993). Memory capacity was assessed by a moderately complex task that required the animal to remember information over short time intervals and to update this information on every trial. Through impressive structure activity comparisons of a series of α_2-adrenergic agonists, two forms of memory-related actions could be defined: hypotensive, sedating, and memory-impairing effects at one type of site and the memory-enhancing effects at another α_2-receptor site. In low doses the α_2-adrenergic agonist guanfacine improved memory without inducing hypotension or sedation, offering hope for the treatment of at least some memory disorders in man.

SELECTED REFERENCES

Alberini, C. M., M. Ghirardi, R. Metz, and E. R. Kandel (1994). C/EBP is an immediate-early gene required for the consolidation of long-term facilitation in *Aplysia*. *Cell 76*, 1099–1114.

Alkon, D. L. (1988). *Memory Traces in the Brain*. Cambridge University Press, Cambridge.

Arnsten, A. F. (1993). Catecholamine mechanisms in age-related cognitive decline. *Neurobiol. Aging 14*, 639–641.

Bailey, C. H., P. Montarolo, M. Chen, E. R. Kandel, and S. Schacher (1992). Inhibitors of protein and RNA synthesis block structural changes that accompany long-term heterosynaptic plasticity in *Aplysia*. *Neuron 9*, 749–758.

Baudry, M. and G. Lynch (1988). Properties and substrates of mammalian memory systems. In *Psychopharmacology: The Third Generation of Progress* (H. Y. Meltzer, ed.), pp. 449–462, Raven Press, New York.

Bear, M. F., and R. C. Malenka (1994). Synaptic plasticity: LTP and LTD. *Curr. Opin. Neurobiol. 4*, 389–399.

Bortolotto, Z. A., Z. I. Bashir, C. H. Davies, and G. L. Collingridge (1994). A molecular switch activated by metabotropic glutamate receptors regulates induction of long-term potentiation. *Nature 368*, 740–743.

Gallagher, M., and P. C. Holland (1994). The amygdala complex: Multiple roles in associative learning. *Proc. Natl. Acad. Sci. U.S.A. 91*, 11771–11776.

Haley, D. A., R. F. Thompson, and J.I.V. Madden (1988). Pharmacological analysis of the magnocellular red nucleus during classical conditioning of the rabbit nictitating membrane response. *Brain Res. 454*, 131–143.

Harvey, J. A. (1987). Effects of drugs on associative learning. In *Psychopharmacology: The Third Generation of Progress* (H. Y. Meltzer, ed.), pp. 1485–1491, Raven Press, New York.

Ito, M. (1987). Long term depression as memory process in the cerebellum. In *Synaptic Function* (G. M. Edelman, W. E. Gall, and W. M. Cowan, eds.), pp. 431–447. Wiley-Interscience, New York.

Kaang, B. K., E. R. Kandel, and S. G. Grant (1993). Activation of cAMP-responsive genes by stimuli that produce long-term facilitation in *Aplysia* sensory neurons. *Neuron 10*, 427–435.

Kuhl, D., T. E. Kennedy, A. Barzilai, and E. R. Kandel (1992). Long-term sensitization training in *Aplysia* leads to an increase in the expression of BiP, the major protein chaperon of the ER. *J. Cell. Biol. 119*, 1069–1076.

Maren, S., J. P. DeCola, R. A. Swain, M. S. Fanselow, and R. F. Thompson (1994). Parallel augmentation of hippocampal long-term potentiation, theta rhythm, and contextual fear conditioning in water-deprived rats. *Behav. Neurosci.* *108*, 44–56.

Markowitsch, H. J. and E. Tulving (1994). Cognitive processes and cerebral cortical fundi: Findings from positron-emission tomography studies. *Proc. Natl. Acad. Sci. U.S.A.* *91*, 10507–10511.

Mintz, M., D. G. Lavond, A. A. Zhang, Y. Yun, and R. F. Thompson (1994). Unilateral inferior olive NMDA lesion leads to unilateral deficit in acquisition and retention of eyelid classical conditioning. *Behav. Neural Biol.* *61*, 218–224.

Posner, M. I., S. E. Petersen, T. Fox, and M. E. Raichle (1988). Localization of cognitive operations in the human brain. *Science 240*, 1627–1631.

Richter-Levin, G., M. L. Errington, H. Maegawa, and T. V. Bliss (1994). Activation of metabotropic glutamate receptors is necessary for long-term potentiation in the dentate gyrus and for spatial learning. *Neuropharmacology 33*, 853–857.

Schacher, S., E. R. Kandel, and P. Montarolo (1993). cAMP and arachidonic acid simulate long-term structural and functional changes produced by neurotransmitters in *Aplysia* sensory neurons. *Neuron 10*, 1079–1088.

Smirnova, T., S. Laroche, M. L. Errington, A. A. Hicks, T. V. Bliss, and J. Mallet (1993). Transsynaptic expression of a presynaptic glutamate receptor during hippocampal long-term potentiation. *Science 262*, 433–436.

Stevens, C. F., S. Tonegawa, and Y. Wang (1994). The role of calcium–calmodulin kinase II in three forms of synaptic plasticity. *Curr. Biol. 4*, 687–693.

Thompson, R. F., and D. J. Krupa (1994). Organization of memory traces in the mammalian brain. *Annu. Rev. Neurosci 17*, 519–549.

Tocco, G., A. J. Annala, M. Baudry, and R. F. Thompson (1992). Learning of a hippocampal-dependent conditioning task changes the binding properties of AMPA receptors in rabbit hippocampus. *Behav. Neural Biol. 58*, 222–231.

Williams, J. H., Y. G. Li, A. Nayak, M. L. Errington, K. P. Murphy, and T. V. Bliss (1993). The suppression of long-term potentiation in rat hippocampus by inhibitors of nitric oxide synthase is temperature and age dependent. *Neuron 11*, 877–884.

13 | Treating Neurological and Psychiatric Diseases

A not-so-subtle theme should by now have been impressed upon the reader, namely, that a full understanding of the molecular mechanisms by which healthy neurons communicate and are sustained will illuminate the basis for using drugs to treat neurological and psychiatric disorders. Except for a few passing references, however, we have not said very much about the nature of the brain's diseases or indicated those for which treatments are now effective and why. In this final chapter, we provide a brief introduction to those important neuropharmacological motivations.

The modern neuropsychiatric therapeutic armamentarium contains some powerful, selective, and highly effective drugs that often save lives and restore function. With the few exceptions boldly noted below, however, most of these treatments came about by refining drugs that were initially developed for dubious disease hypotheses, where potentially useful central nervous system actions were observed as side effects. Often the predecessors for today's drugs told us more about the nature of the brain's normal operations by revealing drug-sensitive functions that had not been previously recognized. Second, the most effective drugs (like those for complex diseases of thinking, emotion, or motoric or appetitive functions) work for reasons that are still not very clear and in which the nature of the illness in molecular terms continues to drive research. The corollary is that the diseases that we understand the most clearly in terms of causative mechanisms are not yet very treatable. One last confession before we begin: New drugs will likely emerge from the mountains of molecular information being compiled, but brain diseases are the most elusive targets because often there are no pertinent animal models of these complex multigenic diseases in which to evaluate new drugs based on far more insightful views of pathogenesis.

WHEN IT'S ALL IN THE GENES

Back in Chapter 3, we went through some of the fundamentals of molecular genetics in order to define some of the basic properties of the cells and their functions. To understand the genetic basis of central nervous system disorders when there is one, we need to revisit those concepts. We refer to the genome as the complete genetic set of a given individual, established at the moment of conception when the half set of genes in a sperm combined with the half set in the lucky ovum. In humans, the cells of the embryo will contain 22 pairs of homologous autosomes and one pair of sex chromosomes that are heterologous (XY) for males or homologous (XX) for females. Each chromosome contains many genes (there are probably 50,000–100,000 in the human genome) each with its own specific location (its "locus") along the chromosome. Since each chromosome in the pair will have one copy (or "allele") of each gene, a person can be either homozygous (if both genes are the same) or heterozygous for any given gene.

Small, nonfunctional differences in nucleotide sequences that arise by mutation allow for some genes to contain polymorphisms. Because these differences also change the vulnerability of the DNA to cleavage by specific restriction endonucleases, a given digestion protocol can produce DNA fragments of different lengths in the alleles for a given gene. This allows us to use the method of restriction fragment length polymorphisms (RFLP) to follow the inheritability of a particular allele by its telltale-sized fragments across family pedigrees.

With that background, we can now categorize disease-producing genetic alterations in individuals in three ways:

1. *Single-gene mutations,* which because they are inherited in recognizable patterns within families are often referred to as

"mendelian" mutations. These can be further categorized as *autosomal dominant* (when having one disease-causing allele is all it takes, as with Huntington's disease or in neurofibromatosis) or *autosomal recessive* (when the disease may be "carried" across generations and not be detected unless the individual inherits the gene mutation from both mother and father and becomes homozygous). Special forms of recessive disorder arise from the X chromosome. Because the Y chromosome carries no somatic traits, and only codes for gender-determining factors, a single X-linked disease-producing allele will cause the disease in males, but nonsymptomatic females will pass it on to half of their sons but none of their daughters. Familial dysautonomia is an autosomal recessive condition affecting Ashkenazi Jews that can be diagnosed by a supersensitivity of the iris to methacholine. Because the gene for phenylalanine hydroxylase is subject to several different forms of inheritable mutation, heterozygote carriers can produce offspring vulnerable to absence of the functional enzyme, leading to phenylketonuria and a progressive mental retardation unless dietary phenylalanine is severely restricted well into early adolescence. Other inheritable metabolic disorders leading to mental retardation are shown in Table 13-1.

2. *Polygenic diseases*, in which multiple genetic factors are responsible for the appearance of the disorder. A good example is the "Westphal" variant form of severe early-onset Huntington's disease when the disease gene is inherited from the father. Most complex diseases that have some inheritable vulnerability, including diabetes mellitus, hypertension, and those we will discuss later in this chapter, arise from an unspecified number of "factors," each of which might easily stem from more than one gene.

3. *Physically abnormal chromosomes* can arise during embryogenesis as cells divide, and the structural abnormality of the chromosomes can be detected by microscopy. Trisomy 21 (a triplication of all or parts of chromosome 21, better known as Down syndrome) and the Fragile X form of mental retardation are examples of this kind of genetic problem.

TABLE 13-1. Hereditary Diseases Associated with Cerebral Impairment

Disorder	Metabolic Defect
Amino acid metabolism	
Arginosuccinic aciduria	Arginosuccinase
Citrullinemia	Arginosuccinic acid synthetase
Cystathionuria	Cystathionine-cleaving enzyme
Hartnup disease	Tryptophan transport
Histidinemia	Histidase
Homocystinuria	Cystathionine synthetase
Hydroxyprolinemia	Hydroxyproline oxidase
Hyperammonemia	Ornithine transcarbamylase
Hyperprolinemia	Proline oxidase
Maple syrup urine disease	Valine, leucine, and isoleucine decarboxylation
Phenylketonuria	Phenylalanine hydroxylase
Lipid metabolism	
A-β-lipoproteinemia (acanthocytosis)	β-Lipoproteins
Cerebrotendonous xanthomatosis	Cholesterol
Gaucher's disease	Cerebrosides
Juvenile amaurotic idiocy	Gangliosides
Krabbe's globoid dystrophy	Cerebrosides
Kuf's disease	Gangliosides
Metachromatic leukodystrophy	Sulfatides
Niemann-Pick disease	Gangliosides and sphingomyelins
Refsum's disease	3, 7, 11, 15-Tetramethylhexadecanoic acid
Tay-Sachs disease	Gangliosides

(continued)

TABLE 13-1. Hereditary Diseases Associated with Cerebral Impairment (*Continued*)

Disorder	Metabolic Defect
Carbohydrate metabolism	
Galactosemia	Galactose-1-phosphate uridyl transferase
Glycogen storage disease (Type 2)	α-Glucosidase
Hurler's disease	Chondroitin sulfuric acid gangliosides
Unverricht's myoclonus epilepsy	Polysaccharides
Lafora's disease	Polyglucosan
Miscellaneous	
Cretinism	Thyroid hormone
Hallevorden-Spaatz disease	Iron deposition in basal ganglia
Intermittent acute porphyria	δ-Aminolevulinic acid
Lesch-Nyhan syndrome	Hypoxanthine-guanine Phosphoribosyl transferase
Methylmalonic acidemia	Methylmalonyl CoA mutase
Wilson's disease	Ceruloplasmin

Inheritable Metabolic Errors and Brain Development

Most of the diseases listed in Table 13-1 are single-gene defects leading to mental retardation or other forms of behavioral abnormality. Almost all of them are untreatable, so the forward-looking emphasis has been on genetic screening. In the *Lesch-Nyhan* syndrome (an X-linked recessive disorder) the affected males are both physically and mentally retarded and exhibit a characteristic self-mutilating behavior. Although the enzyme is expressed in many other cellular systems, it is basically only the formation and function of the brain in which the problems arise, for reasons that are not clear. Mice lacking this enzyme show no behavioral phenotype.

Cretinism, a severe neonatal hypothyroidism, may arise from loss of thyroxin-producing enzymes and can of course be treated by replacement therapy as soon as it is recognized. The same is true of deficiencies of growth hormone and of gonadotropin-releasing hormone (GnRH). The latter hormone is lost in males by a mutation *(Kallman's Syndrome)* in a specific neuronal adhesion molecule required to provide the trail by which the GnRH neurons migrate from their birthplace in the olfactory bulb to their intended final location in the ventral hypothalamus. The *porphyrias* are a group of mostly autosomal dominant mutations of enzymes that synthesize the blood pigment heme; in one form, acute intermittent porphyria, an otherwise asymptomatic case (such as King George III's), can be triggered by infections, or dehydration, or other unknown factors into a variety of incapacitating neurological and behavioral symptoms including depression, mania, and hallucinations.

WHEN GENES ARE INFLUENCED BY ENVIRONMENT

Next we consider diseases in which several unknown genes clearly play a role, but in a manner that is both environmentally and behaviorally influenced. The extent to which specific gene products are expressed in brain is a matter of the demands placed on the neuron or glial cell by the incoming information as we saw in Chapter 3. The environment, both physical and social, is the source of much of this external information. Eventually the information is converted into synaptic signals, which are in turn transduced into intracellular second messengers, and ultimately into altered cytoplasmic and nuclear signals that, through transcriptional regulation, can determine which genes are turned on or kept off. Thus, the degree to which our genotype is reflected in our functional form or phenotype comes under environmental influence. The effects of the environment are likely to be cumulative and perhaps vary with the stages of

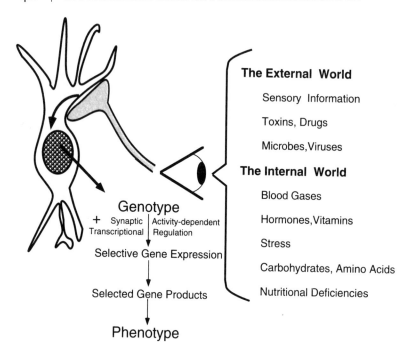

The External World

Sensory Information

Toxins, Drugs

Microbes, Viruses

The Internal World

Blood Gases

Hormones, Vitamins

Stress

Carbohydrates, Amino Acids

Nutritional Deficiencies

Genotype

+ Synaptic | Activity-dependent
Transcriptional | Regulation

Selective Gene Expression

Selected Gene Products

Phenotype

FIGURE 13-1. A neuron sees the world through the information received by its synaptic afferents. The cumulative results of synaptic activity modify gene expression by means of the nuclear actions of metabolic end products (intracellular second messengers, ions, and immediate early genes leading to transcriptional regulation) triggered by synaptic events. Changes in gene expression, resulting from the influence of external events or from changes in the internal environment of the brain, alter the neuron's phenotype (such as the rate of utilization of its own transmitter or receptors) and hence the operation of the circuits in which it participates.

development, and these influences on early brain function may not be apparent until much later in life (see Fig. 13-1). The three examples we will now examine are all relatively common disorders, and many powerful medications have been developed to provide significant alleviation for some but not all of those affected.

Affective Disorders

The family of psychiatric diseases epitomized by changes in mood (emotion, affect) are called the affective disorders, and they are very common. Between 10% and 20% of the population will have a serious (i.e., clinically significant) episode over their lifetime, and they may account for as many as 50% of those hospitalized for psychiatric reasons. There are two major forms with different gender and genetic characteristics. *Bipolar affective disorder* is characterized by recurrent, wide swings in mood from depression to extreme elation, occurs slightly more often in women than men, shows strong familial patterns of inheritability, and is primarily found in young adults 18–45. *Unipolar depression* is characterized by one or more prolonged episodes of depression (measured in weeks), occur far more frequently in the population, and is nearly three times as common in women as in men. These diseases epitomize the possibility for successful biological approaches to the treatment of psychiatric disorders and exemplify current thinking about the study of animal models of human brain diseases.

Catecholamine Theory of Affective Disorder

Animal studies indicate that catecholamines play a role in the periphery with regard to stress and emotional behavior. The catecholamine hypothesis of affective disorder states that, in general, behavioral depression may be related to a deficiency of catecholamine (usually norepinephrine) at functionally important central adrenergic receptors, while mania results from excess catecholamine. While substantial experimental work supports this proposal, it must be kept in mind that most of the experiments on which this hypothesis is based derive from "normal" animals.

The original impetus for developing the catecholamine hypothesis was the finding in patients being treated for tuberculosis that various monoamine oxidase inhibitors administered with other goals in mind, notably iproniazid, acted clinically as mood elevators or antidepressants. Shortly thereafter, it was found that this class of compounds also produced marked increases in brain amine levels. By the same token, reserpine, a potent tranquilizer and an effective antihypertensive, depletes brain amines (serotonin as well as the catecholamines) and produces a serious depressed state (clinically indistinguishable from endogenous depression) in about 20% of those treated, and even suicidal attempts in some people.

Although both classes of drugs alter brain levels of catecholamines and 5-HT quite equally, the fact that a precursor of catecholamine biosynthesis, DOPA, can reverse most of the reserpine-induced symptomatology in animals has tended to bias many researchers in favor of the catecholamine theory. In fact, by no means does the available evidence for the involvement of norepinephrine rule out the participation of dopamine, 5-HT, epinephrine, or other putative transmitters in similar events. The three general classes of drugs used to treat depressive disorders are the monoamine oxidase inhibitors, the tricyclic antidepressants, and the psychomotor stimulants such as amphetamine. All of these pharmacological agents appear to interact with catecholamines in a way that is consistent with the catecholamine hypothesis. Thus, all four of the suggested modes of action of amphetamines (partial agonist, inhibitor of catecholamine reuptake, competitive inhibitor of monoamine oxidase, and displacer of presynaptic norepinephrine and dopamine) would be expected to increase catecholamines temporarily at their receptors. Administration of long-term or high-dose amphetamine produces an eventual depletion of brain norepinephrine and an inhibition of neuronal activity in catecholamine neurons. This chronic depletion of transmitter and the prolonged inactivity of catecholamine neurons may account for the clinical observation of amphetamine tolerance or the well-known poststimulation depression or fatigue observed after chronic administration of this class of drugs.

Many of the tricyclic antidepressants influence catecholamine neurons by inhibiting the catecholamine and 5-HT presynaptic plasma transporters. These drugs were being used to treat depressed people even before the phenomenon of active reuptake had been recognized as a means to terminate synaptic release. While their inhibition of transporters is consistent with the catecholamine hypothesis, the earlier tricyclic reuptake inhibitors did not affect this system exclusively. Some of the tricyclic agents, notably imipramine, potentiate peripheral and central effects of serotonin and also have both central and peripheral anticholinergic actions. More recently, other reuptake inhibitors, not necessarily of the "tricyclic" structural motif, have been made selective for one or another monoamine. For example, mazindol and bupropion are dopamine-selective reuptake inhibitors, and fluoxetine (Prozac) was the first of the serotonin-selective reuptake inhibitors. These drugs are far more selective than was the original tricyclic imipramine, which is now recognized as a mixed norepinephrine/5-HT transporter inhibitor. Although they have proven to be effective in treating subsets of depressed patients, some depressed patients—especially those with childhood or adolescent onset—are quite resistant to such treatment.

The action of the monoamine oxidase inhibitors also supports the catecholamine hypothesis. All these agents inhibit an enzyme responsible for the metabolism of norepinephrine and various other amines (5-hydroxytryptamine, dopamine, tyramine, tryptamine). This inhibition results in a marked increase in the intraneuronal levels of norepinephrine. According to the conceptual model of the adrenergic neuron presented above, this intraneuronal norepinephrine might eventually diffuse out of the neuron and reach receptor cells, thereby overcoming the presumed deficiency. A similar mechanism may also explain the antagonism of reserpine-induced sedation with monoamine oxidase inhibitors, since there will be a replenishment of the norepinephrine deficiency initially caused by reserpine.

Lithium, one of the main agents used to treat manic episodes in bipolar disease, has also been studied with regard to its action on the life cycle of the catecholamines. Interestingly, pretreatment with lithium blocks the stimulus-induced release of norepinephrine from rat brain slices. Other investigators have suggested that lithium may facilitate reuptake of norepinephrine. If the inhibition of release observed in stimulated brain slices is due to a facilitated recapture mechanism, then the mechanism of action of lithium is the exact opposite of that of the antidepressant drugs, as would be expected according to the catecholamine hypothesis. Although a single injection of lithium may antagonize responses to norepinephrine, chronic treatment of rats leads to modest supersensitive norepinephrine responses.

On closer scrutiny of the clinical and experimental pharmacological data cited in support of the catecholamine theory of affective disorder, however, even the keenest enthusiasts recognize significant inconsistencies: (1) cocaine is a very potent inhibitor of catecholamine reuptake, and thus, like the tricyclic antidepressants, it should increase the availability of norepinephrine at central synapses; but cocaine does not possess any significant antidepressant activity. (2) Iprandole, a tricyclic compound without any significant effect on catecholamine uptake or any influence on central noradrenergic neurons, is an effective antidepressant. Furthermore, desensitization of postsynaptic β-receptors after chronic administration of tricyclic antidepressant drugs is not entirely compatible with the proposed theory of a synaptic catecholamine deficit. (3) In laboratory studies the ability of tricyclic antidepressants to inhibit catecholamine uptake and the ability of monoamine oxidase inhibitors to block monoamine oxidase and elevate brain catecholamines are apparent soon after administration. Clinically, antidepressants must be given for several days (10–14) to produce therapeutic effects. Perhaps some novel genes must be turned on or off, or perhaps enhanced catecholamine action at recurrent axons onto monoamine cell bodies further reduces their firing, such that even less catecholamine is released at synapses. (4) On the surface, the pharmacological and clinical effects of the tricyclics and lithium seem to fit nicely with the catecholamine theory of affective disorder. These agents produce opposite effects on norepinephrine disposition, and lithium is effective in treating mania while the tricyclics are useful in treating depression. The water is muddied, however, by the finding that lithium is also effective in the treatment of bipolar depressed patients, where it dampens the emotional oscillations into either mania or depression.

The fact that several questions have been raised about the pharmacological data used to support the catecholamine theory of affective disorder has prompted many investigators to seek direct evidence of the involvement of norepinephrine systems in affective disorders. The most extensive studies have involved analysis of urinary excretion patterns of catecholamine metabolites in patients with affective disorders. The rationale has been that the urinary excretion of a particular metabolite like MHPG may be a useful reflection of central catecholaminergic processes. Despite the inherent problems in urinary catecholamine metabolite measurement, such as complications because of large contributions from peripheral sources, some interesting findings have emerged.

Findings from several clinical studies now indicate that (1) depressed patients as a group excrete less than normal quantities of MHPG; (2) diagnostic subgroups of depressed patients are particularly likely to have low urinary MHPG values; (3) bipolar patients who switch from a depressive to a euthymic or hypomanic state show a corresponding increase in urinary MHPG; and (4) pretreatment urinary MHPG values are predictive of the type of therapeutic response obtained with catecholamine-directed reuptake inhibitors. Although provocative and essentially consistent with the norepinephrine theory of affective disorder, these clinical findings all hinge on the issue of the degree to which urinary MHPG reflects central norepinephrine metabolism.

For the above reasons, the original hypothesis that the mechanism of action of antidepressant drugs is to increase the availability of monoamines in the brain has been updated to include the effects of long-term antidepressant treatment on monoamine receptor sensitivity. A wide array of effects of long-term treatment with antidepressants has been reported in several monoamine systems. The most consistent findings observed following chronic administration of most of the clinically effective antidepressant drugs to experimental animals are (1) a reduction in the number of β-adrenoceptor recognition sites and a down-regulation of β-adrenoceptor functioning; (2) an increase in the sensitivity of central α-adrenoceptors, suggesting an up-regulation in central α-adrenoceptor functioning; and (3) similarly, both behavioral and electrophysiological studies also point to an up-regulation in the sensitivity of central 5-HT receptors. These changes require time to be accomplished, and perhaps that is why the delay in antidepressant action is observed. However, the monoaminergic neurons have a rich recurrent innervation, and an enhancement of this autoinhibition would also be expected from the use of the re-uptake inhibitors. If the monoamine neurons are thus reduced in their excitability, the actual amount of transmitter released from distal axons may be initially reduced, even with re-uptake inhibition. Therefore, another part of the lag in response to treatment may be the requirement for re-equilibration of perikaryal excitability and a restoration of baseline synaptic function before the reuptake inhibition per se can enhance postsynaptic receptor durations of action. Or perhaps new gene products must be made and time is required to ship them in sufficient numbers to the synapses.

Throughout this analysis of antidepressant drug actions, we have noted frequent intrusions by 5-HT-related effects. In fact, 5-HT "deficiency" deserves attention as an independent causative factor in some forms of depression. Diminished plasma levels of the rate-limiting 5-HT precursor, the amino acid L-tryptophan (L-TRP), have been seen in several series of depressed patients and have been linked to either reduced intestinal absorption or increased hepatic catabolism of absorbed L-TRP stimulated by pyrrolase; dietary restriction of L-TRP may also precipitate depression in recently remitted patients. Cerebrospinal fluid levels of the 5-HT catabolite 5-hydroxyindoleacetic acid (5-HIAA) have been studied extensively in depressed subjects, but the best correlations here are with attempts to commit suicide and with impulsive, violent behavioral patterns. Some postmortem studies of depressed patients have shown increased numbers of certain 5-HT receptors, especially in frontal cortical regions. In depressed persons this region is hypometabolic when studied by PET during life. Other pharmacological challenges to depressed subjects, based on changes in sleep EEG patterns and neuroendocrine secretion patterns, further support the concept of a central 5-HT deficiency and postsynaptic receptor up-regulation. Perhaps the greatest support for this concept arose with the major, and to some extent unexpected, antidepressant success of fluoxetine, sertraline, and paroxetine—the serotonin selective re-uptake inhibitors.

More direct studies are necessary, however, before it is possible to conclude whether any or all of the therapeutic actions of antidepressant drugs are indications that the emotional disorder was truly caused by functional deficiencies of one or another monoamine. These studies await the development of appropriate models for monitoring central neurotransmitter functioning in humans, especially those at high risk of developing depression based on their family pedigrees.

Dopamine Hypothesis of Schizophrenia

Schizophrenia is a chronic disorganization of mental function that affects *thinking* (paranoid ideas, inability to maintain a focused thought, easily distracted, loose associations between thoughts), *feeling* (typically referred to as blunted affect and inappropriate responses to social situations), and *movement* (from hyperactivity and excitement to persistent inactivity to the point of maintaining bizarre postures for long periods of time). The disease is most commonly recognized in very young adults, particularly when confronted with severely stressful life events, although evidence of social withdrawal and disorganized thinking may have been noted before the episode. About one-third of those having a single episode may fully recover. Like bipolar depression, schizophrenia shows strong familial patterns of inheritance in some cases. Available medications can be very beneficial, but for most affected individuals, the blunted affect, apathy, lack of volition, and social withdrawal may be intractable.

In analogy to the catecholamine hypothesis of depression, a biochemical explanation for schizophrenia arose from the observations that the only consistent feature among the antipsychotic drugs used to treat the disease was their ability to antagonize D_2 dopamine receptors and that chronic administration of amphetamine and other (mainly) dopamine-mediated psychostimulants could produce a psychotic state loosely resembling some aspects of schizophrenia. This hypothesis in its simplest form states that schizophrenia may be related to a relative excess of central dopaminergic neuronal activity. Attempts to validate this hypothesis in clinical studies have been intense but inconclusive. While D_2 receptors are consistently increased in most postmortem studies of schizophrenic brains, the influence of prior antipsychotic treatments confounds the interpretation, and no direct evidence of increased dopamine synthesis or turnover or function has been obtained. Nor has noninvasive imaging of dopamine receptor densities in previously unmedicated young schizophrenics produced any consistent evidence for an increase in D_2 receptors within the striatum. However, the D_4 receptor subtype, functionally akin to the D_2 receptors in inhibiting adenylate cyclase, has been reported to have greater affinity for the atypical antipsychotic drug clozapine (these drugs were called "atypical" precisely because they were not good D_2 antagonists) and to be increased specifically in postmortem schizophrenic brains.

Unlike depression, where there is no obvious neuropathology, postmortem microscopic examination of the schizophrenic brain has revealed pathological evidence for abnormalities of neuronal density, but without frank neurodegeneration. Together with other findings (enlarged ventricles, thin cortex) that are stable over years of observation, this has led to the interpretation of schizophrenia as a neurodevelopmental disorder. The negative symptoms of cognitive disruption and affect have been correlated with reduced brain metabolic activity in the frontal lobes, and it has recently been suggested that this may be an adaptive down-regulation of dopaminergic activity in this region of cortex. The "atypical" antipsychotic drugs have received increasing attention for three reasons: 1) they can reduce symptoms in patients resistant to other drugs; 2) they produce fewer side effects; and 3) their pharmacology may extend the neurotransmitter base for schizophrenia to 5-HT as well as MA dopamine. Other evidence points to disruption of corticostriatal glutamatergic circuits in the disease. The area is being intensively explored (see Fig. 13-2).

Over the past decade experiments in animals have generated substantial support for the idea that antipsychotic drugs are effective blockers of dopamine receptors. But most of these animal studies (whether behavioral, biochemical, or electrophysiological) have been carried out after acute drug administration. This is a very serious drawback, since, as with the antidepressants, the clinical effects of antipsychotic drugs (both antipsychotic and neurological) take days, weeks, or even months to develop. In nonhuman primates, chronic treatment with antipsychotic drugs causes tolerance to develop to the homovanillic acid (HVA) increase normally observed in the putamen, caudate, and olfactory cortex after a challenge dose of an antipsychotic drug. In cingulate, dorsal frontal, and orbital frontal cortex, however, increased levels of HVA are maintained throughout the time course of chronic treatment. Similar observations have also been made in studies carried out on autopsied human brain specimens. In schizophrenic patients chronically treated with antipsychotic drugs, a significant increase in HVA is found in the cingulate and frontal cortex but not in the putamen and nucleus accumbens, suggesting a possible locus for the therapeutic action of these drugs and providing the first direct experimental evidence that antipsychotic drug treatment increases the metabolism of dopamine in human brain in a regionally specific manner. Establishing with certainty that the elevation in dopamine receptor density is part of the disease process would be very important for both etiological and diagnostic purposes and would serve as a possible basis for treatment strategies. The use of new noninvasive techniques such as positron emission tomography and single photon emission computed tomography to examine dopamine receptor distribution and density in schizophrenia holds promise that this may be accomplished soon.

Drug Abuse

The continued, compulsive obsession with obtaining, consuming, and experiencing self-administered drugs is a major social and med-

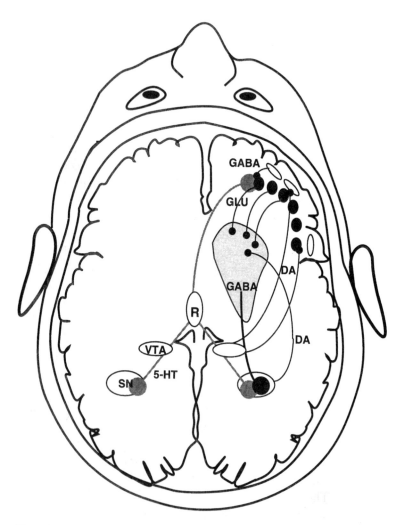

FIGURE 13-2. Schematic diagram of the main neuronal circuits and transmitters dysregulated in schizophrenia. The frontal neocortical projections to the striatum and other subcortical structures are primarily glutamatergic (GLU) efferents. Within the cortex, these units are regulated by GABA-containing interneurons (neuropeptides of those GABA neurons not shown) and by dopamine (DA) projections from the ventral tegmental area (VTA) and serotonergic (5-HT) projections from the raphe (R) nuclei. The latter monoaminergic neurons may project to either the cortical interneurons or

ical problem throughout the world. Specific drugs of abuse have specific patterns of use and dependence. Seven families of drugs have been recognized to be obsessively self-administered by humans. In order of prevalence, they are caffeine (as in coffee or tea), nicotine, alcohol (grouped with benzodiazepines and barbiturates), marijuana (and the congeners hashish and tetrahydrocannabinol), psychostimulants (cocaine and amphetamines), opiates (morphine, heroin, and other agonists), and the hallucinogenic drugs (LSD, phencyclidine, and MDMA [3,4 dioxymethylene methamphetamine] otherwise known as "ecstasy"). Note that the most widely abused substances—caffeine, nicotine, and alcohol—are all legal in the United States, and both the federal and state governments collect substantial "sin taxes" on the latter two while attempts to place taxes on caffeine helped spark the American Revolution. Recreational use of alcohol or nicotine, however, may serve as the gateway to illicit and powerful drugs of abuse.

In a large series of generally replicable studies, strong patterns of genetic transmission for alcoholism, and perhaps other forms of drug abuse, have been observed in humans. Animal models have proved to be extremely effective in examining the circuitry involved in drug-induced reward and in the brain's adaptive responses to specific forms of drug dependence, tolerance, and withdrawal. The genetic analysis of animals bred for their willingness to consume alcohol has been a fertile area of research.

The neurobiological substrate for the acute rewarding effects of the four major classes of abused drugs (alcohol, nicotine, opiates, and psychostimulants) have focused on the area of the ventral fore-

---

to the cortical efferent neurons. The raphe also projects to the substantia nigra (SN), whose DA axons innervate the basal ganglia but not the cortex. Current evidence favors decreased levels of operation in the nigrostriatal projections, but over-activity in the mesolimbic DA systems. "Atypical" antipsychotic drugs may work in part through their capacity to inhibit cortically enriched DA receptors (such as the D_4) as well as 5-HT receptors. (Based on circuitry described by Maes and Meltzer, 1994)

brain that surrounds the nucleus accumbens. This nucleus, with neurons of several forms and efferents, is heavily innervated by cells containing all the monoamines and is considered to be in a continuum with the amygdaloid complex. Increased release of dopamine within the region of the nucleus accumbens has been directly observed by microdialysis in animals trained to self-administer alcohol and nicotine. In the case of cocaine, antagonists of the D_3 dopamine receptor will selectively reduce the reward, that is, the self-administering behavior. However, while the key site for eliciting opiate reinforcement also seems to be localized within the nucleus accumbens, the dopamine afferents are not required for self-administration of these drugs to be maintained.

Certain forms of alcoholism are thought to arise from unknown but genetically transmittable factors that increase vulnerability to alcohol dependence, especially in the male offspring of male alcoholics who may also show antisocial personality disorder. Animal models of genetically transmissible alcohol preference have been achieved several times by constant cross breeding of the few rats that initially show modest alcohol preference. Later generations develop an almost universal predilection to consume alcohol for its pharmacological (as opposed to nutritional or thermal) properties. In the alcohol-preferring animals developed in the United States and in Sardinia, there is diminished dopamine and 5-HT innervation of the nucleus accumbens and increased GABA innervation of that site. Among other manipulations, 5-HT-specific reuptake inhibitors will reduce alcohol self-administration in these rats. Alcohol-accepting rats bred in Finland, however, do not show these neurochemical differences. Alcoholism vulnerability in Oriental human populations is actively suppressed by a prevalent genetic mutation leading to reduced capacity to oxidize the ethanol catabolite acetaldehyde. Those carrying this mutation (half of these populations) react to alcohol as though they were receiving the drug antabuse, which directly inhibits acetaldehyde dehydrogenase. After a brief flurry of excitement over the possibility that certain alleles of certain dopamine receptors might be associated with alcoholism and other forms of dependency, the search for the responsible genes has resumed. While there are major differences in the sensitivity of certain inbred rat strains to the reinforcing actions of opiates and psychostimulants, the evidence for such inherited vulnerability has not been well studied in humans largely because these drugs are universally highly reinforcing once their recreational use has been initiated. Nonetheless, these high- and low-sensitivity animal models of drug response may prove useful in determining the genetic basis for vulnerability to drug dependency in humans.

Efforts to deal with drug abuse have focused on reducing the supplies of the illicit drugs by law enforcement strategies aimed at the prevention of drug smuggling. After spending hundreds of millions of dollars in "supply reduction" only to see the suppliers able to sell more products at lower prices, agencies have devised other strategies directed toward demand reduction. For many years methadone, a mixed agonist–antagonist for morphine with reduced euphorigenic properties, and some of its long-acting congeners have been known to reduce opiate dependency, increasing job performance and reducing criminal behavior, but its wide-scale adoption was delayed because of the belief in some quarters that it was "merely" substituting one drug for another. Only recently has methadone been approved for the long-term treatment of opiate abuse. In animal studies the opiate antagonist nalaxone, also reduces some of the rewarding effects of alcohol self-administration for reasons that are not yet clear. Naltrexone, the orally active congener of naloxone, has recently been approved for the long-term treatment of alcoholism.

WHEN KNOWING THE GENES DOESN'T HELP

Lastly we consider a group of diseases that all fall under the general topic of neurodegenerative disorders, most of which remain untreatable although animal models are beginning to emerge that pose striking challenges to pharmacological development in this area. Some of the more creative avenues to restoring brain function, from transplantation of neurons to the transfer of autosomal genes, are being explored here.

Alzheimer's Disease

This chronic, progressive, degenerative disorder, which is in some rare cases familial, is becoming a greater and greater health care problem as the population survives to older and older ages. The nearly linear incidence of Alzheimer's detection with age makes this a very prevalent disorder, perhaps being detected in as many as 50% of those older than age 85, which is no longer a unique occurrence in our society. Alzheimer's disease is characterized behaviorally by a severe impairment in cognitive function, including memory and the ability to recognize objects and orient oneself in time and space. Neuropathologically, the disease is epitomized by the appearance of neuritic plaques and neurofibrillary tangles containing β-amyloid protein within restricted cortical regions (hippocampus, frontal, and temporal, more than parietal and far more than occipital lobes) and layers (especially prominent in layers III, V, and VI, where the large neurons are located), and within a few of the subcortical structures (ventral pallidum, locus ceruleus) projecting to the cortical areas of involvement.

A cholinergic dysfunction has been implicated in Alzheimer's disease based on the findings that (1) the administration of centrally acting muscarinic-blocking agents to normal individuals induces a loss of recent memory; (2) in the cerebral cortex and hippocampus of patients with Alzheimer's disease there is a dramatic reduction of ACh, choline acetyltransferase, and high-affinity choline uptake; (3) in Alzheimer patients there is a severe reduction of neurons in the nucleus basalis of Meynert, the primary cholinergic input to the cortex; and (4) in some but not all studies, a decrease in muscarinic and nicotinic receptors has been noted. It should also be noted, however, that patients with Alzheimer's disease have decreased levels of somatostatin, neuropeptide Y, and CRF; increased numbers of CRF and NMDA receptors; as well as reduced numbers of locus ceruleus neurons. A decrease in the neuronal content of the nucleus basalis of Meynert has also been observed in some patients with Down syndrome (trisomy 21, the only other condition known to express neuritic plaques) and with Parkinson's disease.

Thus, the marriage of Alzheimer's disease to a cholinergic dysfunction may involve some extramarital relationships. Little success has been achieved to date by treating patients with choline, lecithin, physostigmine, or the muscarinic agonist arecoline. Current efforts focus on the development of drugs that can block acetylcholinesterase centrally without hepatoxicity and peripheral autonomic side effects. Future therapy will perhaps require the development of a specific M_2-muscarinic antagonist, or an M_1-muscarinic or some other nicotinic agonist, or the transplantation of fetal cholinergic tissue to the brain with the hope that appropriate synaptic contacts or trophic factors can be restored. A potentially direct assault on the causes of the disease has been mounted by analyzing the proteins that are aggregated in the neuritic plaques and tangles and the variations in the processing of β-amyloid precursor protein in healthy aged brains compared with those of Alzheimer's patients. Combinations of genetic predispositions (e.g., inheritance of the apoliprotein E-ε4) with environmental conditions or events such as head trauma may initiate the process of pathological change. The creation of a transgenic mouse overexpressing one form of the β-APP gene in cortical neurons and producing plaques within 18 months offers the possibility of screening for drugs that may reduce amyloid depositing, and perhaps may reveal reasons why the aged brain overproduces this and other proteins in the first place.

Parkinson's Disease

Parkinson's disease (PD) is a progressive neurodegenerative disorder of the basal ganglia characterized by tremor, muscular rigidity, difficulty in initiating motor activity, and loss of postural reflexes. It is observed in approximately 1% of the population over age 55. For over 70 years it has been known that PD is characterized pathologically by loss of pigmented cells in the substantia nigra, but only since 1960 has the substantial loss of dopamine in the striatum been documented. It is now clear that PD can be defined in biochemical terms as primarily a dopamine-deficiency state resulting from degeneration or injury to dopamine neurons. The most striking degenerative loss of dopamine neurons is observed in the nigrostriatal system. Even in patients with mild symptoms, a striatal dopamine loss of 70%–80% occurs while severely impaired subjects have striatal dopamine depletions in excess of 90%. Since the dopamine transporter is heavily expressed in the terminals of dopamine neurons that are lost in PD, it is not surprising that striatal binding of agents that label this site (cocaine, nomifensine, GBR, and mazindol) is lost in the Parkinsonian striatum. This alteration corresponds well with the loss of functional dopamine uptake visualized in vivo by positron emission tomography (using ^{18}F-L-DOPA uptake or ^{18}F-nomifensine) or by single photon emission computed tomography (SPECT) using other presynaptic dopamine labels. Although striatal dopamine loss represents the primary neurochemical abnormality in the PD brain, typical parkinsonism is accompanied by the loss of other dopamine systems and other monoamine neurons as well. Some degeneration of dopamine-containing neurons is also apparent in the mesolimbic, mesocortical, and hypothalamic systems, as is the loss of norepinephrine-containing neurons in the locus ceruleus, and of serotonin neurons. Non-monoamine systems are also affected, with depletions observed in somatostatin, neurotensin, Substance P, enkephalin, and cholecystokinin-8. Since many of these nondopamine systems indirectly interact with mesotelencephalic dopamine systems, it is obvious that changes in some of them are bound to influence the function of dopamine neurons in a complex way. Before it was known that DA is severely depleted, for example, treatments with anticholinergic drugs had been viewed as moderately effective.

Since the earliest and most substantial neurochemical abnormality in PD is the loss of dopamine, the modern strategy for treatment has concentrated on restoring the dopamine deficit. The theoretical strategies here include substrate supplementation, direct and indirect dopamine agonists, metabolic inhibitors (monoamine oxidase inhibitors), and uptake inhibitors. The most successful treatment has been the use of L-DOPA, now usually combined with an inhibitor of DOPA decarboxylase that acts only outside the brain.

This enhances the amount of the absorbed L-DOPA that can enter the brain, and alleviates the gastrointestinal symptoms that arose from the higher doses required previously. As the degeneration of dopamine (and other) neurons progresses, the requirements for L-DOPA increase and are accompanied by interruptions of its effectiveness that are poorly understood ("wearing-off" and "on/off" responses). Direct dopamine agonists have some benefit for patients whose responsiveness to L-DOPA is greatly reduced or erratic. So far, the only direct-acting dopamine agonist that has found extensive use in PD is bromocriptine, primarily a D_2 agonist. However, selective D_1 agonists such as ropinirole and cabergoline have been developed which exhibit longer durations of action without the absorption or blood–brain barrier permeation problems that have been suggested as causes for the variations in L-DOPA effectiveness. Other agents belonging to this class will no doubt prove useful in the future as supplements or alternatives to L-DOPA.

In view of the behavioral and electrophysiological studies that suggest that D_1 receptor activation is necessary for the effects of D_2 receptor stimulation to be maximally expressed in normal animals as well as in animals with supersensitive dopamine receptors, the functional interaction between D_1 and D_2 receptors could have important implications in PD, where stimulation of postsynaptic dopamine receptors confers symptomatic benefit. Knowledge of the optimal ratio of relative drug activity at D_1 and D_2 receptors that is required to elicit effective stimulation of dopamine-mediated function may provide a basis for the design of new drugs. Also, more knowledge of the distribution and function of various dopamine receptor subtypes should facilitate the development of new agents to treat dopamine deficiency states.

Primate Model of Parkinson's Disease

In 1983, researchers at Stanford University identified a contaminant in a locally produced "synthetic heroin" that induced a Parkinson-like syndrome in some of the individuals who self-administered this street drug. The contaminant identified in this preparation, 1-methyl, 4-phenyl 1, 2,3,6-tetrahydropyridine (MPTP), exhibits a high degree of anatomical and species-specific toxicity. MPTP administered systemically in low doses to nonhuman primates produces parkinsonian symptoms and destroys nigrostriatal dopamine neurons while sparing several other brain dopamine systems. The discovery of the selective neurotoxic properties of MPTP and the development of a primate model of parkinsonism stimulated a strong resurgence of inquiry into the causes and treatment of PD. Still, the mechanism responsible for selective dopamine neurotoxic features of MPTP in primates has not been conclusively established.

MPTP appears to act as a pro-toxin, and monoamine oxidase-B–mediated bioactivation of MPTP to MPP^+ plays a critical role in the ultimate neurotoxic action. The administration of monoamine oxidase-B inhibitors, including L-deprenyl, affords full protection against the neurotoxic action of MPTP. Since dopamine uptake inhibitors such as mazindol and GBR-12909 are also able to protect against MPTP-induced neurotoxicity, it has been suggested that MPTP is oxidized to MPP^+ outside the dopamine neurons, perhaps by monoamine oxidase in astrocytes. MPP^+ is then transported and concentrated by the dopamine uptake system in dopamine neurons. Once inside the dopamine neuron, MPP^+ exerts its neurotoxic effect by acting as a mitochondrial poison through inhibition of respiration at site **i** of the electron transport chain. This postulated mechanism for MPTP's neurotoxicity in dopamine neurons is illustrated in Fig. 13-3.

Attempts have already been made to apply these animal data on the mechanisms of MPTP killing of dopamine neurons in the effort to delay the progression of early PD: large clinical trials combining monoamine oxidase inhibitors and antioxidants have been interpreted as showing some effectiveness. Future treatments may also profit from studies now actively being pursued in the animal model: surgical transplantation of fetal substentia nigra neurons, transplantation of autologous (that is, from the same patient) adrenal medullary cells or skin fibroblasts genetically transfected to express tyrosine hydroxylase, and even special virus constructs to transfer

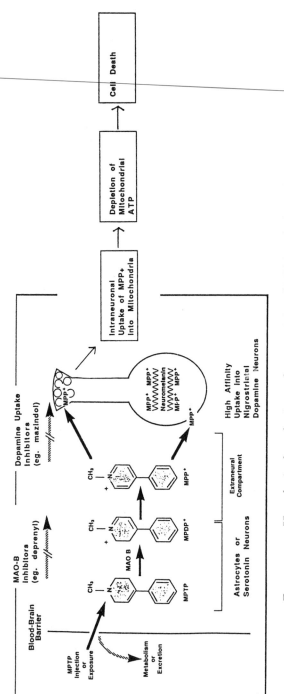

FIGURE 13-3. Hypothesized mechanisms of neurotoxicity of MPTP, which produces parkinsonism in primate species.

the tyrosine hydroxylase gene into surviving striatal neurons, with or without some of the neurotrophic factors which may be required.

FUTURE PROSPECTS

There is no shortage of treatment opportunities for neuropharmacologists, and we have only surveyed some of the diseases in which causes and treatments are under intensive investigation. The interested reader may find additional targets of opportunity in the literature of other neurological or psychiatric disorders that follow the dictum of the moment—that treatment success seems inversely proportional to the molecular understanding of the etiology or pathophysiology. Along this spectrum from treatable, not-understood to untreatable, almost-understood diseases, we find:

> *The hypertensions:* A large collection of disorders identified by the medical finding, often without symptoms, of elevated systolic and diastolic blood pressures. Depending upon cause, a variety of medications may restore blood pressure to normal levels: diuretics, peripheral sympatholytics, ganglionic blocking agents, central α_2-adrenergic agonists (clonidine) and α_1-antagonists (prazosin), and quite unexpectedly drugs that inhibit angiotensin III, either by inhibiting its formation through molecules engineered as angiotensin-converting enzyme inhibitors or by receptor antagonism.
>
> *Obsessive-compulsive disorder* and *anxiety disorder:* Highly prevalent conditions characterized by chronic, unrealistic anxiety and compulsive behavior patterns, in some cases chemically elicitable, exhibiting semispecific cerebral metabolic alterations. For unknown reasons these disorders often respond well to certain medications such as fluoxetine; inbred animal model systems may provide clues some day.
>
> *The epilepsies:* Convulsive disorders arising from trauma, infections, tumors, or unknown causes. Postencephalitic or posttraumatic scarring and acute fever are known causes, but many

others are suspected. Several drugs are partially effective for reasons that are poorly understood, and the only useful animal models are those that simulate the disease condition (electroshock or drug treatments).

Multiple sclerosis: Chronic or recurrent demyelination leads to progressive, focal neurodegenerative loss of sensory and motor function, often accompanied by depression and cognitive deterioration. Drugs that slow down the immune system (interferon-β) or reduce cell division in the macrophage–monocyte lineage (2-Cl-deoxyadenosine) have recently begun to show some potential.

Human immunodeficiency virus-associated cognitive and motor dysfunction: The HIV virus, known not to infect neurons, nevertheless produces progressive neurodegeneration, perhaps mediated by either inflammatory cytokines or neurotoxins of either central (e.g., NO or GLU) or peripheral (e.g., quinolinic acid) origins. No current treatments are effective.

Amyotrophic lateral sclerosis: An untreatable progressive neurodegenerative disorder restricted to the motor neurons of the spinal cord, corticospinal tract, and the medulla and recently linked to a mutation in the enzyme Cu–Zn superoxide dismutase. In a mouse mutant animal model this enzyme has already been knocked out.

Before leaving student readers to find their own path forward into this daunting field of major untreated diseases, let us make one final observation. The present-day armamentarium of drugs useful in the treatment of central nervous system disorders arose—for the most part—without good insight into the nature of the diseases, by making fortuitously clever structural modifications of drugs often developed initially for the wrong reasons and by refining animal models that only very loosely simulated the human condition or predicted the results of structure–activity modifications in the leading compound. Future treatments may be accomplished most directly when one or more disease-causing genes are understood and then replaced by gene transfer technology. Several promising animal models have been reported in which cells transfected with gene constructs are stereotaxically inserted into the brain or a cerebral ventricle; these cells then express the desired gene product (e.g., tyrosine hydroxylase or a neurotrophic growth factor) and provide enduring replacement at least in experimental Parkinson's or Alzheimer's disease models. Much work is being done to develop viruses as packaging vectors to deliver next-generation gene constructs to specific target neurons, without the need for neurosurgical manipulation to transfer the genes, although this remains an unrealized goal for now.

It is not surprising that drug development in this field is a multi-billion dollar annual investment since untreatable chronic dysfunctions of the brain represent one of the largest pockets of cost in health care. Investors in this process are firmly convinced that logical drug development in the future will be successfully guided by the "tools of modern drug discovery," as exemplified by the angiotensin-converting enzyme inhibitors and the angiotensin III antagonists. These tools are (1) the genetic understanding of diseases and their environmentally sensitive regulatory factors, (2) animal models that closely simulate the pathophysiology, (3) cellular models in which to express cloned transmitter receptors, and (4) detailed three-dimensional, atomic-level understanding of the target molecules where the drugs are intended to act.

Given these tools, the imaginative drug design team will then rely upon the capacity of synthetic chemistry to compute potential steric requirements of receptor pockets for manipulation and to create molecular dynamic simulations of putative drug candidates. In the next-to-final step before selecting the color of the tablets, the ideal candidate will then be synthesized within the computer work station from chiral molecular building blocks in the archives and then matched against the world's database of previously made molecules for initial validation of activity. After that, it's off to the patent office, the FDA, and a few million dollars' worth of clinical trials. Piece of cake, eh? Maybe there are better ways, but no one knows them yet. Students, start your engines.

SELECTED REFERENCES

Baxter, L.R.J. (1994). Neuroimaging studies of human anxiety disorders: Cutting paths of knowledge through the field of neurotic phenomena. In *Psychopharmacology: The Fourth Generation of Progress* (F. E. Bloom and D. J. Kupfer, eds.), pp. 1287–1300, Raven Press, New York.

Bohacek, R. S., and C. McMartin (1995). Exploring the universe of molecules for new drugs. *Nature Medicine 1*, 177–179.

Kahn, R. S., and K. L. Davis (1994). New developments in dopamine and schizophrenia. In *Psychopharmacology: The Fourth Generation of Progress* (F. E. Bloom and D. J. Kupfer, eds.), pp. 1193–1204, Raven Press, New York.

Korczyn, A. D. (1994). Parkinson's disease. In *Psychopharmacology: The Fourth Generation of Progress* (F. E. Bloom and D. J. Kupfer, eds.), pp. 1479–1484, Raven Press, New York.

Kosten, T. A., M. J. Miserendino, S. Chi, and E. J. Nestler (1994). Fischer and Lewis rat strains show differential cocaine effects in conditioned place preference and behavioral sensitization but not in locomotor activity or conditioned taste aversion. *J. Pharmacol. Exp. Ther. 269*, 137–144.

Kupfer, D. J. (1994). Introduction to clinical neuropsychopharmacology. In *Psychopharmacology: The Fourth Generation of Progress* (F. E. Bloom and D. J. Kupfer, eds.), pp. 813–822. Raven Press, New York.

Maes, M., and H. Y. Meltzer (1994). The serotonin hypothesis of major depression. In *Psychopharmacology: The Fourth Generation of Progress* (F. E. Bloom and D. J. Kupfer, eds.), pp. 933–944. Raven Press, New York.

Martin, J. B. (1993). Molecular genetics in neurology. *Ann. Neurol. 34*, 757–773.

Phillips, M. I., E. A. Speakman, and B. Kimura (1993). Levels of angiotensin and molecular biology of the tissue renin angiotensin systems. *Regul Pept 43*, 1–20.

Plaitakis, A., and P. Shashidharan (1994). Amyotrophic lateral sclerosis, glutamate, and oxidative stress. In *Psychopharmacology: The Fourth Generation of Progress* (F. E. Bloom and D. J. Kupfer, eds.), pp. 1531–1544. Raven Press, New York.

Price, D. L., D. W. Cleveland, and V. E. Koliatsos (1994). Motor neurone disease and animal models. *Neurobiol. Dis. 1*, 3–11.

Schatzberg, A. F., and J. J. Schildkraut (1994). Recent studies on norepinephrine systems in mood disorders. In *Psychopharmacology: The*

Fourth Generation of Progress (F. E. Bloom and D. J. Kupfer, eds.), pp. 911–920. Raven Press, New York.

Sipe, J. C., J. S. Romine, J. A. Koziol, R. McMillan, J. Zyroff, and E. Beutler (1994). Cladribine in treatment of chronic progressive multiple sclerosis [see comments]. *Lancet 344*, 9–13.

Stern, R. A., D. O. Perkins, and D. L. Evans (1994). Neuropsychiatric manifestations of HIV-1 infection and AIDS. In *Psychopharmacology: The Fourth Generation of Progress* (F. E. Bloom and D. J. Kupfer, eds.), pp. 1545–1558. Raven Press, New York.

Tallman, J. F., and S. G. Dahl (1994). New drug design in psychopharmacology: The impact of molecular biology. In *Psychopharmacology: The Fourth Generation of Progress* (F. E. Bloom and D. J. Kupfer, eds.), pp. 1861–1874. Raven Press, New York.

The Interferon-β Multiple Sclerosis Study Group (1993). Interferon-β-1 is effective in relapsing-remitting multiple sclerosis. I. Clinical results of a multicenter, randomized, double-blind, placebo-controlled trial. *Neurology 43*, 655–661.

Timmermans, P. B., P. Benfield, A. T. Chiu, W. F. Herblin, P. C. Wong, and R. D. Smith (1992). Angiotensin II receptors and functional correlates. *Am. J. Hypertens. 5*, 221S–235S.

Weinberger, D. R. (1994). Neurodevelopmental perspectives on schizophrenia. In *Psychopharmacology: The Fourth Generation of Progress* (F. E. Bloom and D. J. Kupfer, eds.), pp. 1171–1184. Raven Press, New York.

Index